Bartosz Tylkowski, Marta Giamberini, Susana Fernandez Prieto (Eds.)

Microencapsulation

Also of interest

Chemical Synergies – From the Lab to In Silico Modelling
Bandeira, Tylkowski (Eds.), 2018
ISBN 978-3-11-048135-8, e-ISBN 978-3-11-048206-5

Metals in Wastes
Wieszczycka, Tylkowski, Staszak (Eds.), 2018
ISBN 978-3-11-054628-6, e-ISBN 978-3-11-054706-1

Polymer Engineering
Tylkowski, Wieszczycka, Jastrzab (Eds.), 2017
ISBN 978-3-11-046828-1, e-ISBN 978-3-11-046974-5

Medical Physics – Models and Technologies in Cancer Research
Bajek, Tylkowski (Eds.), 2020
ISBN 978-3-11-066229-0, e-ISBN 978-3-11-066230-6

Chemical Technologies and Processes
Staszak, Wieszczycka, Tylkowski (Eds.), 2020
ISBN 978-3-11-065627-5, e-ISBN 978-3-11-065636-7

Microencapsulation

Edited by
Bartosz Tylkowski, Marta Giamberini
and Susana Fernandez Prieto

2nd Edition

DE GRUYTER

Editors

Dr. Bartosz Tylkowski
Chemical Technologies Unit, Eurecat
Carrer Marcelli Domingo s/n
43007 Tarragona
Spain
bartosz.tylkowski@eurecat.org

Prof. Dr. Marta Giamberini
Rovira i Virgili University
Department of Chemical Engineering
Av. Paisos Catalans 26
43007 Tarragona
Spain
marta.giamberini@urv.cat

Susana Fernandez Prieto
Procter & Gamble
Services Company
Temselaan 100
1853 Strombeek-Bever
Belgium
fernandezprieto.s@pg.com

ISBN 978-3-11-064176-9
e-ISBN (PDF) 978-3-11-064207-0
e-ISBN (EPUB) 978-3-11-064195-0

Library of Congress Control Number: 2019955048

Bibliographic information published by the Deutsche Nationalbibliothek
The Deutsche Nationalbibliothek lists this publication in the Deutsche Nationalbibliografie; detailed bibliographic data are available on the Internet at http://dnb.dnb.de.

Typesetting: Integra Software Services Pvt. Ltd.
Printing and binding: CPI books GmbH, Leck
Cover image: Science Photo Library / Scharf, David

www.degruyter.com

Preface

According to *Science*, microencapsulation, developed approximately 65 years ago, is defined as a major interdisciplinary research technology. Microencapsulation is being used to deliver everything from advanced drugs to unique consumer sensory experiences. It's rapidly becoming one of the most important opportunities for expanding brand potential. Recent advances in polymer science and inorganic chemistry have further enhanced this growth. We decided to edit this book for those who see the potential benefit of using microencapsulation but need practical insight into using the know-how. The book aims to review the art of microencapsulation and to provide the readers with a comprehensive and in-depth understanding of recent developments and innovative applications of this leading-edge technology. Controlled release from microcapsules can be activated as a response to different triggering events: biological/chemical conditions (pH, reactive oxygen species) or external stimuli (thermal, magnetic, electrical). However; these mechanisms allow only limited control over the time and location of delivery. This can be overcome when the encapsulated material is protected in a photo-sensitive microcapsule shell. The topics detailed in Chapters 1 and 2 provide an overview on different materials and strategies used for photo-triggered microcapsules preparation and applications. Chapters 3 to 6 describe the growth and applications of microencapsulation in textile, food, medical and pharmaceutical industries. Developments of encapsulation technologies in the field of functional coatings are outlined in Chapter 7, while Chapter 8 provides the state of the art of encapsulation advances in extraction processes. Chapter 9 emphasizes micro and nanocapsules as supports for Surface Enhanced Raman Spectroscopy (SERS). Investigation on Si-based inorganic microencapsulation is given in Chapter 10 while application of encapsulation technology in agriculture field is detailed in Chapter 11.

We would like to express we gratitude to the contributing authors in making this project a success, and to Mrs. Vivien Schubert from DeGruyter for her assistance and encouragement in this venture.

Editors

https://doi.org/10.1515/9783110642070-202

Contents

Contributing authors

Domenico Pirone
The Procter & Gamble Company
Temselaan 100
1853 Strombeek-Bever,
Belgium
pirone.d@pg.com

Rita Del Pezzo
The Procter & Gamble Company
Temselaan 100
1853 Strombeek-Bever,
Belgium
delpezzo.r@pg.com

Todd L Underiner
The Procter & Gamble Company
Temselaan 100
1853 Strombeek-Bever,
Belgium
underiner.tl@pg.com

Susana Fernandez Prieto
The Procter & Gamble Company
Temselaan 100
1853 Strombeek-Bever
Belgium
fernandoprieto.s@pg.com

Anna Trojanowska
Universitat Rovira i Virgili
Departament d'Enginyeria Quimica
Av. Pais Catalans, 26
43007 Tarragona, Spain
anna.trojanowska@urv.cat

Marta Giamberini
Universitat Rovira i Virgili
Departament d'Enginyeria Quimica
Av. Pais Catalans, 26
43007 Tarragona, Spain
marta.giamberini@urv.cat

Nuno A.G. Bandeira
BioISI – Biosystems & Integrative Sciences Institute
C8, Faculty of Sciences, University of Lisbon
Campo Grande
1749-016 Lisboa, Portugal
nuno.bandeira@ciencias.ulisboa.pt

Krzysztof Artur Bogdanowicz
Professor Józef Kosacki's Military Institute of Engineer Technology
ul. Obornicka 136,
50-961 Wrocław, Poland
bogdanowicz@witi.wroc.pl

Bojana Boh Podgornik
University of Ljubljana
Faculty of Natural Sciences and Engineering
Department of Textiles, Graphic Arts and Design
Snežniška ulica 5,
SI-1000 Ljubljana, Slovenia
bojana.boh@ntf.uni-lj.si

Marica Starešinič
University of Ljubljana
Faculty of Natural Sciences and Engineering
Department of Textiles, Graphic Arts and Design
Snežniška ulica 5,
SI-1000 Ljubljana, Slovenia
marica.staresinic@a.ntf.uni-lj.si

Cinta Panisello Ilatje
Universitat Rovira i Virgili, Departament d'Enginyeria Quimica,
Av. Pais Catalans, 26
43007 Tarragona, Spain

Tania Gumi
Universitat Rovira i Virgili
Departament d'Enginyeria Quimica,
Av. Pais Catalans, 26
43007 Tarragona, Spain
tania.gumi@urv.cat

Ricard Garcilla Valls
Universitat Rovira i Virgili
Departament d'Enginyeria Quimica,
Av. Pais Catalans, 26, 43007 Tarragona
Spain
ricard.garcia@urv.cat

Monika Haponska
Eurecat, Centre Tecnològic de Catalunya, C Marcellí Domingo s/n,
43007 Tarragona, Spain
monika.haponska@eurecat.org

Anna Bajek
The Ludwik Rydygier Collegium Medicum
Nicolaus Copernicus University in Torun
Department of Tissue Engineering, Ul
Karłowicza 24, 85-092, Bydgoszcz, Poland
a_bajek@wp.pl

Irene Tsibranska
Institute of Chemical Engineering
Bulgarian Academy of Sciences
1113 Sofia, Bulgaria
tsibranska@yahoo.com

Justyna Kozlowska
Faculty of Chemistry
Nicolaus Copernicus University in Torun,
Gagarina 7
87-100 Torun, Poland
justynak@umk.pl

Natalia Stachowiak
Faculty of Chemistry
Nicolaus Copernicus University in Torun
Gagarina 7, 87-100 Torun
Poland
naat.staa@gmail.com

Weronika Prus-Walendziak
Faculty of Chemistry
Nicolaus Copernicus University in Torun
Gagarina 7, 87-100 Torun
Poland
279502@stud.umk.pl

Bartosz Tylkowski
Eurecat, Centre Tecnològic de Catalunya, C
Marcellí Domingo s/n
43007 Tarragona, Spain
bartosz.tylkowski@eurecat.org

Gaetano Palumbo
AGH University of Science and Technology of
Krakow
Department of Chemistry and Corrosion of
Metals Reymonta
23 30-059 Krakow, Poland
gpalumbo@agh.edu.pl

Karolina Wieszczycka
Poznan University of Technology
Institute of Chemical Technology and
Engineering
Berdychowo, St. 4
60-965 Poznan
Poland
karolina.wieszczycka@put.poznan.pl

Katarzyna Staszak
Poznan University of Technology
Institute of Chemical Technology and
Engineering
Berdychowo, St. 4
60-965 Poznan, Poland
katarzyna.staszak@put.poznan.pl

Renata Jastrzab
A. Mickiewicz University
Faculty of Chemistry
Umultowska 89b
61-614 Poznan, Poland
renatad@amu.edu.pl

Magdalena Olkiewicz
Eurecat, Centre Tecnològic de Catalunya, C
Marcellí Domingo s/n
43007 Tarragona, Spain
magda.olkiewicz@eurecat.org

Adrianna Nogalska
Eurecat, Centre Tecnològic de Catalunya, C
Marcellí Domingo s/n
43007 Tarragona, Spain
adrianna.nogalska@eurecat.org

Joanna Kowalska
Institute of Plant Protection – NRI,
Department of Biological Pest Control and
Organic Agriculture,
Wegorka street 20,
60-318 Poznan, Poland
J.Kowalska@iorpib.poznan.pl

Eligio Malusa
CREA – Research Centre for Engineering and Agro-Food Processing
Strada delle Cacce 73
10135 Turin, Italy
malusa@inrete.it

Marcin Luczak
Samorządowa Szkoła Podstawowa nr 1 im. 68 Wrzesińskiego Pułku Piechoty,
Szkolna 1, 62-300 Września, Poland
luczak.marcin@wp.pl

Patryk Nowak
Stary Rynek 3/2
62-230 Witkowo,
Poland
patryknowak@windowslive.com

Xavier Montane
Universitat Rovira i Virgili
Av. Pais Catalans, 26, 43007 Tarragona
Spain
xaviermontanemontanem@gmail.com

Josep M. Montornes
Eurecat, Centre Tecnologic de Catalunya,
C Marcellí Domingo s/n
43007 Tarragona, Spain
josep.montornes@eurecat.org

Leon Marteaux
Dow Corning S.A. 83 Av. Mounier
1200 Brussels, Belgium
e-mail: leon.marteaux@dowcorning.com

Domenico Pirone, Rita Del Pezzo, Todd L. Underiner, Susana Fernandez Prieto, Anna Trojanowska, Marta Giamberini and Bartosz Tylkowski

1 Photo-triggered microcapsules

1.1 Introduction

The encapsulation of materials for protection, by isolating them in a separate phase, has evolved into a major interdisciplinary research focus [1]. One of the most challenging tasks, and the ultimate purpose of developing delivery systems, is to modulate the release of encapsulated cargo substances. Strategies such as heat treatment, ionic strength, magnetic fields and light-induced morphology change have been used to alter shell density and integrity and, thereby, to influence capsule permeability [2]. As one of the most unique arts of stimuliresponsive capsules, photo-stimuli-responsive capsules can affect their micro-/nanostructures in the form of remote control triggered by external light, e.g. sunlight, without the need for direct contact or interaction. Moreover, triggering the release of microcapsules using light has several advantages over other external stimuli:

- In contrast to chemicals, photons do not contaminate the reaction systems and they have very low or negligible toxicity.
- The excitation wavelength can be controlled through the design of the photo-responsive molecule.
- It is easy to control the time and/or local excitation [3].

Triggered release of microcapsule contents using light is appealing for several applications. Nanoparticles and chromophores absorb light over a range of wavelengths, and their absorption cross-sections can be tuned both for one-photon and multiphoton absorption. For applications in cosmetics and agriculture, ultraviolet (UV)- and visible light-sensitive capsules are used because of the abundance of near-UV light. Near-infrared (NIR)-absorbing capsules are of greater interest in biological systems because of decreased light scattering in tissues at those wavelengths [4]. Moreover, light absorption can increase the vibrational energy (heat) of a substrate. Therefore, light can be used to activate thermal release mechanisms involving phase transitions and changes in polymer morphology.

1.2 Microcapsules containing carbon nanotubes

Since their discovery by Iijima in 1991, carbon nanotubes are becoming one of the most promising types of materials for use in nanotechnology [5]. They are unique materials that absorb infrared (IR) radiation, particularly between 700 and 1,100 nm.

https://doi.org/10.1515/9783110642070-001

Absorbed IR promotes molecular oscillation, leading to efficient heating of the surrounding environment. During the last 10 years, several patents and scientific publications have been dedicated to developing this phenomenon and to investigating a new class of photosensitive microcapsules containing carbon nanotubes in their structures. Recently, Pastine and coworkers [6] designed remotely triggered microcapsules in which the encapsulated liquid contained opto-thermally active species. The authors mixed the active material to be encapsulated with a small amount of carbon nanotubes and the precursors for making nylon. All reagents were continuously stirred, which caused the nylon to form spheres that captured the nanotubes and the active material. The obtained results clearly showed that when the laser was aimed at a capsule, the carbon nanotubes absorbed the light, heating up the liquid inside and causing it to expand until it exploded, releasing the contents. By varying the stir rate, Pastine at al. were able to influence the diameter of the microcapsules and vary their size from approximately 100 to 1,000 μm. Figure 1.1 shows an optical image of the prepared capsules.

Other examples of photo-triggered capsules containing carbon nanotubes as one of the fillers are described in US Patent Application US20120253000 (submitted by Okawa and coinventors) and US Patent Application US20100215724 (submitted by Prakash and coinventors). Okawa et al. reported the escape of phenylacetylene droplets, under irradiation with a 40 mW laser, from polyamide microcapsules prepared using an interfacial polymerization of triamines and diacid or triacid chlorides in an oil-in-water emulsion. Prakash et al. disclosed the release of a pharmaceutically active ingredient, such as a small molecule drug (organic molecules with molecular weight less than about 1,000 D – i.e. a peptide or a DNA molecule), under light irradiation, from microcapsules formed by alginate-poly-L-lysine-alginate and containing a low concentration of carbon nanotubes as a secondary active material. Caruso et al. [7] incorporated a suspension of single-walled nanotubes in chlorobenzene/ethyl phenylacetate into microcapsules using *in situ* emulsification polymerization of urea-formaldehyde, while Paunov and Panhuis [8] fabricated nanotube-based microcapsules using a colloid templating technique. Recently, Yashchenok et al. [9] prepared and characterized polyelectrolyte multilayer microcapsules containing carbon nanotubes and glutaraldehyde treatment in their shells. Glutaraldehyde treatment is typically used in biology for fixing cells in order to preserve their morphology on drying, and in the present case, this leads to cross-linking of amino groups of cationic polyelectrolytes. The authors demonstrated that incorporation of carbon nanotubes and subsequent treatment with glutaraldehyde result in free-standing structures, and these structures do not collapse after drying. In order to open the microcapsule shell and release encapsulated material, the authors irradiated the microcapsule walls using lasers operating at 473 nm and 830 nm. The Raman spectra and atomic force microscopy images of microcapsule before and after laser irradiation unequivocally confirmed that remote laser activation was accompanied by selective local destruction

Fig. 1.1: Toluene-filled polyamide microcapsules containing 1 wt.% carbon nanotubes. (a) Optical image of microcapsules in a scintillation vial. (b) Optical image of microcapsules in oil. (c) Scanning electron micrograph of microcapsules. (d) Scanning electron micrograph of crushed microcapsules. Carbon nanotubes (white bundles) are visible in the interior and exterior and incorporated into the wall [6]. Copyright 2009 American Chemical Society.

of carbon nanotubes and polyelectrolyte matrix in the microcapsule shell. Thus, the microcapsule structures described by the researchers provide new opportunities for biomedical applications owing to their improved mechanical stability and a pointwise manipulation with laser beams.

Saito and Kato [10] developed a method in which combining layer-by-layer (LbL) technique and templated-assisted technologies, various kinds of polyelectrolyte-based hollow capsules have been produced for anticancer treatments. To prepare the hollow capsules with shells that contain single-wall carbon nanotubes (SWCNTs), partially oxidized SWCNTs were used as polyanions and were embedded into one of the polyelectrolyte layers on the silica core. By dissolving the silica core, well-dispersed capsules of the polyelectrolyte/SWCNT composite were obtained. The anticancer drug was loaded into the capsules, and the release rates with and without the irradiation of the NIR laser beam at the wavelength of 800 nm were compared. Distinct release was confirmed in the latter case, whereas almost no release was detected in the former case, indicating that the SWCNT molecule is a suitable light absorber for use with optically addressable drug carriers. Notwithstanding, it is important to highlight that the toxicology of carbon nanomaterials, as well as their application in medicine and pharmacology, could be closely dependent on their surface chemistry, sizes, doses, and administration routes [11].

1.3 Microcapsules containing silver and gold particles

It is well known that nanoparticles of silver and gold, like carbon nanotubes, absorb light in the visible spectrum and release this energy as heat in their surroundings, which can be harvested to release encapsulated substances from microcapsules either destructively or nondestructively. This is due to surface plasmon resonance absorption of the nanoparticles, whose absorption cross-section is drastically more intense than for a typical dye. In the case of silver and gold, these plasmon oscillations have frequencies of approximately 400 nm and 525 nm, respectively. Like gold nanoparticles (AuNPs), gold nanorods (AuNRs) absorb light, but the mean absorbance region of nanorods is situated in the NIR region. As for colloidal gold particles, this absorption peak is largely dependent on surface properties, but additionally in the case of nanorods, on their length-to-width ratio (i.e. aspect ratio). If selecting the synthetic conditions, one can produce AuNRs with the dimensions that correspond to a specific absorption window [12]. The use of noble metal nanoparticles to optically induce changes in the shell of microcapsules was first demonstrated by Skirtach et al. in 2004 [13]. The authors doped silver nanoparticles into poly(sodium 4-styrenesulfonate)/poly(allylamine hydrochloride) micro shells and studied their response to light irradiation. Figure 1.2 shows transmission electron microscopy images of the shell of the capsule with Ag nanoparticles: (a) whole capsule and (b) a section with larger magnification.

Published results show that the obtained composite shell was sensitive to laser light with a wavelength of 830 nm, and local heating of nanoparticles resulted in the deformation and rupture of the shell. This effect allowed the realization of the remote release of the encapsulated material from microcapsules inside the living tissue. A consequence of the doping method applied by Skirtach et al. is difficult nanoparticle characterization due to their polydispersity. Indeed, the authors reported that it was not possible to estimate the exact amount of reduced silver on the capsule surface and they did not provide detailed analysis of the distribution of silver nanoparticles in the capsule wall. According to the authors, residual absorption of silver nanoparticles with a diameter of 20 nm in the NIR was shown to be sufficient to produce heat and break microcapsules [13] but insufficient for microcapsules containing 4 nm silver nanoparticles [14, 15]. Based on the published

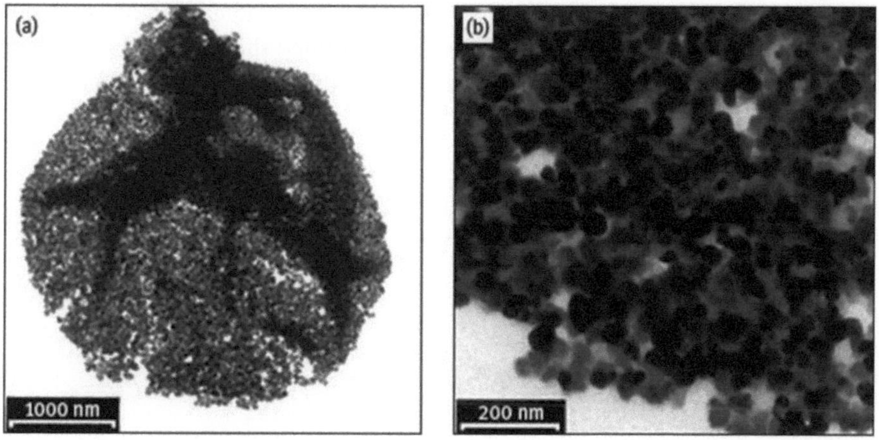

Fig. 1.2: Transmission electron microscopy (TEM) images of the shell of the capsule with Ag nanoparticles: (a) whole capsule and (b) a section with larger magnification. Reprinted with permission from [12]. Copyright 2004 American Chemical Society.

results, it can be concluded that the irradiation time necessary to break microcapsules strongly depends on the size of the nanoparticles. Because living tissues are sensitive to monochromatic coherent radiation, the selection of the minimal values of laser intensity and its time duration is an important issue for remote opening of the drug-loaded capsules inside tissues. Using a 532 nm continuous wave (CW) laser, Radzuik and coworkers [16] showed that capsules impregnated with 10 nm silver nanoparticles took, on average, 22 s to open, whereas it took 10 s of laser exposure when using 18 nm nanoparticles. Surprisingly, the combination of the 10 and 18 nm nanoparticles in capsule shells decreased the laser exposure requirements to 5 s or less. The authors showed that the microcapsules' diameter size depends on the volume and molar ratios of the precursor materials (Figure 1.3).

One of the first examples of AuNP exploitation for photo-triggered microcapsules fabrication was published by the Caruso research group at The University of Melbourne. In an early study [17], the group reported that the enzyme lysozyme can be encapsulated within the polyelectrolyte/AuNP shell via LbL assembly on the surface of lysozyme crystals and that the enzyme can be released on demand without significant loss of bioactivity following irradiation with short pulses of NIR laser light. Then, in subsequent investigations, they described the loading/release of macromolecules into/from light-responsive polyelectrolyte/AuNP microcapsules formed by a polyelectrolyte multilayer shell with lipids and ligands, respectively. Distinguished from their previous study, where the enzyme was used as a template to achieve encapsulation, in this investigation, fluorescein isothiocyanate (FITC)-labeled dextran, a convenient model of high-molecular-weight biomaterials such as DNA, was postloaded into preformed capsules by switching the capsule permeability between the "closed" and "open" states through variation of the bulk solution pH. Electron microscopy confirmed that the laser irradiation had no apparent effect on capsules without the light-absorbing shell component (Figures 1.4 and 1.5) [18].

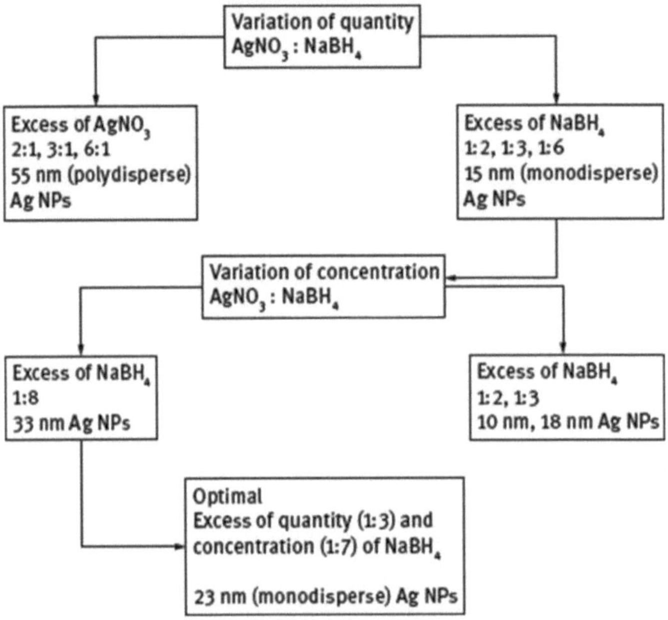

Fig. 1.3: Size of resulting nanoparticles prepared at various volumes and molar ratios of the precursor materials. Reprinted with permission from [15]. Copyright 2007 American Chemical Society.

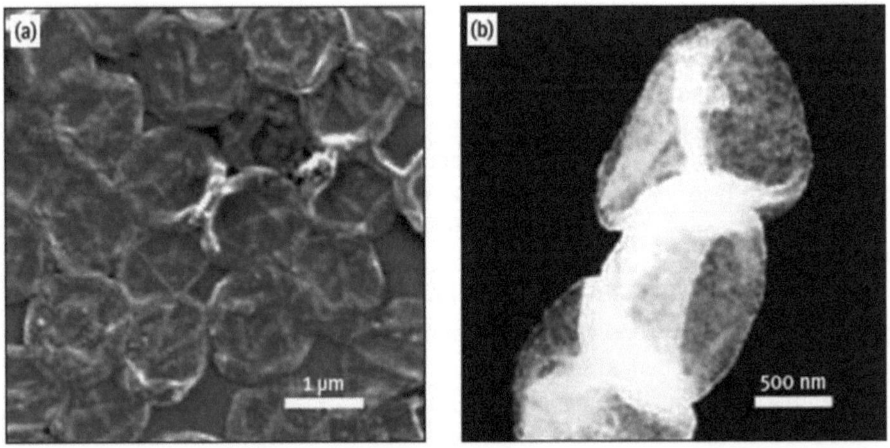

Fig. 1.4: (a) Scanning electron microscope (SEM) and (b) transmission electron microscopy (TEM) images of poly(styrene sulfonate) (PSS) and poly(allylamine hydrochloride) (PAH), before laser irradiation. Reprinted with permission from [17]. Copyright 2005 American Chemical Society.

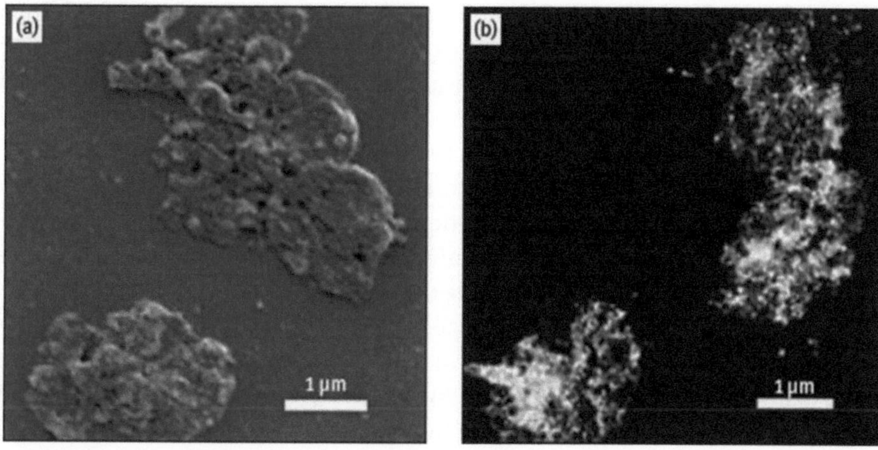

Fig. 1.5: (a) Scanning electron microscope (SEM) and (b) transmission electron microscopy (TEM) images of PSS/PAH/gold nanoparticle capsules loaded with FITC-dextran after laser irradiation (radiant exposure per pulse 700 mJ/cm^2). Reprinted with permission from [17]. Copyright 2005 American Chemical Society.

Based on these data, Prof. Caruso proposed that laser induced release involves:
- heating of the capsule shell to high temperatures above the spinodal point of water upon nanoparticles light absorption;
- development of thermal stresses within the capsule shell because of the variations in thermal expansion coefficients of shell wall materials; and
- capsule rupture.

A similar outcome was reported by Skirtach et al. [14], who incorporated the AuNPs inside the walls of polyelectrolyte multilayer capsules. In this investigation, an active material – AF-488 dextran – was successfully incorporated into the capsules using a heat-shrinking method. According to the authors, the capsules obtained by such a method exhibit improved mechanical stability: properties important for the *in vivo* and *in vitro* delivery of encapsulated material. Upon illumination by laser light, the encapsulated dextran leaves the interior of a capsule inside a living cancer cell. This study has served as a significant step toward the use of polyelectrolyte multilayer capsules for the delivery of medicine into biological cells. Furthermore, release from polyelectrolyte microcapsules functionalized with AuNPs, by burst opening and deformation, has also been demonstrated by Volodkin et al. [19]. The authors reported photo/temperature-triggered release of a liposome cargo from surface-supported vesicles embedded inside biocompatible polyelectrolyte multilayers. In an effort to enlarge the scope of applications of remote release and to extend it further to other surface-supported drug delivery vesicles, the same research team applied remote release to liposome-AuNPs, referred to as Lip-NP assemblies or complexes [20]. The goals of this work were to show that Lip-NP assemblies could be prepared in a controlled manner in terms of size and

NP state and then to use NIR light to selectively release encapsulated dye from the assemblies. In the remote-release experiments, the micron-sized aggregate of Lip-NP was illuminated by a focused laser beam, which resulted in release of the encapsulated fluorescent dye. According to the authors, the functionalized liposome-nanoparticle assemblies can be used for transdermal applications in which an active compound is delivered through the skin, easily accessible by light. Due to quite deep IR-light tissue penetration, the light-responsive liposome assemblies could serve as active constituents of implanted devices. Yang and coworkers [21] reported a novel NIR light-responsive drug delivery capsule platform based on AuNRs integrated within a mesoporous silica framework that was surface functionalized with an anticancer aptamer DNA. The authors observed that upon exposure to NIR light, the photothermal effect of the AuNRs led to a rapid rise in the local temperature, resulting in the dehybrization of the linkage DNA duplex, allowing the release of the entrapped guest. *In vitro* studies showed the feasibility of using this nanocarrier as a targeted and noninvasive remote-controlled drug delivery system in cancer cells with high spatial/temporal resolution. According to the authors, this multifunctional capsule platform could integrate chemotherapy, photothermotherapy and imaging into one system. The good biocompatibility, cancer cell recognition ability and efficient intracellular drug release provided a basis for *in vivo* controlled-release biomedical applications and cancer therapy. Incorporation of AuNPs into microcapsule walls has also been used to trigger increased microcapsule permeability upon temperature increase as an effect of light irradiation. Kim et al. [22] designed hydrogel capsules with membranes composed of poly (N-isopropylacrylamide) containing AuNRs and achieved reversible control of permeability. However, these polymeric or hydrogel microcapsules exhibited relatively poor mechanical properties, frequently leading to undesired rupture or deformation of the shell membrane in physiological environments. In addition, drug diffusion through the polymer matrix of the shell was limited to small molecules with molecular weight (MW) cutoff below 200. West and coworkers [23] suggested that the permeability changes of the N-isopropylacrylamide/acrylamide capsule walls containing AuNPs, upon laser irradiation, could be useful in insulin therapy. Recently, Lee et al. [24] prepared microcapsules of poly(dL-lactic-co-glycolic acid) (PLGA) containing AuNRs by using water-oil-water (W/O/W) double emulsions. Upon laser irradiation, the nanorods generated heat via a photothermal effect, and as a result, the membrane became more permeable above the glass transition temperature of the PLGA, thereby providing remote control release of the active material. Very recently, Jeong and coworkers [25] obtained photo- and thermo-responsive microcapsules whose porous membranes contain reversible valves. In a similar manner to Lee et al., the authors prepared W/O/W double-emulsion drops, which served as templates to produce microcapsules. The ethyl cellulose capsule membrane, made from the middle oil phase, was designed to contain interconnected pores that could be filled by a thermo-responsive hydrogel. To provide photo-responsiveness, AuNRs were embedded in the hydrogel matrix to act as nanoheaters under NIR irradiation. The microcapsules were shown to exhibit photo-triggered release of a model drug with high molecular weight, with the release event successfully occurring repeatedly over extended periods by a series of NIR exposures. In addition, a porous shell made of the biocompatible polymer, ethyl cellulose, provided high mechanical stability and membrane integrity. The system developed by Chen et al. [26], called FA-MC@GNR, exhibits high stability, no obvious toxicity, and remarkably improved tumor-targeting capabilities *in vivo*. That system is composed by a multilayered hollow microcapsule of folic acid (FA) modified and loaded with AuNRs (GNR) that can undergo thermal degradation under NIR light. The composite capsules were loaded via physical adsorption with a NIR dye mercaptopropionic acid (MPA) or anticancer drug (doxorubicin, DOX), yielding FA-MC@GNRs/MPA or FA-MC@GNRs/DOX, respectively. In Figure 1.6, the scheme of the synthesis of the microcapsules is presented. The group of Chen demonstrated the system's ability to simultaneously elicit photothermal therapy and controlled chemotherapy, achieving synergistic cancer treatment both *in vitro* cellular and *in vivo* animal experiments by utilizing the strong NIR absorption of FA-MC@GNRs/DOX.

The method developed by Wang et al. [27] is also very innovative and demonstrates that gas-filled capsules (GFCs) can be used as pressure sensors. The production method developed is based on a combination of a microfluidic technique and laser-triggered reactions with which it is possible to fabricate functional GFCs on-demand. The method involves (i) the generation of monodispersed alginate microcapsules containing ammonium bicarbonate (AB) as gas resource and AuNRs as a heating resource, in a microfluidic device, and (ii) the NIR light-triggered generation

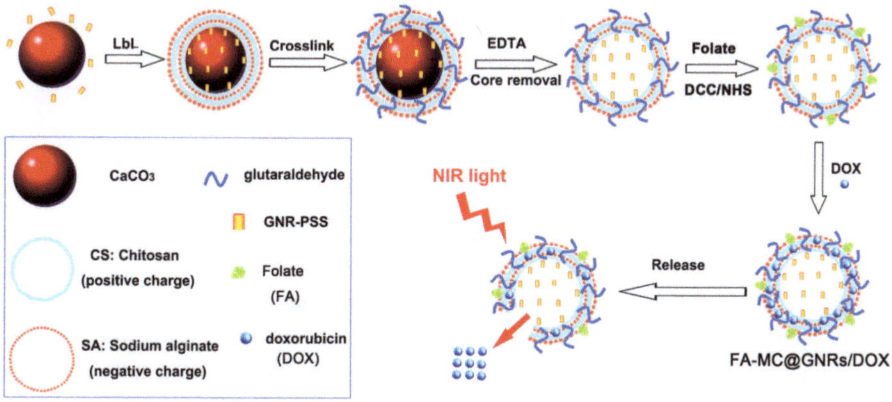

Fig. 1.6: The synthesis routine of the multifunctional microcapsules (FA-MA@GNRs/DOX). Reprinted with permission from [26]. Copyright 2019 from Royal Society of Chemistry.

of gases from the AB-containing microcapsules and simultaneous encapsulation of the gases in an alginate shell to produce GFCs. The GFCs can be readily functionalized with inorganic nanoparticles in the alginate membranes. The functional GFCs may find applications in various fields such as bioimaging, drug delivery, and sensors. Peralta et al. [28], studied a desolvation and cross-linking method that was used to successfully encapsulate AuNRs into human serum albumin nanoparticles (HSAPs) simultaneously with the chemotherapeutic drug paclitaxel (PAC). They were able to form particles with overall size of 299 ± 6 nm and a loading efficiency of PAC up to 3 µg PAC/mg HSAP. Thanks to the incorporation of AuNRs, they can go through a photothermal heating when irradiated by NIR; the bulk reaches up to 46 °C after 15 min of NIR laser exposure, causing severe cellular hyperthermia and necrosis. The hybrid particle system could be also customized by external functionalities via conjugated targeting ligands, such as antibodies. The encasement strategy developed facilitated (i) a colloidal hybrid treatment system capable of enhanced permeability and retention effects, (ii) photothermal ablation of cancer cells, and (iii) release of the active up to 188 ng in a single 15-min irradiation session. Xu et al. [29] studied multi stimuli responsive capsules made of thermal-/pH-dual sensitive aliphatic poly(urethane-amine) (PUA), sodium poly(styrenesulfonate) (PSS), and AuNPs. The results indicate that the prepared hollow microcapsules have the great potential to be used as a novel smart drug carrier for the remotely controllable drug delivery. Xu et al. made interdependent multiresponsive drug delivery with by the LbL technique. The electrostatic interactions among PUA, PSS, and AuNPs contribute to the successful self-assembly of hollow multilayer microcapsules. Thanks to the shrinkage of PUA above its lower critical solution temperature and the interaction variation between PUA and PSS at different pH conditions, hollow microcapsules exhibit distinct pH and thermal-sensitive properties. Moreover, AuNP aggregates can effectively convert light into heat upon irradiation with NIR laser and endow the hollow microcapsules with distinct NIR-responsiveness. More importantly, the NIR-responsive study also demonstrates that the microcapsule morphology and the corresponding NIR-responsive drug release are strongly dependent on the pH value and temperature of the media.

1.4 Microcapsules containing titanium dioxide particles

Titanium dioxide particles and related composites are well known for their catalytic activity and oxidative potential. They have low production costs and a strong absorbance in the UV region, making such nanoparticles a smart choice of compound to sensitize thin films to UV light [30]. TiO_2 nanoparticles have been used in the last years in many applications as filler in transparent, flame-retardant, thermally insulating polymer composite films; conferee UV-blocking properties to such materials make them the best candidates to be used in fields as construction,

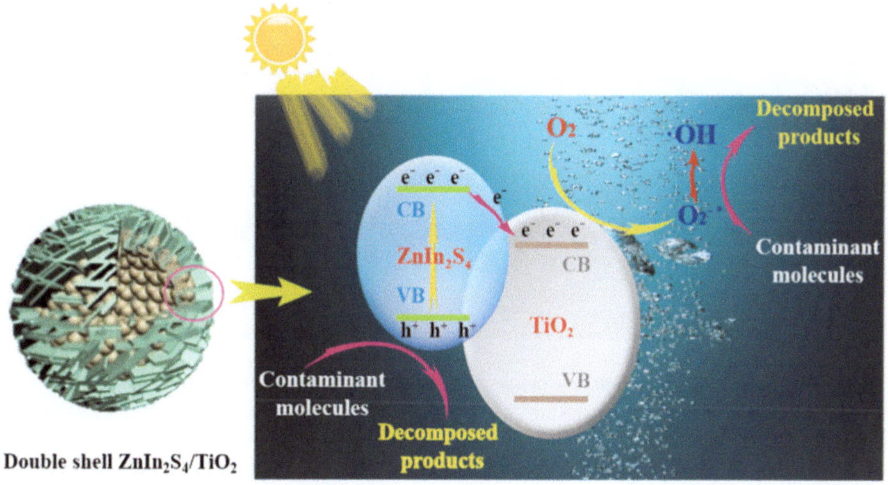

Fig. 1.7: Schematic illustration of degradation mechanism by double shell ZIS/TiO_2-x composite. Reprinted with permission from [31]. Copyright 2019 from Elsevier.

transportation, electronics, aérospace, and medicine [31]. Composite materials based on TiO_2 have also been realized in the field of nanospheres. Double shell $ZnIn_2S_4$ nanosheets/TiO_2 (ZIS/TiO_2) hollow composite nanospheres (Figure 1.7) demonstrated enhanced photocatalytic degradation activities of tetracycline hydrochloride, levofloxacin, and rhodamine B under visible light irradiation as well as excellent stability due to the synergic effects between $ZnIn_2S_4$ and TiO_2 [31, 32].

TiO_2 nanoparticles are also deeply investigated, thanks to their excellent physiochemical properties [33]. TiO_2-based hybrid, nanostructured catalysts are more and more explored for efficient photocatalytic degradation of gaseous and volatile organic pollutants in water or air. Coupling TiO_2 with nanocarbons or polymers having suitable bandgaps, in fact, it is possible to form composites with desirable bandgaps and specific light-photocatalytic activity ranges. Since the photocatalytic activity of these materials depends on their light-responsive range and carrier-separation capacity, rational design of TiO_2-based nanostructures that meet the requirement is crucial [34]. Several reports have been published on the incorporation of TiO_2 nanoparticles into microcapsules [34, 36, 37]. Hu et al. [35] developed a self-templated approach for the synthesis of TiO_2 microcapsules (Figure 1.8) with tunable size and wall by heating sol-gel-derived TiO_2 microspheres with poly(acrylic acid) (PAA) in a diethylene glycol (DEG) solution. According to the authors, PAA plays a crucial role in the formation of microcapsules by crosslinking the surface TiO_2 nanoparticles and preventing them from dissolution by DEG. Moreover, the investigators showed that hollow microcapsules were formed when DEG molecules penetrated the outer layer and removed the core materials by forming soluble titanium glycolate.

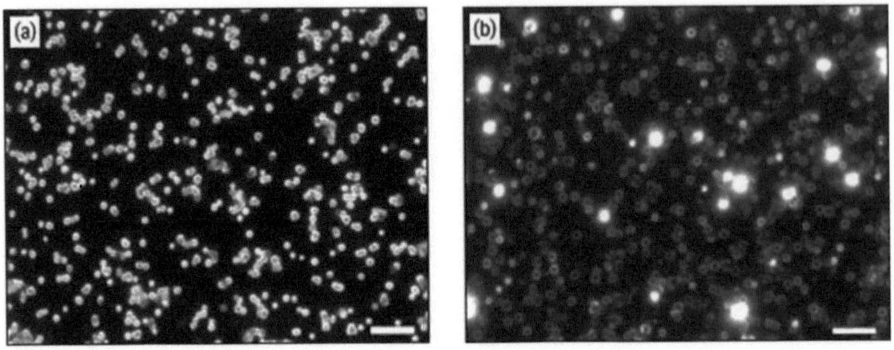

Fig. 1.8: Dark-field optical microscopy images of (a) the primary TiO_2 microspheres and (b) the corresponding microcapsules. The particles were dispersed in DEG and then deposited as a thin liquid film on a silicon substrate to reduce their movement during imaging. All scale bars are 10 µm. Reprinted with permission from [35]. Copyright 2007 American Chemical Society.

Very interesting results were presented by Katagiri and coworkers [38]. These authors showed that the UV irradiation of polyelectrolyte capsules coated with SiO_2/TiO_2 resulted in capsule obliteration due to the UV absorption of TiO_2. Prior to UV irradiation, the particles have a spherical structure for both SiO_2-polymer and SiO_2/TiO_2-polymer capsules. After UV irradiation, the SiO_2-polymer capsules remained spherical and intact, whereas the SiO_2/TiO_2-polymer capsules completely decomposed. In TiO_2, electron-hole pairs are generated through excitation with UV light and the holes residing in the valence band of TiO_2 have high oxidative activity. In addition to their triggering capabilities, these capsules are unique compared to other polyelectrolyte capsules because of their improved mechanical integrity resulting from the metal oxide coating, which may allow them to better withstand the mechanical loadings of solid-state applications. In the work from Wang et al., TiO_2 nanospheres were employed as typical photo-initiators, for the synthesis of various inorganic/polymer nanocomposites via photocatalytic surface-initiated polymerization. The excitation of TiO_2, deposited on the surface, by UV-light irradiation, produces electrons and holes that drive the free radical polymerization at the surface, producing core/shell composite nanospheres (Figure 1.9). This approach opens up new opportunities in the design and synthesis of inorganic and polymeric composite nanomaterials for a variety of applications [39].

1.5 Microcapsules containing photosensitive chromophores in the shell

The exposure of some photoactive groups to light can generate reversible structural changes, thereby directly changing the hydrophilic-hydrophobic balance without addition of other reagents. Typical groups that display photochemically induced

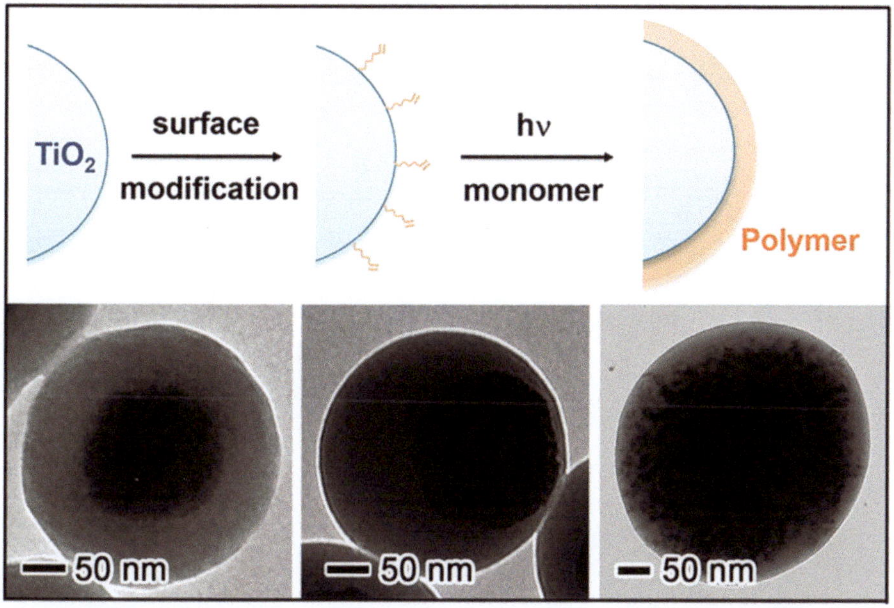

Fig 1.9: Schematic illustration showing the photocatalytic surface-initiated polymerization with TiO_2 nanospheres as the photo-initiator. Reprinted with permission from [39]. Copyright 2019 American Chemical Society.

transitions include azobenzene (change of dipole moment, size, and shape), spyrobenzopyran (ring opening formation of zwitterionic species), triphenylmethane leucohydroxide (generation of charges), and cinnamoyl (photodimerization), as shown in Figure 1.10. These transitions can further induce changes in the optical, mechanical, and chemical properties of the system containing the chromophore [40].

Currently, these chromophores are intensively investigated to implement light sensitivity in polymers as well as in microcapsule shells formed by them. Azobenzene derivatives are the most studied photoactive groups, and microcapsules based on these compounds are a subject of Chapter 2. The azo group can undergo reversible isomerization between the trans and cis configurations by light and heat (see Figure 1.10a) The photoisomerization of azobenzene chromophores has no side reactions, and the wavelength to induce the transformation can be tuned by incorporating substituents into the chromophores. Tao et al. [41] showed that an azo-based dye, when incorporated into the shell walls, altered the permeability of capsules exposed to visible light. Bédard and coworkers [42] also demonstrated that when the azo dye was incorporated into the polyelectrolyte backbone, the permeability of polyelectrolyte capsules could be altered upon light absorption by azobenzene moieties. Spirobenzopyran derivatives can also be used to create water-soluble polymers that associate into aggregates under UV irradiation. Upon UV irradiation, the neutral spirobenzopyran undergoes reversible isomerization into a zwitterionic merocyanine

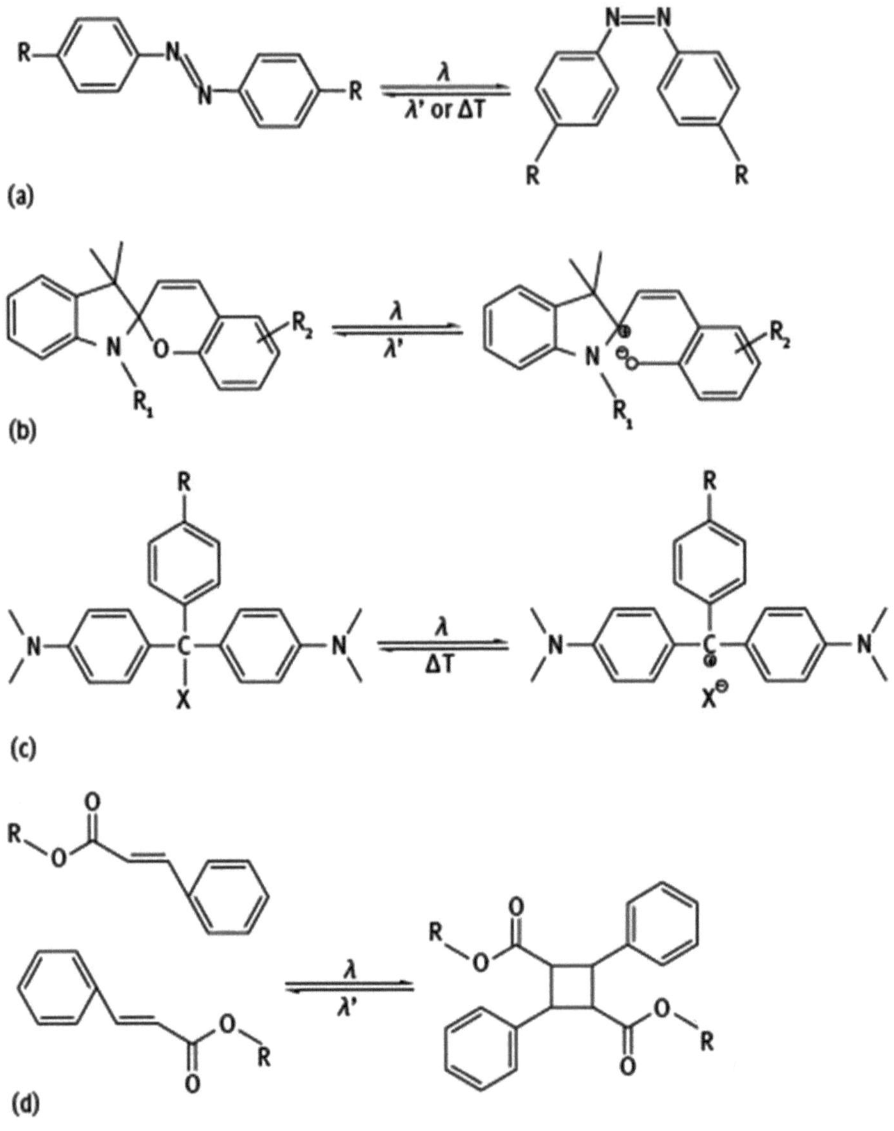

Fig. 1.10: Examples of chromophores that display photochemically induced transitions. (a) Reversible trans (left) and cis (right) photoisomerization of azobenzene. (b) Reversible photoisomerization of spirobenzopyran derivatives. (c) Dissociation of triphenylmethane leucoderivatives into an ion pair under ultraviolet irradiation. (d) Reversible photodimerization of the cinnamoyl group.

(see Figure 1.10b). For example, a water-soluble random copolymer formed from N-(2-hydroxypropyl)methacrylamide (HPMA) and a methacrylate monomer containing a spirobenzopyran side chain was soluble in demineralized water. However, UV exposure (20 m) induced the aggregation of large polymer clusters (400 nm) due to zwitterionic merocyanine formation. In a nearly nonionic solution, the attractive dipole-dipole interaction between these zwitterionic merocyanines provided intra- and inter-chain associations between the polymer chains. Upon exposure to visible light (20 min), the clusters disintegrated as the spirobenzopyran returned to its neutral form. Interestingly, this photoreversible cluster formation was reversed in a 1 M NaCl solution. For this high-ionic-strength solution, the neutral form of the copolymer had low solubility, while its charged form with zwitterionic merocyanine was highly soluble [43]. The following examples concern microcapsules containing photoactive polymers bearing triphenylmethane leucohydroxide or nitrocinnamate. Kono's group [35] prepared polyelectrolyte complex capsules from partly crosslinked PAA, polyethylenimine, and a copolymer of acrylic acid and bis(4-(dimethylamino) phenyl) (4-vinylphenyl)methyl leucohydroxide. Upon UV irradiation, the triphenylmethane derivative dissociated into an ion pair, thereby generating charges (see Figure 1.10c). While the active material was released at a low rate in the dark, upon light irradiation, the release rate of the encapsulated material increased 10-fold after a time lag of several minutes. When the capsules were put in the dark again, the release rate decreased immediately. Kono et al. [44] showed that this permeability control responding to the light can be achieved over several cycles and, moreover, that the photo-response was diminished after 75 min. According to the authors, release of encapsulated p-toluenesulfonate from studied capsules was caused by degradation of the photosensitive groups on the capsule shell. The main diameter of the studied capsules was 3 mm. Microcapsules were also built with a shell composed of nanoparticles of poly(orga-nosiloxane) functionalized with nitrocinnamate [45]. Photo-crosslinking the shell by 2 + 2 photocycloaddition of nitrocinnamate ($\lambda > 275$ nm) (Figure 1.10d) generated stable microcapsules. Destruction of the microcapsules was achieved by photocleavage of the dimer linkages via illumination at $\lambda = 254$ nm. Cyclodextrin was encapsulated and subsequently released by this reversible photocrosslinking reaction. However, the process was rather slow and several hours were required for photoaddition and photodissociation. Chang and coworkers [46] developed photosensitive microcapsules based on α-phenylcinnamylideneacetylated poly(L-lysine) polymer containing α-phenylcinnamylideneacetate moieties in its structure, which are derivatives of cinnamate. The authors reported that a very low concentration (0.5%) of photosensitive moieties is sufficient to achieve a less than 20% breaking frequency in 48 h under UV-light irradiation. Recently, self-immolative polymers containing photosensitive functional groups have been reported in the literature in order to trigger release of core contents from polymeric microcapsules. Fomina et al. [47] developed a light-sensitive self-immolative polymer [poly(lactic-co-glycolic acid)] containing a quinine methide backbone and photocleavable nitrobenzyl alcohol groups as the triggers. The cleavage of the triggering group by light at 350 nm induces cyclization of the diamine spacer, which in turn unmasks an unstable quinine methide moiety. As illustrated in Figure 1.11, incorporation of these moieties into a polymer chain causes degradation of the polymer backbone upon

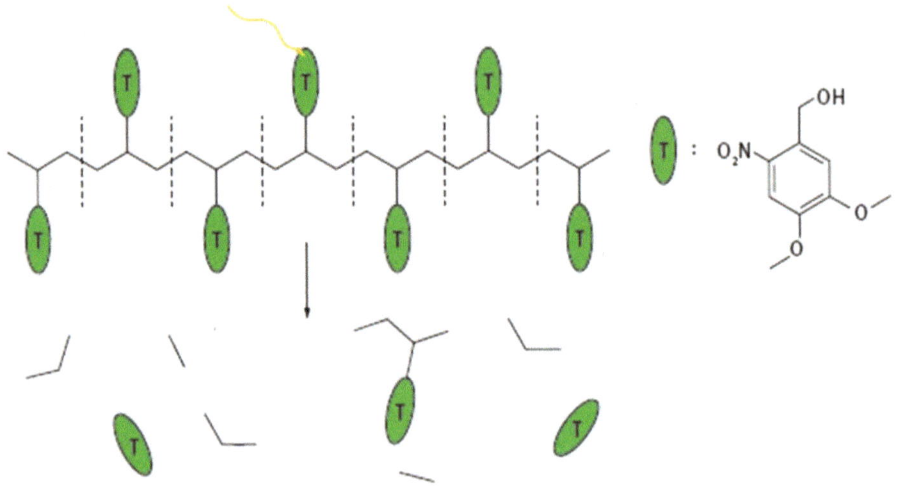

Fig. 1.11: Degradation of light-sensitive polymer containing quinine methide backbone upon irradiation. Reprinted with permission from [47]. Copyright 2009 American Chemical Society.

irradiation. In a preliminary study, the authors prepared nanocapsules based on this polymer and demonstrated that release of encapsulated dye occurs only in presence of UV light. Unfortunately, the obtained nanocapules possessed limited application in biological systems due to the long irradiation time required for polymer degradation and release of the active.

In order to overcome this limitation and to create microcapsules whose shell is more sensitive to brief irradiation, the same research group [48] designed a photodegradable polymer based on a well-established quinonemethide self-immolative disassembly mechanism using o-nitrobenzyl and 4-bromo-7-hydroxycoumarin as end-cap moieties instead of nitrobenzyl alcohol. The authors encapsulated hydrophobic dye Nile red and demonstrated 40% release of the dye over 2 h of UV irradiation. Moreover, the obtained results showed that the NIR irradiation is less efficient than the photocleavage obtained with UV irradiation. According to the authors, no bulk response or dye release were observed upon NIR irradiation, likely because the hydrophobic environment created around this photosensitive moiety strongly affects the absorption properties and quantum yields of this triggering group. Wang and coworkers [49] intended to design a series of photosensitive expandable particles for nanocarriers based on photolabile monomers (see Figure 1.12). The nanoparticles, with a main diameter size ca. 120 nm, synthesized through miniemulsion polymerization, were stable in aqueous solutions, although light at 350 nm irradiation led to particle uncaging and further particle expansion up to ca 800 nm. The authors used Coumarin 6 (C6), a hydrophobic fluorescence dye, as a model to evaluate photo-controlled release in aqueous solution and reported that the release efficiency reached 85% based on the fluorescence intensity after 5 min of irradiation with UV light. This new light-triggered expandable particle system

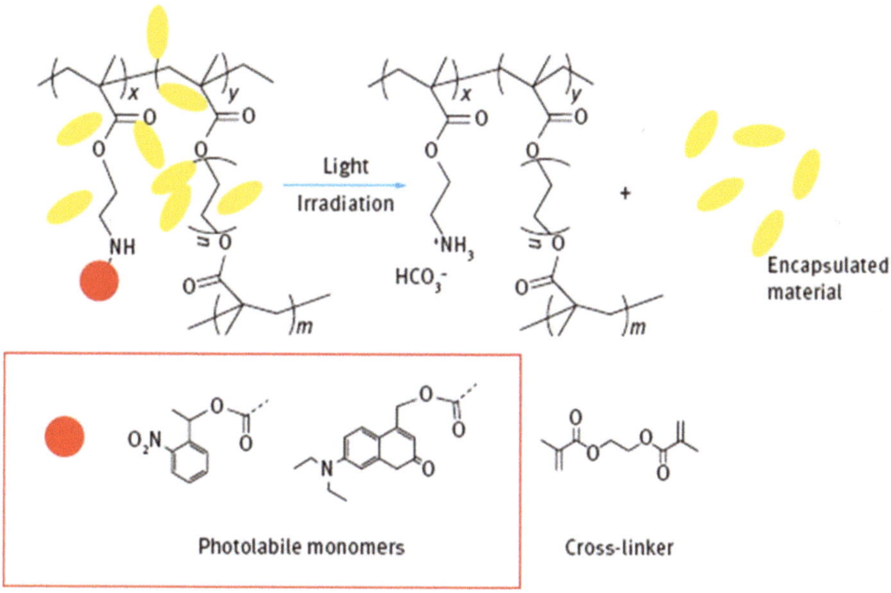

Fig. 1.12: Strategy for controllable release from light-induced expandable nanoparticles and structures of photolabile monomers as well as a cross-linker for particles prepared by miniemulsion polymerization.

may afford a new option for the photocontrollable treatment of cancers and other diseases with spatiotemporal resolution.

Lv et al. [50] simplified the syntheses of photolabile polymers with direct formation of polyesters by the coupling of 2-nitrophenylethylene glycol and dioyl dichlorides with different lengths. Obtained polymers were able to assemble photodegradable nanoparticles via a simple oil/water emulsion. The published results showed that particles formed with these polymers are quite stable in aqueous solutions with different pH buffers and at elevated temperatures. High stability of polymeric nanoparticles before triggering is greatly favored for an ideal drug carrier, for preventing drugs from leaking during circulation in blood. The authors demonstrated that UV light can trigger the cleavage of particles and the release of encapsulated substances with the efficiencies up to 8-8 based on Nile red fluorescence intensity upon 15 min light irradiation. The photo-degradation behaviors of nanoparticles were confirmed by a scanning electron microscope. Without light irradiation, particles were in spherical shapes; however, their shape was lost upon light activation. Azagarsamy and coworkers [51], to develop the photodegradable nanoparticles, designed and synthesized a short photodegradable cross-linker (see Figure 1.13). The authors selected a methoxy-nitrobenzyl ether derivative as the photodegradable moiety, because of its well-established chemistry and greater utilization in numerous biological applications (due to its absorbance at longer wavelengths of light). Moreover, in this study,

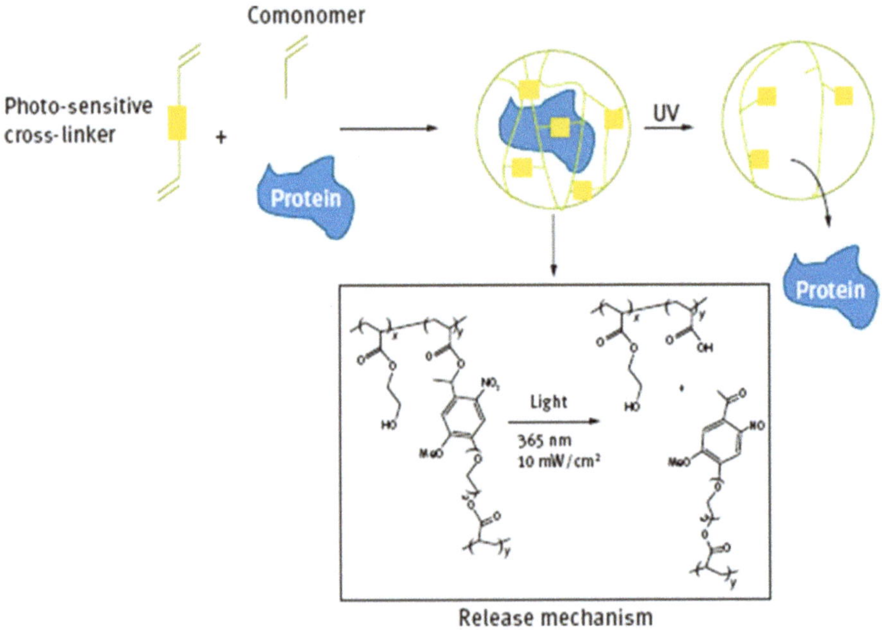

Fig. 1.13: Protein release from the microcapsules based on de-crosslinking of the polymer under illumination with UV light.

the authors used alkaline phosphatase (ALP) as a model encapsulated protein. ALP is a biologically important enzyme that is readily available and known to specifically cleave only phosphate esters. Figure 1.13 shows a mechanism of protein release from the microcapsules based on de-crosslinking of the polymer under illumination with UV light.

A different approach was described by Koo and coworkers [52]. The authors have focused their research on microcapsules whose walls contain photoacid generators (PAGs), which makes them optically addressable. The authors reported that by exposure to UV light with a wavelength of 254 nm, the triphenylsulfonium triflate as a PAG within the capsule walls was activated, which caused the release of protons from the capsules. This release of protons caused a decrease in the pH of the capsule solution, which then triggered a swelling of the microcapsules. The authors reported that these microcapsules could be opened and closed via alternate repetition of exposure to UV light and washing with neutral water. Moreover, prolonged exposure led to a breakage of the capsules, which promoted a rapid release of the entrapped substances. Jiang et al. [53] propose photo-switchable crosslinked coumarin-modified microcapsules. A wall material was based on coumarin and isocyanate group, while a core material was a dye, 2-anilino-6-dibutylamino-3-methylfluoran. The capsules have been obtained via interfacial polymerization followed by photo-crosslinking reaction. As-prepared capsules present low leakage due to crosslinked structure. Designed crosslinking is reversible, and the authors exploit coumarin property to

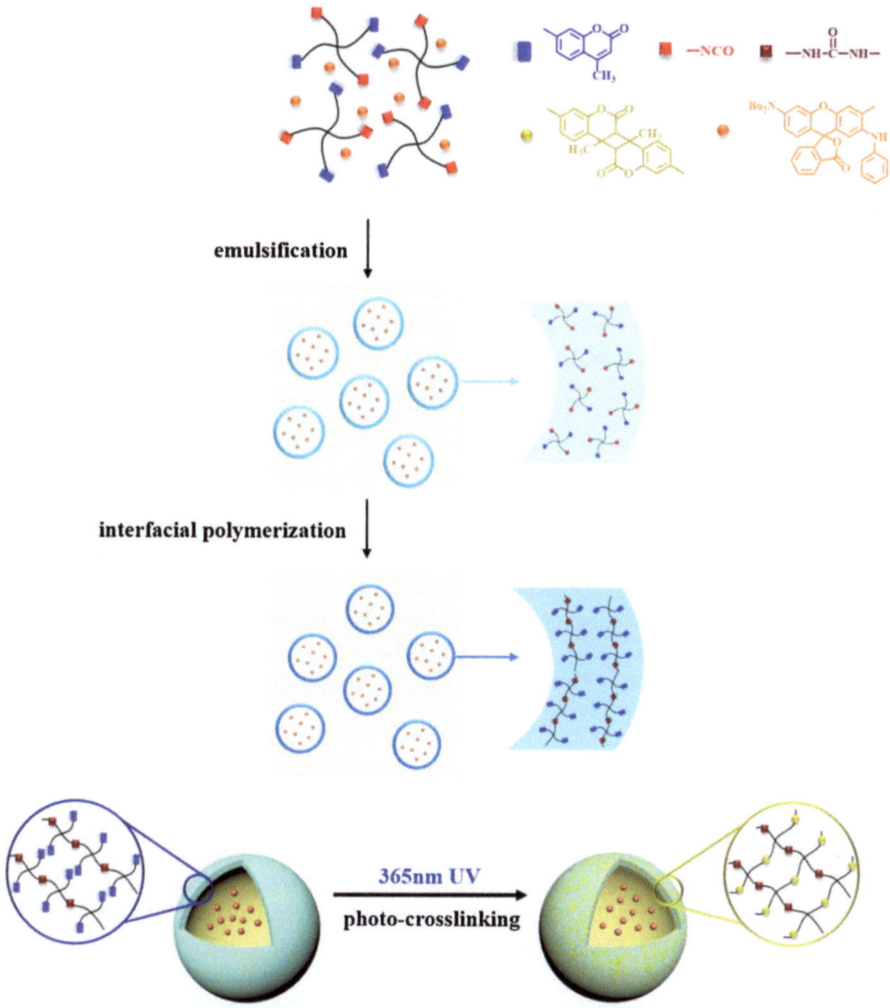

Fig. 1.14: Preparation of coumarin-modified microcapsules via interfacial polymerization and photo-crosslinking reaction. Reprinted with permission from [53]. Copyright 2019 from Elsevier.

photo-dimerize while exposed to 365 nm light and photo-cleavage upon 254 nm irradiation via the [2 + 2] cyclization reaction (Figure 1.14). This property allows release of an encapsulated dye to be controlled by capsules irradiation upon specific UV light.

1.6 Acknowledgements

Financial Support for Smartmem-Stimuli-Responsive Membranes for consumer goods sustainability project under European Community's Horizon 2020 ITN Marie Curie grant agreement no. 675624 is gratefully acknowledged.

1.7 References

[1] Zhang, J., Coulston, R. J., Jones, S. T., Geng, J., Scherman, O. A., Abell, C., One-step fabrication of supramolecular microcapsules from microfluidic droplets, Science **335** (2012) 690–694.
[2] Abbaspourrad, A., Carroll, N. J., Kim, S.-H., Weitz, D. A., Polymer microcapsules with programmable active release, J Am Chem Soc **135** (2013) 7744–7750.
[3] Asanuma, H., Liang, X., Nishioka, H., Matsunaga, D., Liu, M., Komiyama, M., Synthesis of azobenzene-tethered DNA for reversible photo-regulation of DNA functions: hybridization and transcription, Nat Protoc **2** (2007) 203–212.
[4] Vogel, A., Venugopalan, V., Mechanisms of pulsed laser ablation of biological tissues, Chem Rev **103** (2003) 577–644.
[5] Iijima, S., Helical microtubules of graphitic carbon, Nature **354** (1991) p. 56.
[6] Pastine, S. J., Okawa, D., Zettl, A., Fréchet, J. M. J., Chemicals on demand with phototriggerable microcapsules, J Am Chem Soc **131** (2009) 13586–13587.
[7] Caruso, M. M., Schelkopf, S. R., Jackson, A. C., Landry, A. M., Braun, P. V., Moore, J. S., Microcapsules containing suspensions of carbon nanotubes, J Mater Chem **19** (2009) 6093–6096.
[8] Paunov, V. N., in het Panhuis, M., Fabrication of carbon nanotube-based microcapsules by a colloid templating technique, Nanotechnology **16** (2005) 1522–1525.
[9] Yashchenok, A., Masic, A., Gorin, D., Inozemtseva, O., Sup Shim, B., Kotov, N., Skirtach, A., Möhwald, H., Optical heating and temperature determination of core-shell gold nanoparticles and single-walled carbon nanotube microparticles, Small **11** (2015) 1320–1327.
[10] Saito, H., Kato, N., Polyelectrolyte/carbon nanotube composite microcapsules and drug release triggered by laser irradiation, Jpn J Appl Phys **55** (2016) 1–4.
[11] Cheng, L., Wang, C., Feng, L., Yang, K., Liu, Z., Functional nanomaterials for phototherapies of cancer, Chem Rev **114** (2014) 10869–10939.
[12] Song, Y., Liu, G., Wang, J., Dong, X., Yu, W., Synthesis and luminescence resonance energy transfer based on noble metal nanoparticles and the NaYF4:Tb3+ shell, Phys Chem Chem Phys **16** (2014) pp. 15139–15145.
[13] Skirtach, A. G., Antipov, A. A., Shchukin, D. G., Sukhorukov, G. B., Remote activation of capsules containing Ag nanoparticles and IR dye by laser light, Langmuir **20** (2004) 6988–6992.
[14] Skirtach, A. G., Muñoz Javier, A., Kreft, O., Köhler, K., Piera Alberola, A., Möhwald, H., Parak, W. J., Sukhorukov, G. B., Laser-induced release of encapsulated materials inside living cells, Angew Chem Int Ed **45** (2006) 4612–4617.
[15] Bukreeva, T. V., Parakhonsky, B. V., Skirtach, A. G., Susha, A. S., Sukhorukov, G. B., Preparation of polyelectrolyte microcapsules with silver and gold nanoparticles in a shell and the remote destruction of microcapsules under laser irradiation, Crystallogr. Rep. **51** (2006) 863–869.
[16] Radziuk, D., Shchukin, D. G., Skirtach, A., Möhwald, H., Sukhorukov, G., Synthesis of silver nanoparticles for remote opening of polyelectrolyte microcapsules, Langmuir **23** (2007) 4612–4617.
[17] Radt, B., Smith, T. A., Caruso, F., Optically addressable nanostructured capsules, Adv Mater **16** (2004) 2184–2189.
[18] Angelatos, A. S., Radt, B., Caruso, F., Light-responsive polyelectrolyte/gold nanoparticle microcapsules, J Phys Chem B **109** (2005) 3071–3076.
[19] Volodkin, D. V., Delcea, M., Möhwald, H., Skirtach, A. G., Remote near-IR light activation of a hyaluronic acid/poly(L-lysine) multilayered film and film-entrapped microcapsules, ACS Appl Mater Interfaces **1** (2009) 1705–1710.
[20] Volodkin, D. V., Skirtach, A. G., Möhwald, H., Near-IR remote release from assemblies of liposomes and nanoparticles, Angew Chem Int Ed **48** (2009) 1807–1809.
[21] Yang, X., Liu, X., Liu, Z., Pu, F., Ren, J., Qu, X., Near-infrared light-triggered, targeted drug delivery to cancer cells by aptamer gated nanovehicles, Adv Mater **21** (2012) pp. 2890–2895.

[22] Kim, R., *et al.*, Extremely deep blue and highly efficient non-doped organic light emitting diodes using an asymmetric anthracene derivative with a xylene unit, Chem Commun **49** (2013) pp. 4664–4666.
[23] Sershen, S. R., Westcott, S. L., Halas, N. J., West, J. L., Temperature-sensitive polymer-nanoshell composites for photothermally modulated drug delivery, J Biomed Mater Res **51** (2000) pp. 293–298, 2000.
[24] Lee, M. H., Hribar, K. C., Brugarolas, T., Kamat, N. P., Burdick, J. A., Lee, D., Harnessing interfacial phenomena to program the release properties of hollow microcapsules, Adv Funct Mater **22** (2012) pp. 131–138.
[25] Jeong, W.-C., Kim, S.-H., Yang, S.-M., Photothermal control of membrane permeability of microcapsules for on-demand release, ACS Appl Mater Interfaces **6** (2014) pp. 826–832.
[26] Chen, H., *et al.*, Combined chemo- and photo-thermal therapy delivered by multifunctional theranostic gold nanorod-loaded microcapsules, Nanoscale **7** (2015) pp. 8884–8897.
[27] Wang, L., *et al.*, Light-triggered generation of multifunctional gas-filled capsules on-demand, J Mater Chem C **4** (2015) pp. 652–658.
[28] Peralta, D. V., Heidari, Z., Dash, S., Tarr, M. A., Hybrid paclitaxel and gold nanorod-loaded human serum albumin nanoparticles for simultaneous chemotherapeutic and photothermal therapy on 4T1 breast cancer cells, ACS Appl Mater Interfaces **7** (2015) pp. 7101–7111.
[29] Xu, S., Shi, J., Yang, L., Wu, Q., Cao, S., Hollow PUA/PSS/Au microcapsules with interdependent near-infrared/pH/temperature multiresponsiveness, J Appl Polym Sci **133** (2016).
[30] Dahl, M., Liu, Y., Yin, Y., Composite titanium dioxide nanomaterials, Chem Rev **114** (2014) pp. 9853–9889.
[31] Jiang, Y., *et al.*, Facile in-situ solvothermal method to synthesize double shell $ZnIn_2S_4$ nanosheets/TiO_2 hollow nanosphere with enhanced photocatalytic activities, Ceram Int **44** (2018) pp. 6115–6126.
[32] Jiang, Y., Li, F., Liu, Y., Hong, Y., Liu, P., Ni, L., Construction of TiO_2 hollow nanosphere/g-C3N4 composites with superior visible-light photocatalytic activity and mechanism insight, J Ind Eng Chem **41** (2016) pp. 130–140.
[33] Jiang, Y., *et al.*, Facile in-situ solvothermal method to synthesize double shell $ZnIn_2S_4$ nanosheets/TiO_2 hollow nanosphere with enhanced photocatalytic activities, Ceram Int **44** (2018) pp. 6115–6126.
[34] Reddy, K. R., Hassan, M., Gomes, V. G., Hybrid nanostructures based on titanium dioxide for enhanced photocatalysis, Appl Catal Gen **489** (2015) pp. 1–16.
[35] Hu, Y., Ge, J., Sun, Y., Zhang, T., Yin, Y., A self-templated approach to TiO_2 microcapsules, Nano Lett **7** (2007) pp. 1832–1836.
[36] Guo, X.-F., Kim, Y.-S., Kim, G.-J., Fabrication of SiO_2, Al_2O_3, and TiO_2 microcapsules with hollow core and mesoporous shell structure, J Phys Chem C **113** (2009) pp. 8313–8319.
[37] Buonomenna, M. G., Figoli, A., Spezzano, I., Morelli, R., and Drioli, E., Combined emulsion and phase inversion techniques for the preparation of catalytic PVDF microcapsules, J Phys Chem B **112** (2008) pp. 11264–11269.
[38] Katagiri, K., Koumoto, K., Iseya, S., Sakai, M., Matsuda, A., and Caruso, F., Tunable UV-responsive organic–inorganic hybrid capsules, Chem Mater **21** (2009) pp. 195–197.
[39] Wang, X., *et al.*, Photocatalytic surface-initiated polymerization on TiO_2 toward well-defined composite nanostructures, ACS Appl Mater Interfaces **8** (2016) pp. 538–546.
[40] Li, M.-H., Keller, P., Stimuli-responsive polymer vesicles, Soft Matter **5** (2009) pp. 927–937.
[41] Tao, X., Li, J., and Möhwald, H., Self-assembly, optical behavior, and permeability of a novel capsule based on an azo dye and polyelectrolytes, Chem – Eur J **10** (2004) pp. 3397–3403.
[42] Bédard, M. F., De Geest, B. G., Skirtach, A. G., Möhwald, H., Sukhorukov, G. B., Polymeric microcapsules with light responsive properties for encapsulation and release, Adv Colloid Interface Sci **158** (2010) pp. 2–14.

[43] Koňák, Č., Rathi, R. C., Kopečková, P., and Kopeček, J., Photoregulated association of water-soluble copolymers with spirobenzopyran-containing side chains, Macromolecules **30** (1997) pp. 5553–5556.
[44] Kono, K., Nishihara, Y., Takagishi, T., Photoresponsive permeability of polyelectrolyte complex capsule membrane containing triphenylmethane leucohydroxide residues, J Appl Polym Sci **56** (1995) pp. 707–713.
[45] Yuan, X., Fischer, K., Schärtl, W., Photocleavable microcapsules built from photoreactive nanospheres, Langmuir **21** (2005) pp. 9374–9380.
[46] Chang, S. J., Lee, C. H., Wang, Y. J., Microcapsules prepared from alginate and a photosensitive poly(L-lysine), J Biomater Sci Polym Ed **10** (1999) pp. 531–542.
[47] Fomina, N., McFearin, C., Sermsakdi, M., Edigin, O., Almutairi, A., UV and near-IR triggered Release from polymeric nanoparticles, J Am Chem Soc **132** (2010) pp. 9540–9542.
[48] Fomina, N., Sankaranarayanan, J., Almutairi, A., Photochemical mechanisms of light-triggered release from nanocarriers, Adv Drug Deliv Rev **64** (2012) pp. 1005–1020.
[49] Wang, Z., Wang, P., and Tang, X., Synthesis of light-induced expandable photoresponsive polymeric nanoparticles for triggered release, ChemPlusChem **78** (2013) pp. 1273–1281.
[50] Lv, J., Liu, Y., Wei, J., Chen, E., Qin, L., Yu, Y., Photocontrol of fluid slugs in liquid crystal polymer microactuators, Nature **537** (2016) pp. 179–184.
[51] Azagarsamy, M. A., Alge, D. L., Radhakrishnan, S. J., Tibbitt, M. W., Anseth, K. S., Photocontrolled nanoparticles for on-demand release of proteins, Biomacromolecules **13** (2012) pp. 2219–2224.
[52] Koo, H. Y., Lee, H.-J., Kim, J. K., Choi, W. S., UV-triggered encapsulation and release from polyelectrolyte microcapsules decorated with photoacid generators, J Mater Chem **20** (2010) pp. 3932–3937.
[53] Jiang, N., Cheng, Y., Wei, J., Coumarin-modified fluorescent microcapsules and their photo-switchable release property, Colloids Surf Physicochem Eng Asp **522** (2017) pp. 28–37.

Rita Del Pezzo, Anna Trojanowska, Dominico Pirone, Nuno A.G. Bandeira, Krzysztof Artur Bogdanowicz, Marta Giamberini and Bartosz Tylkowski

2 Light-sensitive microcapsules based on modified and un-modified azobenzene moieties

2.1 Introduction

Environmentally responsive materials have been the subject of great interest in the last two decades due to their versatile applications. Such materials are sometimes called "smart" because their properties allow them to react in a specific way to external stimuli, such as temperature, pH, light, ionic strength, and magnetic fields.

The previous chapter of this book was dedicated to an overview of different types of photo-stimuli materials that have been used for light-triggered release of encapsulated actives. In this chapter, we focus on photosensitive microcapsules, whose shells are based on azobenzene moieties. It is well known that aromatic azobenzenes are excellent candidates as molecular switches because they can exist in two forms: namely the *cis* (*Z*) and *trans* (*E*) isomers, which can interconvert both photochemically and thermally. This transformation induces a molecular movement and a significant geometric change; therefore, the azobenzene units are excellent candidates to build dynamic molecular devices. This strategy is very attractive in microencapsulation technology because it allows control over the conformation and, consequently, the release of the encapsulated active, such as drugs, perfume, etc., not only in required time but also in a reversible way without the addition of any reagent or different stimuli. According to results presented in literature and patents, development and testing of photo-control release microcapsules have had a significant impact on:

1. environmentally friendly production methods:
   - encapsulation and smart controlled release of crop protection agents, i.e. pesticides;
2. health protection:
   - encapsulation and smart release of protective substances (i.e. used as a main component of sun protection creams) only at the appropriate time – during sunlight illumination – in order to minimize their side effects;
   - encapsulation and smart release of pharmaceutics or supplements in order to protect their properties from the external environmental;
   - encapsulation and smart release of pharmaceutics or supplements in order to mask their taste;
   - increase of standard of life while decreasing the costs;
   - encapsulation and smart release of different active agents with future potential applications in electronics, textiles, catalysis, graphics and printing, the chemical industry, etc.

https://doi.org/10.1515/9783110642070-002

This chapter is divided into two sections. In the first two paragraphs, we discuss the photoisomerization processes of unmodified and modified azobenzene molecules, while in the second two, we describe selected examples of microcapsules whose shells are formed by materials containing ultraviolet (UV)-sensitive or white-light-sensitive moieties in their structures.

2.2 Photoisomerization of azobenzene

Azobenzenes are organic molecules that have two aromatic rings linked by an azo group (N=N). They have properties that have led to some applications of great importance, mainly for the chemical industry. The azobenzenes are highly colored compounds and belong to the group of so-called FD&C (food, drug, and cosmetics) dyes. Azobenzene was described for the first time in 1834 [1], and in 1937 – one century later – G.S. Hartley published a study on the influence of light on the configuration of N=N double bonds [2]. The exposure of a solution of azobenzene in acetone to light allowed the discovery of the *cis* isomer. This was the starting point of the development of one of the best organic molecular switches described so far. Nowadays, azobenzene dyes represent approximately 60% of the world production of industrial dyes [3]. Like a C=C double bond, azobenzenes have two geometric isomers (*Z*/*E*) around the N=N double bond, and the *trans* isomer (*E*) is ~12 kcal/mol more stable than the *cis* isomer (*Z*) [4]. The energy barrier of the photo-excited state is ~23 kcal/mol, such that the *trans* isomer is predominant in the dark at room temperature [5].

The *trans*-azobenzene easily isomerizes to the *cis* isomer by irradiation of the *trans* isomer with a wavelength between 320 and 350 nm. The reaction is reversible, and the *trans* isomer is recovered when the *cis* isomer is irradiated with light of 400–450 nm, or heated. For many azobenzenes, the two photochemical conversions occur on the scale of picoseconds, while the thermal relaxation of the *cis* isomer to the *trans* isomer is much slower (milliseconds to days). The photo-induced isomerization of the azobenzenes leads to a remarkable change in their physical properties, such as molecular geometry, dipole moment, or absorption spectrum [6, 7]. The isomerization process involves a decrease in the distance between the two carbon atoms in position 4 of the aromatic rings of azobenzene, from 9.0 Å in the *trans* form to 5.5 Å in the *cis* form (Figure 2.1) [8]. The *trans*-azobenzene is almost flat and has no dipole moment, whereas the *cis* isomer presents an angular geometry and a dipole moment of 3.0 D. One of the rings rotates to avoid steric repulsions caused by the facing of one of the π clouds of one aromatic ring to the other. The free volume requirement is that the *cis* is larger than the *trans*, with estimates of approximately 0.12 nm^3 required for isomerization to proceed via an inversion of the azo bond, and 0.28 nm^3 for a rotation about the azo bond [9]. The UV-Visible (vis) absorption spectrum of azobenzene presents two characteristics absorption bands corresponding to $\pi \rightarrow \pi^*$ and $n \rightarrow \pi^*$ electronic transitions. The transition $\pi \rightarrow \pi^*$ is usually in the

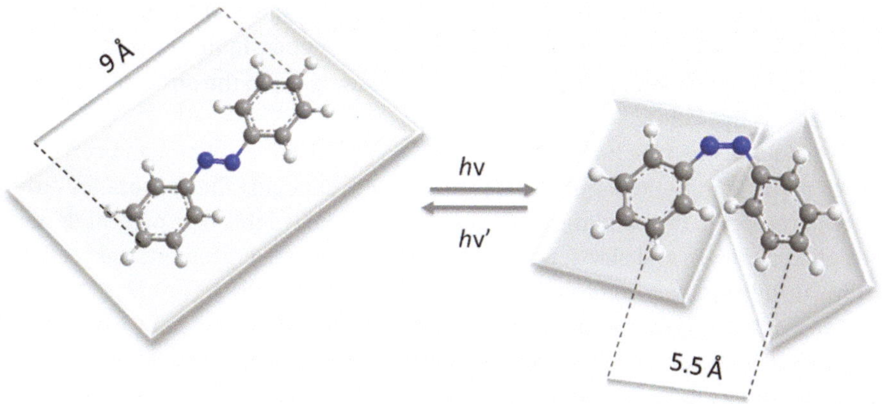

Fig. 2.1: Azobenzene photoisomerization. The *trans* form (left) can be converted to the *cis* form (right) using an appropriate wavelength (UV at 300–350 nm) of light. A different wavelength (visible blue light >400 nm) can be used to convert the molecule back to the *trans* form. Alternately, the molecule will thermally relax to the stable *trans* form.

near-UV region and is common to carbonate systems, such as stilbene [10]. The electronic transition n→π* is usually located in the visible region and is due to the presence of lone electron pairs of nitrogen atoms [11].

Due to this second electronic transition, the dynamic photoisomerization process of azobenzenes is different to the carbonate compounds [12]. Azobenzene undergoes *trans-cis* isomerization by S1←S0 and S2←S0 excitations and *cis-trans* isomerization by exciting into the S1 or S2 state. The sum of the quantum yields (QYs) is different to unity, which indicates multiple pathways for isomerization [9].

In the last years, the QYs of azobenzene photoisomerization in methanol solution as a function of temperature and at various wavelengths were redetermined using newly obtained molar absorption coefficients of its *cis* and *trans* isomers. The results were found to be not comparable with those published previously, especially in the range of the n→π* absorption band (from 250 to 550 nm), where the difference goes up to 44%; this disclosure lead to a full review of all the data collected in the past years. The method used in the work from Ladanyi et al. for the statistical analysis of spectrophotometric QY measurements is generally applicable [13].

The aromatic azo-compounds are classified into three types based on the order of their energetic electronic states π→π* and n→π* [5]. This order depends on the electronic nature of the aromatic rings of azobenzene.

Each type of azobenzene also has a predominant color defined by the wavelength of the maximum absorption band (λ_{max}) (indicated in brackets in each case):

- Azobenzene type: the π→π* band is very intense in the UV region and there is one n→π* weaker in the visible region (yellow color). The electronic nature of the aromatic rings is very similar to simple azobenzene (Ph–N=N–Ph).

- Aminoazobenzene type (*o*- or *p*-(X)–C_6H_4–N=N-Ar): the π→π* and n→π* bands are very close or collapsing in the UV-vis region. In this case, the azo-compounds have electron-donor substituents (X) in the *ortho* or *para* positions (orange color).
- Pseudo-stilbene type [(X)–C_6H_4–N=N–C_6H_4–(Y)]: the absorption band corresponding with π→π* *transi*tion is shifted to red, changing the appearance order with respect to the band n→π*. The azo-compounds of this type present donor substituents (X) and electron acceptors (Y) at the 4 and 4′ positions, respectively (push/pull system) (red color).
- The isomerization process normally involves a color change to more intense colors. The absorption spectra of both isomers differ mainly in the following aspects (see Figure 2.2) [14]:
- *Trans* isomer: the absorption band π→π* is very intense, with a molar extinction coefficient (ε) ~ 2–3 × 104 $M^{-1}·cm^{-1}$. The second band (n→π*) is much weaker (ε ~ 400 $M^{-1}·cm^{-1}$) as this transition is not allowed in the *trans* isomer by symmetry rules.
- *Cis* isomer: the absorption band π→π* is shifted to shorter wavelengths (hypochromic effect), decreasing significantly in intensity (ε ~ 7–10 × 103 $M^{-1}·cm^{-1}$). The electronic *transi*tion π→π* (380–520 nm) is allowed in the *cis* isomer, resulting in an increase in the intensity (ε ~ 1500 M–1·cm–1) with respect to the *trans* isomer.

These differences allow a photochemical interconversion by irradiation with light of a certain wavelength, obtaining different proportions of the *cis*- and *trans*-photostationary states. The excitation caused by the wavelength is dependent on the nature of the substituents of the aryl groups. In most cases, *trans* → *cis* isomerization

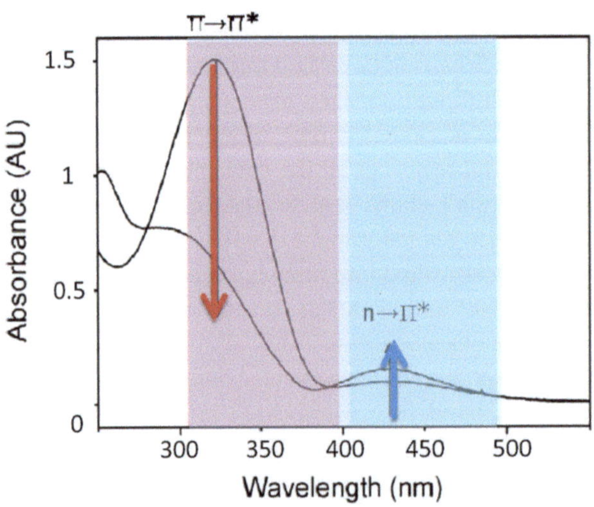

Fig. 2.2: Respective example of a UV spectrum of azobenzene compounds.

is promoted by irradiation with wavelengths between 320 and 380 nm, while exposures to $\lambda \sim 400$–450 nm favor the *cis* → *trans* photoreversion. The mechanism is not well established. Several mechanistic studies have been performed on the isomerization reversal route *cis* → *trans* of azobenzene to investigate the effect of the substituents on the benzene rings as well as the influence of several parameters. The available data suggest that the isomerization of azo-compounds can proceed through the reversal of one of the N–C bonds or by the rotation of the N=N bond. The nonbonding electron pair of each nitrogen atom may lead to one $n \rightarrow \pi^*$ electronic *trans*ition (S0→S1) with inversion at the nitrogen atom (inversion mechanism). Conversely, the isomerization can also occur through a rotation mechanism, which involves a $\pi \rightarrow \pi^*$ *trans*ition (S0→S2) (Figure 2.3) [9, 15, 16].

Taking into account that in the next paragraphs we will focus on the modification of the azobenzene molecule and on the incorporation into polymeric chains and shells, it is important to underline that the photoisomerization mechanism is highly affected from the environment in which the molecule is inserted. While azobenzenes readily photoswitch in solutions and thus in liquid state, where the molecular switch is unconstrained, their isomerization can be obstructed in electronically coupled and sterically hindered systems as in densely packed self-assembled monolayers (SAMs). By means of nonadiabatic molecular dynamics with trajectory surface hopping calculations, they found a very similar *trans* → *cis* QY for the free dimer ($\Phi = 0.19 \pm 0.05$) in comparison to the monomeric reference ($\Phi = 0.21 \pm 0.05$) and a decrease in the case of the SAM ($\Phi = 0.11 \pm 0.03$), concluding that the steric effects are the main reason why switching is suppressed [17].

2.3 Photoisomerization of modified azobenzene

As described in the first paragraph of this book, the class of reversible light-switchable molecules is very wide and, besides molecules which photo dimerize or which allow intramolecular photo-induced bond formation, more attention will

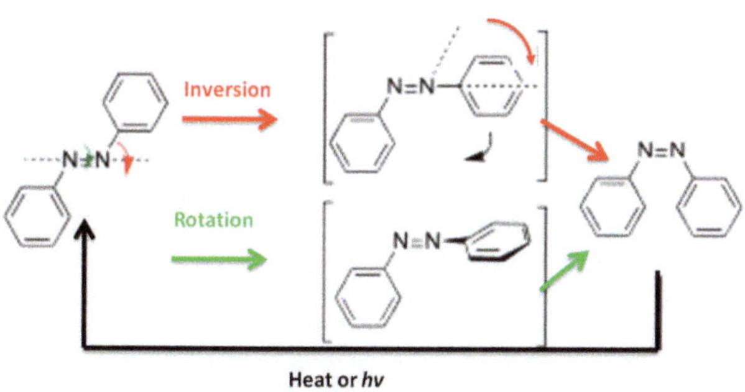

Fig. 2.3: Mechanistic proposals for the isomerization of azobenzenes.

be focused, in this book, on those that exhibit photo-isomerization; this family is further divided into the azobenzene derivative, the amino-azobenzene derivatives, and the push-pull azobenzene, according to the Rau classification of azobenzenes (Figure 2.4) [4].

Those compounds all exhibit the central azobenzene structure while they differ for the ring substitution patterns; this helps to hold the common properties of azobenzene such as the strong electronic absorption, while the ring substitution leads to a modification of the absorption spectrum with a shift in the absorption wavelength [18]; moreover, depending on the nature of the substituents on the aromatic group, the equilibria between the *trans* and *cis* isomers are variable.

The UV light photoisomerization of most azobenzenes can be considered as a limit for a wide range of applications; for example, in applications *in vivo* or comprising the biological systems, the UV light is considered unhealthy and can trigger unwanted responses, including cellular apoptosis.

This is the main reason that the attention is particularly focused nowadays on the modification of these molecules and on the shift of the isomerization at different (higher) wavelengths. Photoisomerization that can occur entirely in the visible region would therefore be desirable and considered as a benefit in many fields [19].

Many approaches for achieving long-wavelength switching have been tested as, for example, the coupling of azobenzene derivatives with upconverting nanophosphors [20] or the incorporation of electron-donating groups [21], which can dramatically red-shift the photoswitching wavelength.

The review from Hamon et al. provides a comprehensive overview of the synthesis of azobenzene derivatives and of azo-linked carbohydrate moieties in the period from 1998 to 2008 [22]. One of those approaches, the reduction reactions of aromatic compounds having nitro groups, has been used by Tylkowski et al. [23] to obtain an azobenzene modified with two electron donor groups. In this work, the synthesis of azobenzene derivative has been conducted via reduction of nitro compounds using glucose as a carbohydrate reducing agent; the reduction furnished an hydroxylamine as a reaction intermediate, as proposed

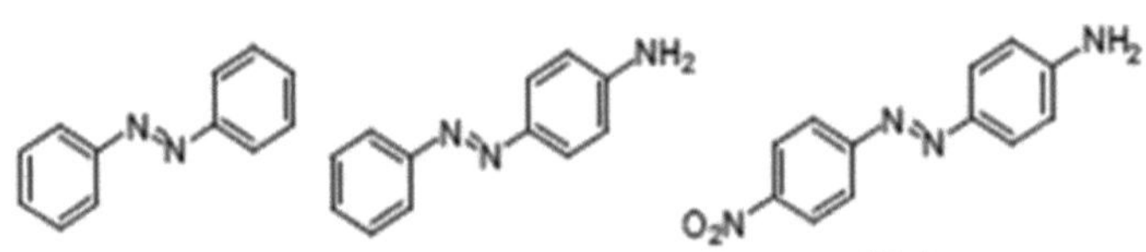

(*E*)-1 azobenzene (*E*)-2 aminoazobenzene (*E*)-3 push-pullazobenzene

Fig. 2.4: Classification of azobenzenes compounds: unmodified azobenzene, azobenzene functionalized with one electron-donor group (es. -NH_2) (aminoazobenzenes), and azobenzene functionalized with both electron-donor and electron aceptor groups (es. -NH_2 and NO_2) (push-pull azobenzenes). Reprinted with permission from [19]. Copyright 2019 American Chemical Society.

by Gowda et al. in 2003 [24]. The hydroxylamine is then modified into an azo-dicarboxylic acid via further oxidation. Its subsequent conversion into the acylic chloride is propaedeutic for the preparation of photosensitive microcapsules, which will be analyzed in the next paragraph [23].

A large number of azobenzene derivatives are known in which enhanced electron-donating nature of ring substituents increases both the wavelength of absorption of the *trans* isomer and the rate of thermal back-isomerization from *cis* to *trans*, so that the most electron donor groups substituents exhibit the longest wavelength absorption, up to 500 nm (Figure 2.5) [25].

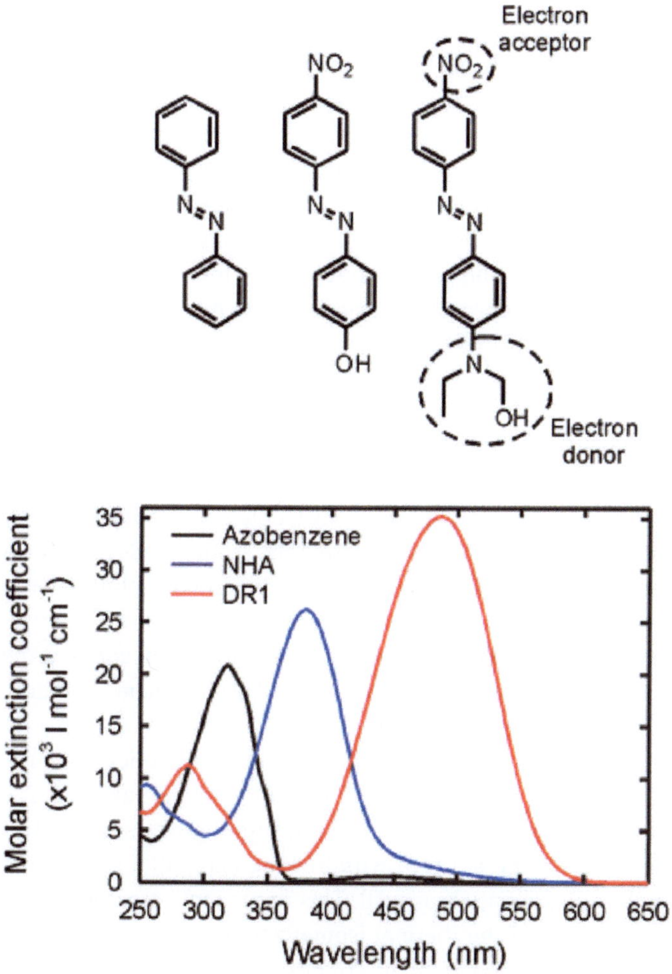

Fig. 2.5: Examples of azobenzenes and their absorption spectra measured from dilute tetrahydrofuran solution. The compounds are classified according to the weak or strong electron-donor behavior of the substituents and, hence, to the absorption spectrum. Reprinted with permission from [18].

This phenomenon has been attributed to similarities between the electronic excited state of the *trans* isomer and the thermal transition state of back-relaxation; both species, in fact, have strong dipolar character. The dipolar nature of the transition state also results in strong solvent sensitivity of the half-life for thermal relaxation. In general, azobenzene photoswitches that absorb at long wavelengths in water relax very quickly back to the *trans* state so that only vanishingly small amounts of the *cis* isomer can be produced under low power steady-state illumination [26].

Every switchable sample in fact will reach, under illumination, an equilibrium that is called photo stationary state associated with a steady-state *trans-cis* composition. Such composition depends on the competing effects of photoisomerization, thermal relaxation, and *cis* reconversion upon light absorption. The steady-state composition is unique to each system, as it depends on the QYs for the two processes and the thermal relaxation rate constant. The kinetics of the isomerization and thermal relaxation processes, as well as the percentages of *cis* or *trans* isomers, can be conveniently studied via absorption spectroscopy or Nuclear Magnetic Resonance (NMR) spectroscopy can also be used [27, 28]. Under moderate irradiation, the composition of the photo stationary state is predominantly *cis* for azobenzenes, mixed for aminoazobenzenes, and predominantly *trans* for pseudo-stilbenes, but for certain patterns, where the absorption spectra of the two isomers overlap significantly, a single wavelength of light activates both the forward and reverse reactions, leading to a mixed stationary state and continual interconversion of the molecules, notably for the pseudo-stilbenes [18].

All the interesting applications of azobenzenes are linked to its facile and reversible isomerization around the azo bond, between the more thermally stable E configuration and the metastable Z form; thus, it is important to deeply understand and study the two processes and to design the molecule accordingly. In the work from Jerca et al. [29], a detailed isomerization study of a series of substituted azobenzenes is conducted by UV-vis spectroscopy in different solvents. They showed that isomerization and relaxation speeds strongly depend on the polarity of the solvent and how the mechanisms of isomerization (rotation or inversion) depend on the chemical nature of the substituents and on the solvent polarity.

Through spectroscopic measurements, they developed a theoretical calculation for the study of the kinetics of photoisomerization and thermal relaxation. The kinetic parameters determined for the thermal Z to E relaxation processes confirms that the mechanism depends on the electron withdrawing capability of the substituent groups and on the solvent nature as well. This approach was used for the calculation of the energy barrier of the thermal isomerization process, via UV-vis spectroscopy or NMR study (Figure 2.6), in many studies, and its value always falls in the range of 90–150 kJ/mol [18, 25, 30].

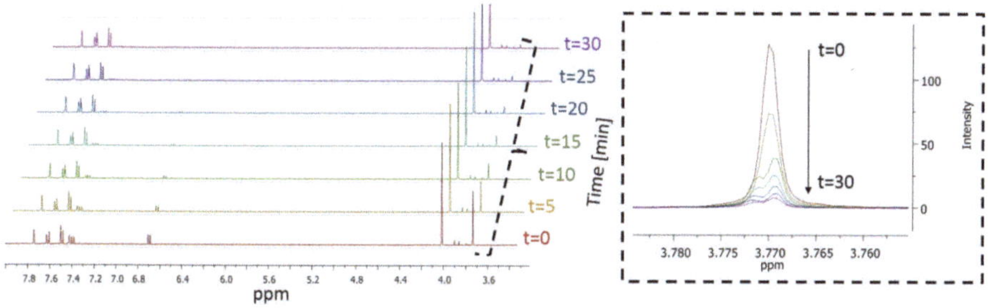

Fig. 2.6: Kinetic of thermal back isomerization calculated via 1H-NMR at 70 °C starting from a mixture of *trans* and *cis* isomers of an ortho-substituted azobenzene. The activation energy E_a was found to be equal to 105.6 KJ/mol.

2.4 UV-sensitive microcapsules based on azobenzene moieties

Using the isomerization of azobenzene and its derivatives to modulate the structure of delivery systems, and thus to trigger the light-induced release of various compounds, has been reported in the literature. Different methods and concepts have been used for microcapsules preparation.

2.4.1 Liposome microcapsules

The first study of incorporating azobenzene moieties in a photoresponsive system to affect release was published by Kano et al. in 1980 [31]. The authors incorporated an amphiphilic azobenzene moiety, along with dipalmitoylphosphatidylcholine (DPPC), at various molar ratios and were able to modulate the release profiles of liposomes based on the azo moiety of choice, the composition of photo stationary state, and the degree of incorporation in the liposome. The authors characterized the photoisomerization process using a UV spectroscopy by illuminating the *trans*-azo compound at 366 nm for 10 s. The *trans*-compound formed a photo-stationary state with 80% *cis* isomer, which reverts back to *trans* when irradiated at >420 nm. The authors also studied the resulting osmotic shrinkage of the vesicles upon incorporation of azo compound by measuring the optical density of the solutions. They encapsulated bromothymol, a blue dye, in the lipid bilayer of liposomes formulated from DPPC and subsequently showed that the permeation of the dye into water increased with greater incorporation of the *cis* azobenzene moiety (formed by irradiation). Unfortunately, in these pioneering studies, percent release and duration of release upon pulsing were not entirely characterized.

Since this seminal study, there have been numerous publications utilizing this concept. Many systems developed since have incorporated azobenzenes in lipid backbones and formulated liposomes that are photoresponsive [32–34].

The photo-responsiveness of the liposomes arises from the fact that in the *trans* configuration, the molecules pack tightly in the bilayer. When irradiated with UV light, they undergo *trans-cis* isomerization, which leads to distortions in the packing of the bilayer and causes the liposomes to become "leaky," allowing the encapsulated drugs to be released. Irradiation of azobenzene results in the formation of a photostationary state, and the composition of this state determines the release rate of the drug. More recently, Smith et al. have used phototriggerable liposomes to trigger gelation of an alginate solution by releasing calcium chloride upon irradiation with 385 nm light for 1 min. Such on-demand gelation is important in tissue engineering applications [35].

The photoisomerization concept has also been successfully utilized in the preparation of photoresponsive micelles. These systems take advantage of the change in net dipole moment upon switching from the *trans* orientation (no net dipole moment) to the *cis* orientation. This leads to disruption in the hydrophobichydrophilic balance of the self-assembled micelles and causes reorganization and subsequent release of encapsulated contents [35].

2.4.2 Self-assembly microcapsules

Self-assembly of amphiphilic block copolymers induces the formation of nanosized polymeric microcapsules or micelles, which have been widely explored as carriers for enzymes or nonbiological catalysts as well as containers for drug or gene delivery. As amphiphilicity is the principal basis of such self-assembly, some approaches have been developed to modulate the polymeric assemblies for controlled drug delivery through tuning the amphiphilicity of the block copolymers. Generally, the drug species encapsulated in or attached to the polymeric assemblies can be released via reversible or irreversible disassembly of the hydrophobic core-forming segments. However, the self-assembly of amphiphilic block copolymers usually involves the use of organic solvents and suffers from complicated preparation processes. More than a decade ago, Kataoka's group [36] developed a method for preparing block copolymer assemblies on the basis of electrostatic interactions. This new family of polymeric assemblies is formed by double-hydrophilic block copolymers, containing ionic and nonionic water-soluble segments (block ionomers), and can incorporate many charged polymers, including synthetic polyions, enzymes, DNA, RNA, and others [37]. One great advantage of this approach is that such assemblies are formed in water, and no organic solvent is required for their preparation. Moreover, the block ionomers with appropriate molecular weight and composition can also form micelle-like or vesicle like aggregates.

The basic mechanism of the formation of such polymeric assemblies involves the core precipitation of the charged blocks of block ionomers with the oppositely charged polyions. Besides the block ionomer-polyion systems, Kabanov et al. [38] proposed a simple method to prepare block ionomer complexes by electrostatic complexation of block ionomers with oppositely charged surfactants. Such a block ionomer complex can be depicted as an amphiphilic supramolecular block copolymer, in which the nonionic block functions as the hydrophilic part while the electrostatic complex of the ionic block and aggregated surfactant counterions serves as the hydrophobic part. It is known that by introducing stimuli-responsive moieties such as azobenzene onto the surfactants, the surfactant aggregates can be tuned toward controllable disassembly.

Wang et al. [37] for the first time demonstrated the possibility of controlled self-assembly and disassembly of the block ionomer and surfactant microcapsules through tuning the amphiphilicity of the surfactants. The authors fabricated the UV-responsive microcapsules through an electrostatic association between an azo-containing surfactant and a double-hydrophilic block ionomer, poly-(ethylene glycol)-b-poly(acrylic acid) (PEG43-PAA153), as shown in Figure 2.7. They found that loading and release of fluorescent molecules included in the microcapsules could be achieved by reversible self-assembly and disassembly under UV light irradiation. By introducing stimuli-responsive surfactants into the block ionomer microcapsule complex, one may broaden the application of this new class of supramolecular materials. Such novel materials containing azobenzene moieties in their structures are of both basic and practical significance, especially as prospective nanocontainers for drug delivery.

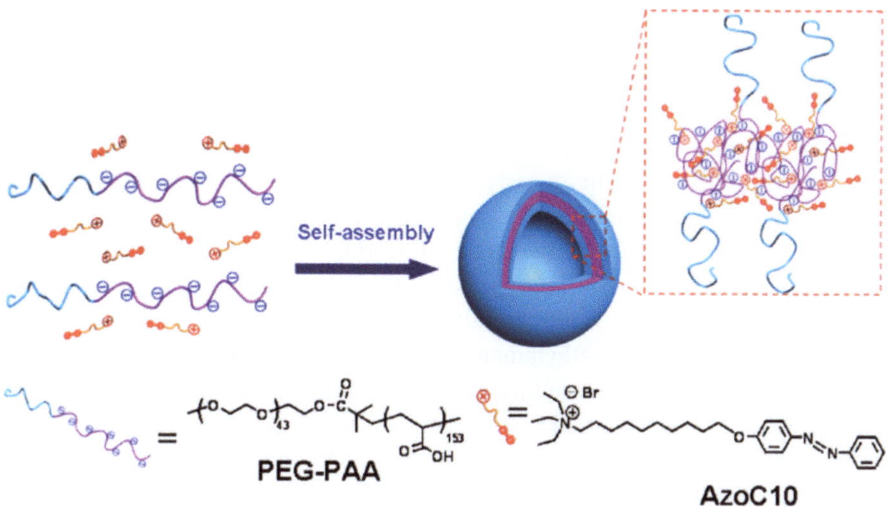

Fig. 2.7: Schematic illustration of the self-assembly of the block ionomer microcapsules. Reprinted with permission from [37].

Self-assembling processes to form azo polyurethane microspheres have been employed by Zhou et al. [39] Fabrication of the beads has been induced by slowly increasing the water content in DMF-H_2O mixture where azo polyurethanes have been dispersed. Obtained structures have been thoroughly investigated via laser light scattering and transmission electron microscopy (TEM) to prove their photo-induced deformation behavior. For this reason, the spheres have been irradiated with linearly polarized light such as Ar^+ laser (488 nm) and diode solid state laser (532 nm), thus allowing the azo polyurethane microspheres to elongate in the light polarization direction. The authors investigated four types of azo polyurethane polymers, each one containing different substituent (-CF_3, -COOH, -COOEt, and -NO_2). The obtained results suggested that the carboxyl and trifluoromethane groups showed the largest deformation degrees when irradiated with 488 and 532 nm light, respectively. Y. Zhou et al. concluded that the microspheres of polyurethanes functionalized with azo chromophores can be used as a new type of switchable photoresponsive materials.

2.4.3 Layer-by-layer microcapsules

As shown in this book, numerous strategies have been used to develop microcapsules with different functionalities. One of the simple but functional methods is the so-called layer-by-layer (LbL) assembly technique. By using the electrostatic interactions between oppositely charged polyelectrolytes, the LbL approach offers diversified multilayer capsule systems with controllable architectures and properties. Moreover, the stepwise polymer deposition procedure facilitates the functionalization of the capsule formations; a typical example is cargo substance encapsulation [40]. Polyelectrolytes are basic components for LbL capsule fabrication. Diverse polyelectrolytes have been used to build up capsules. Different combinations of the oppositely charged polyelectrolytes with active functional groups endow their capsules with unique properties, which would affect their further applications. In an early example, Möhwald and coworkers reported that LbL capsules containing an azo dye in their shell allowed photochemical control of the permeability of the shell [41]. In this study, the microcapsule shell was composed of an azo dye – Congo red (CR) – and different polymers, including poly(styrenesulfonate, sodium salt) (PSS), poly-(allylamine hydrochloride) (PAH), and poly(diallyldimethylammonium chloride) (PDDA). Figure 2.8 shows the general procedure of LbL self-assembly of PDDA/CR onto the (PSS/PAH)3/PSS shells templated on MF latex particles.

In order to observe morphological changes in the microcapsules shell before and after irradiation, the authors used scanning force microscopy (SFM). Moreover, the optical changes in the capsules were verified by using confocal laser scanning

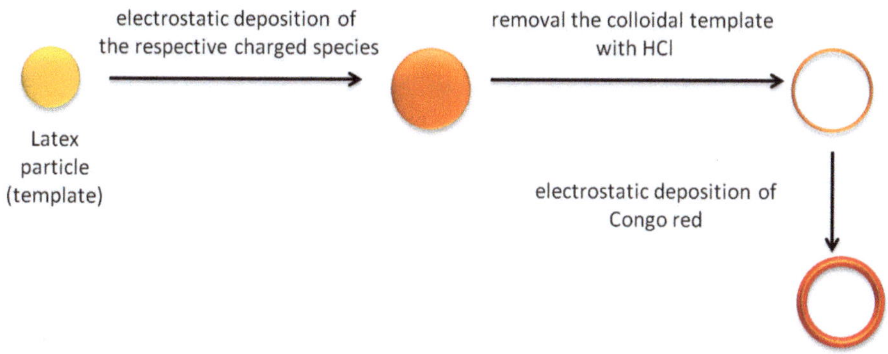

Fig. 2.8: General procedures for the fabrication of hollow capsules composed of CR and polyelectrolytes [41].

microscopy and SFM. All results obtained by the authors provide useful insights into the photochemical reaction mechanisms on the self-assembled PDDA/CR composite capsules and release of encapsulated material.

This kind of capsule with photocontrolled permeability could be of particular interest for applications in drug delivery, photocatalysis, optical materials, and related medical areas such as photodynamic therapy or skin care.

Bédard et al. [11] also constructed the microcapsules containing azobenzene moieties through LbL self-assembly of sodium salt of azobenzene, poly(vinylsulfonate) (PVS), and poly(allyamine hydrochloride); however, contrary to Möhwald and coworkers, they investigated how *trans-cis* isomerization of the azo moieties influences the permeability changes of the shell and on encapsulation of the active material instead of its release during light irradiation. According to the authors, incorporation of azobenzene groups can cause shrinking of the microcapsule wall, increase their permeability, and as a consequence encapsulate required materials.

More recent examples using stimuli-responsive capsules based on azobenzene moieties in the capsule wall comprise the work of Yi and Sukhorukov [40–42], Lin et al. [43], and Xiao et al. [44].

Yi and Sukhorukov fabricated UV-responsive microcapsules containing azobenzene by sequential deposition of oppositely charged poly[1-[4-(3-carboxy-4-hydroxyphenylazo)benzenesulfonamido]-1,2-ethanediyl, sodium salt] (PAZO) and poly(diallyldimethyl ammonium) chloride (PDADMAC). The authors showed that the combination of PDADMAC and PAZO led to aggregation of PAZO segments in the progress of polymer deposition, which facilitated the large extent of J aggregates when the capsules were exposed to UV light.

Moreover, the same authors [42] developed a multifunctional capsule system using an electrostatic attraction LbL assembly, which can integrate both encapsulation and release processes in one system, simply triggered by only one external stimulus. Dual-functional complex microcapsules (PDADMAC/PAZO-)4-(DAR/Nafion)2 containing both diazonium and aozbenzene groups were proposed to realize a time-dependent UV response for successive encapsulation and release. Upon exposure to UV light, the DAR/Nafion layers underwent a rapid *in situ* crosslinking and hence sealed the capsule shells through diazonium-related photolysis. Then, further gradual shell swelling was followed by realignment of azobenzene molecules in PDADMAC/PAZO layers, as it is shown in Figure 2.9. Fluorescent polymers were consequently investigated as cargo substances. Results indicated continuous UV light-triggered rapid cargo encapsulation over a time scale of minutes and gradual release with continuous irradiation over hours. Morphological changes in the capsule shell during UV irradiation were investigated by scanning electron microscope (SEM) and are reported in Figure 2.10.

It is well known that the driving force for the LbL assembly of microcapsules mainly uses electrostatic attraction; however, this limits the building blocks to a narrow range of oppositely charged and water-soluble polymers.

Contrary to Yi and Sukhorukov or Möhwald and coworkers, both Xiao et al. and Lin et al. designed photoswitchable LbL capsules that use a supramolecular interaction as the driving force of LbL assembling and layer drug loading. The capsules were assembled by host polymeric layers containing α-cyclodextrin (CD) and guest polymeric layers containing azo. Using the supramolecular interaction instead of electrostatic interaction as the driving force of LbL, the authors have been able to enhance the stability of microcapsules in various pH conditions.

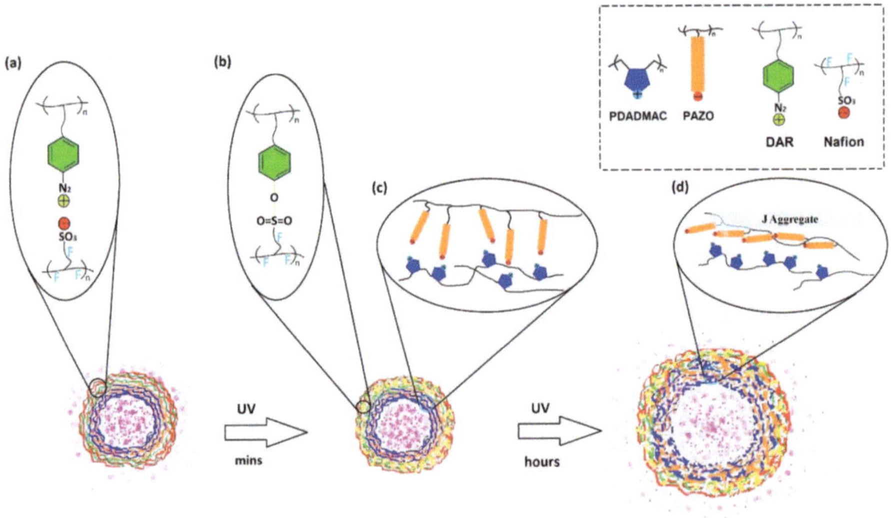

Fig. 2.9: Schematic illustration of UV-induced complex capsule shell sealing and further swelling. Reprinted with permission from [42].

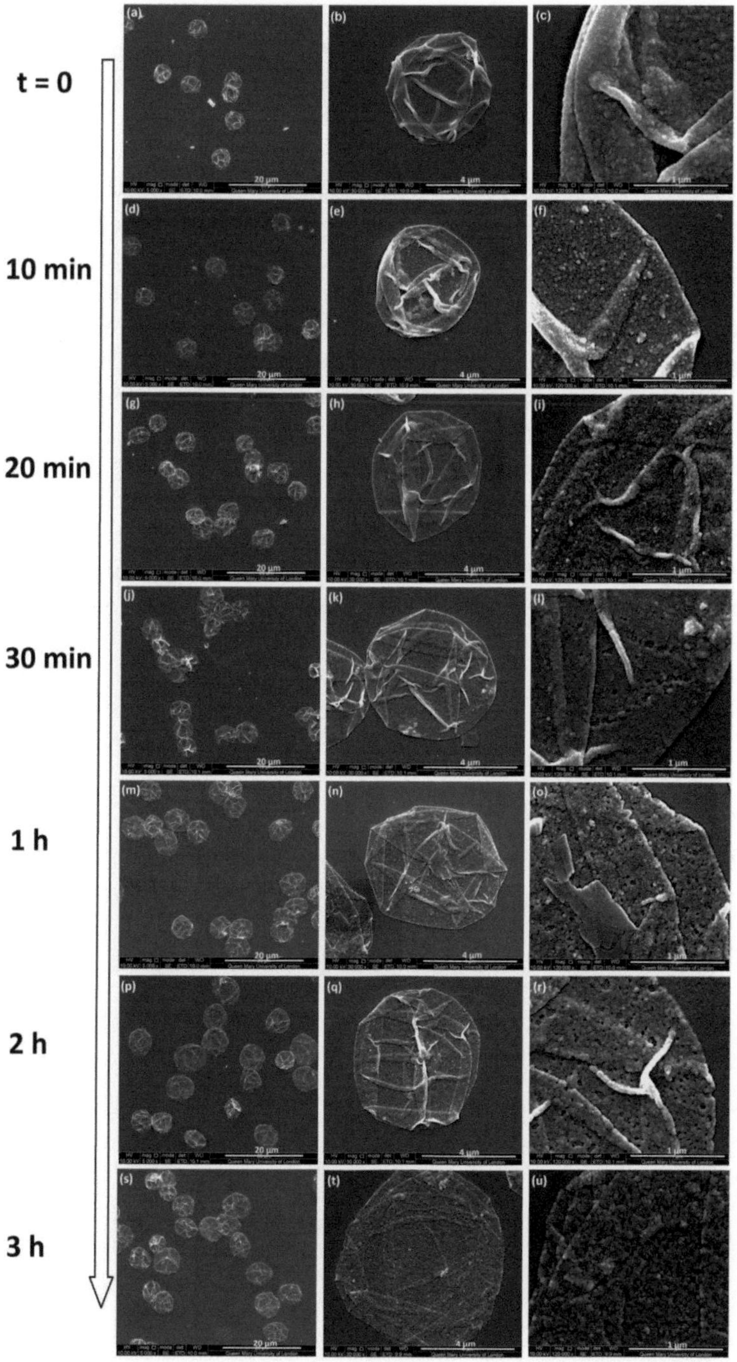

Fig. 2.10: Scanning electron microscope (SEM) images of complex microcapsules before (first row) and after UV irradiation of 10 min (second row), 20 min (third row), 30 min (fourth row), 1 h (fifth row), 2 h (sixth row), and 3 h (last row) at different magnifications. Reprinted with permission from [42].

Xiao et al. investigated UV-sensitive microcapsules based on host-guest interactions between carboxymethyl dextran-graft-α-CD (CMD-g-α-CD) and poly(acrylic acid) naminododecane p-azobenzeneaminosuc-cinic acid (PAA-C12-azo), which were assembled LbL on $CaCO_3$ particles. They used an antineoplastic drug modified with α-CD α-CD-rhodamine B (α-CD-RhB) as a model drug, which was loaded onto PAA-C12-azo layers by host-guest interaction. After removal of $CaCO_3$ particles by ethylenediaminetetraacetic acid, hollow microcapsules loaded with α-CD-RhB were obtained. Because the interactions between α-CD and azo were photosensitive, obtained capsules were dissociated upon irradiation by a UV lamp (λ = 365 nm) due to the transformation of *trans*-azo to *cis*-azo (see Figure 2.11). As a result, more than 60% of the drug was released from the microcapsules within 300 min of irradiation (Figure 2.12), while in a dark environment, the drug release was very slow and less than 5% of the drug was released in 300 min.

Using a similar method, Lin and coworkers designed microcapsules; however, these were much more "advanced" compared to those prepared by Xiao and coworkers. In this investigation the host-guest interactions were based not only between β-CD and azobenzene (like in Xiao's study) but also between β-CD and adamantane (AD). Prepared microcapsules were able to be controllably switched between the "on"

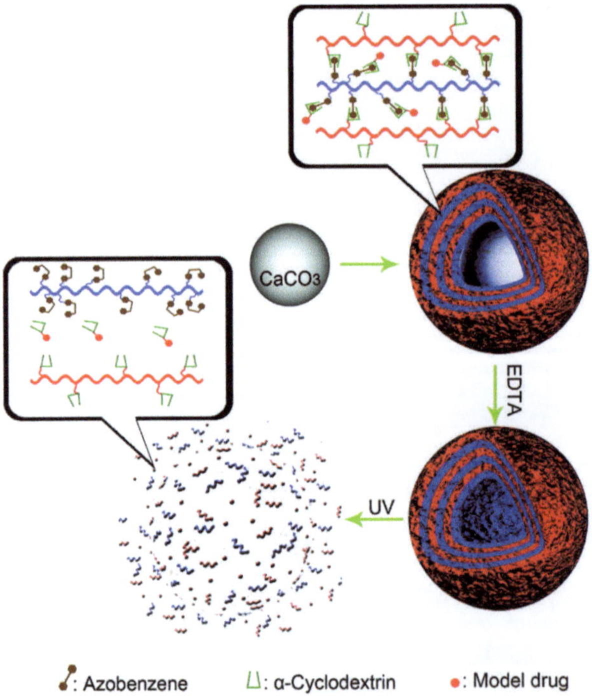

Fig. 2.11: LbL microcapsules prepared by Xiao et al. Reprinted with permission from [44].

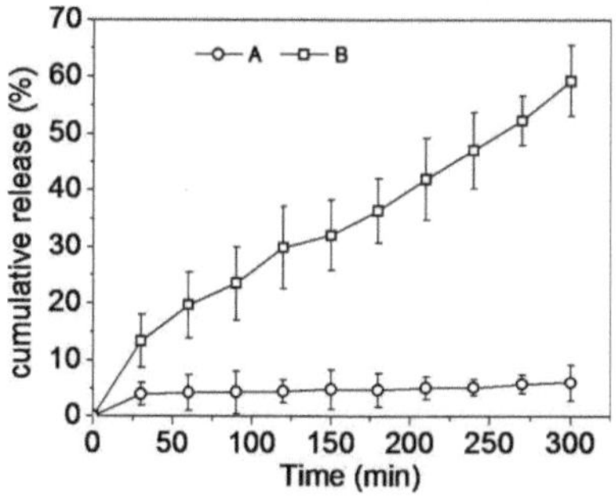

Fig. 2.12: Drug release behaviors of (PAA-C12-azo)5/(CMD-*g*-α-CD&α-CD-RhB)5 microcapsules in the dark (A) and under 365 nm UV light irradiation (B). Reprinted with permission from [44].

and "off" state. In this study, Lin et al. showed that the stable host-guest interactions between β-CD and AD maintained the structure as a permanent frame, while the reversible UV-sensitive ones between azo and β-CD could form a denser membrane to keep the drug inside. In order to prepare this type of microcapsules, the authors used two specific polymer chains: poly(acrylic acid-graft-azobenzene-graft-AD) (PAA-g-ADg-azo) and poly(aspartic acidgraft-b-CD) (PASP-g-β-CD). Poly(ethylene glycol)5000-graft-fluorescein isothiocyanate (PEG5000-FTIC) was loaded inside the microcapsules as a model drug. In the "off" state, *trans*-azo participates in the host-guest interaction with β-CD, and the model drug cannot pass. When the UV ray switches azobenzene moieties to the *cis*-state, the interaction between azobenzene and β-CD diminishes. The electrostatic repulsion between the negatively charged polyelectrolyte chains makes the membrane no longer dense enough to keep the model drug inside. As the photoisomerization of azobenzene moiety is reversible, the release process could be controlled by UV irradiation reversibly. Once the "on" state microcapsule is stimulated by visible light, it would switch back to the "off" state; i.e., the release could be ceased and recommenced controllably. These photosensitive microcapsules exhibit great potential in biomedical applications.

2.4.4 Interfacial polymerization microcapsules

To date, self-assembly techniques have been mainly used in the fabrication of microcapsules containing azobenzene units; however, the self-assembled microcapsules are generally not robust. Therefore, the stable, reversible photoresponsive microcapsules are highly desirable but little reported to the best of our knowledge. Tylkowski et al. [45] prepared new lightly crosslinked liquid crystalline polyamide microcapsules that contained azobenzene moieties in the main chain by using an interfacial polymerization method. The preparation procedures for this technology, also known as interfacial condensation, have been thoroughly described in the literature [46, 47].

Briefly, the microcapsule wall is formed from monomers that are dissolved in the two separate phases (oil and water phase) and they polymerize at the interface of the emulsion droplets. For example, monomers such as diamine can be dissolved in the water and the aqueous phase is dispersed in the oil phase. The second monomer that is oil-soluble, e.g. diacryl chloride, is then added and reacts with the first monomer at the interface forming the wall material. Different types of polymers may be produced by selecting different monomers, but most publications refer to polyamide membrane. Tylkowski et al. prepared microcapsules whose shell was constituted by liquid crystalline polyamide and contained either toluene as the core or concentrated solutions of naphthalene or β-carotene. According to the authors, obtained results were the first published example of microcapsules whose shell is completely constituted by a liquid crystalline lightly crosslinked polymer. They published results concerning the characterization of these microcapsules, which show that release could be easily triggered by irradiating with UV light at 364 nm for a few minutes. This suggested that microcapsules that meet the target of specific applications can be designed by optimizing characteristics such as the state of order of the shell, its range of thermal stability, and the structural changes that occur upon irradiation.

Figure 2.13 shows the release of β-carotene from these polyamide microcapsules in water at 20 °C, in the time range 0–5.5 min, in the absence (▲) and in the presence (■) of continuous irradiation with UV light, measured as described in the experimental part. The difference between the two curves is straightforward: in the absence of

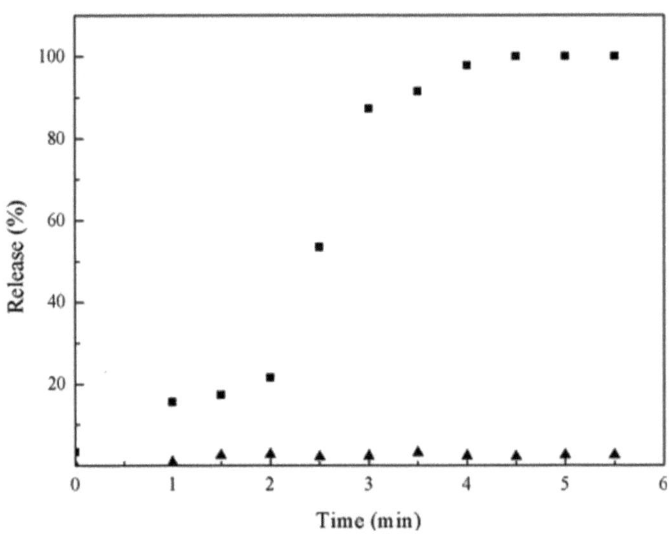

Fig. 2.13: Release kinetics of β-carotene from polyamide microcapsules in water at 20 °C in the absence (▲) and in the presence (■) of continuous irradiation with UV light. Reprinted with permission from Elsevier [45].

irradiation, release was practically negligible and the plateau value of about 2.5% reached in the first minutes remained constant even after 120 min of observation. Differently, when microcapsules were submitted to continuous irradiation with UV light, after an induction period of about 2 min, β-carotene was quickly released and reached its highest concentration value after 5 min. It is important to underline that the initial time corresponds to the UV lamp switch-on; the observed induction time can be therefore reasonably ascribed to the time needed by the polymer for rearranging in the isotropic structure, as a consequence of photoisomerization. Moreover, optical micrograph during the release of β-carotene from these polyamide microcapsules at room temperature after suspending 20 min in water (Figure 2.14a,b) clearly confirmed the occurrence of release: in the case of the sample suspended in water without irradiation, a trend toward the formation of large agglomerates of β-carotene (dark spots) into the core of microcapsules was observed (Figure 2.14a). However, when the sample was irradiated, microcapsules looked almost empty: moreover, they did not look broken or damaged, thus confirming that β-carotene release was due to a change in the barrier properties of the shell material as a consequence of UV irradiation.

Marturano et al. [48], by using miniemulsion interfacial polymerization, synthesized nanosized capsules also based on a lightly crosslinked polyamide containing azobenzene moieties in the main chain. The obtained nanocapsules were loaded either with toluene or with the fluorescent probe Coumarin-6 (dissolved in toluene) as a core. Under continuous UV irradiation, the polymer underwent *E-Z* photoisomerization, allowing the release of the encapsulated material. In this study, variation in diameter of the nanocapsules with the time of UV irradiation was detected through dynamic light scattering analysis. Between 10% and 30% growth was observed, depending on the sample.

In 2019, Trojanowska et al. [49] published an article describing a mechanism of active material release from photo-sensitive microcapsules based on cross-linked liquid crystalline polyamide containing azobenzene moieties incorporated in the main chain. Performed quantum chemical calculations of this phenomenon have been described in detail. Atomistic scale modeling of the polymeric building blocks of

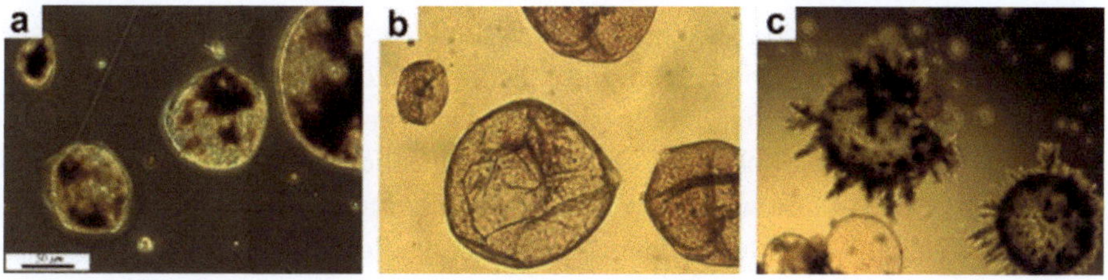

Fig. 2.14: Optical micrographs during the release of β-carotene from the polyamide microcapsules at room temperature after suspending 20 min in (a) water, (b) water under irradiation with UV light, and (c) tetrahydrofuran. Reprinted with permission from Elsevier [45].

azobenzene modified capsules was performed, and obtained information was relative to the structural transformations of the *trans-cis* isomerization process. In addition, experimental data, such as Fourier-transform infrared spectroscopy (FTIR) studies and atomic force microscopy (AFM) analysis were presented. FTIR spectra recorded before and after capsules exposure to UV light shows of the *trans* isomer band at 1409 cm^{-1} in dark, and band corresponded to *cis* isomer at 1491 cm^{-1} after irradiation, which reveals a successful transformation with UV light. AFM studies report a significant increase in the root mean square roughness after irradiation with the light and a surface skewness by 20 times the initial value taking the *trans* isomer as a reference. Those results were in good agreement with the described computational investigation. Therefore, the authors proposed, based on theoretical and experimental data, a squeezing mechanism of encapsulated active release from capsule shells.

2.5 Visible-light sensitive microcapsules based on modified azobenzene

In recent years, high interest has been devoted to the development of optically addressable, micron-sized delivery systems, thanks to their great potential as tools in bioengineering or in biomedical applications. Hollow capsules containing photoactive materials such as azobenzene moieties constitute a good approach to this challenge. As already stated in the previous section of this chapter, what makes the azobenzene highly interesting for many applications is its facile and reversible photo-induced isomerization [50–54]. The azobenzene switching around the N=N double bound has been used to influence conformational, optical, or surface properties of polymeric materials that incorporate it [55, 56]; it reflects in a length reduction of this group from 9.0 to 5.5 A° [57]. Moreover, isomerization of this class of dye is accompanied by molecular changes, which leads to the modification of physical properties such as polarity, viscosity, and absorbance. When the azobenzene moieties are incorporated in a polymeric structure (into the backbone or grafted onto it), the isomerization may affect the film's properties, such as thickness, wettability, and geometry [58, 59]. The possibility of incorporating such photo-switchable molecules in capsule polymeric shells seems a promising approach to create a remote-controlled release system triggered by photo irradiation; the effects of photo-isomerization of the incorporated azobenzene could potentially change the wall permeability of capsules.

As reported in paragraph 3, the chemical modification of azobenzene structures to switch their absorption wavelength in the visible range have been highly investigated; visible-light sensitive azobenzene can be successfully applied to produce microcapsules that could release their cargo when irradiated at a wavelength between 400 and 700 nm. Light triggered microcapsules open new horizons spanning both industrial and academic research due to their numerous benefits over other external stimuli. Light irradiation is durable and easy to control; moreover, the molecular response to light does not contaminate the reaction environment and can be tuned trough appropriate molecular design. In the following section, we report some interesting examples of light-sensitive microcapsules.

Bédard et al. [11] reported in 2007 the light-induced shrinking and encapsulation process of fluorescent dextran in a novel type of light-responsive micron-sized polymeric shells. This system comprises a hollow photosensitive shell made out of PAH, poly{1-[4-(carboxy-4-hydroxyphenylazo)benzenesulfoamido]-1,2-ethanediyl, sodium salt} (PAzo), and PVS. Upon exposure to a light source, these shells start to shrink rapidly, encapsulating fluorescently labeled dextran molecules. This was attributed to changes in the permeability of the shell due to light exposure. The group observed that the effectiveness of this method increases with the duration of irradiation time, as well as for larger fluorescent molecules. The new system was shown to possess a high thermal stability; therefore, the size and shape of capsules remained unaltered at high temperatures. Robust monodisperse microcapsules, made by Wang at al. [60], were obtained via distillation precipitation polymerization of an azobenzene polymeric coating on silica particles. After a selective removal of the silica templates by hydrofluoric acid etching, the hollow structures were formed. These microcapsules, confirmed by TEM investigation, had excellent reversible photoisomerization under both UV and visible lights. Acetonitrile was trapped in the network of the shell during polymerization due to its compatibility with the azobenzene polymer. The subsequently acetonitrile evaporation was used to produce pore channels in the shell. The presence of such channels was subsequently confirmed by nitrogen adsorption-desorption test. Loading and release of RhB molecules in the microcapsules core were carried out. The release profiles obeyed pure Fickian diffusion with a power law of $t^{0.42}$. The diffusion coefficient of RhB from azobenzene polymer microcapsules under visible light (1.47×10^{-12} cm^2/s) was lower than that under UV light (2.12×10^{-12} cm^2/s).

In the work made by Kamiya et al. [61], a biocompatible photo-responsive microcapsule with potential application as a photo-controlled drug-release system was realized. The promising system consists of spherical photoresponsive microcapsules composed of three photo-switchable DNA strands. These strands formed a three-way junction motif that further self-assembled to form microspheres through hybridization of the sticky-end regions of each branch. Multiple unmodified azobenzene (Azo) or 2,6-dimethyl-4-(methylthio)azobenzene (SDM-Azo) was introduced into the sticky-end regions via a d-threoninol linker serving as the photo-switch. A specific light irradiation induced *trans*-to-*cis* isomerization of Azo or SDM-Azo deforming, as a consequence, the DNA capsule structure. Thus, photo-triggered release of encapsulated small molecules from the DNA microcapsule was successfully achieved. Moreover, cytotoxicity to cultured cells caused by the photo-triggered release of doxorubicin was demonstrated.

Recently, Tylkowski et al. [23] developed visible light-triggered microcapsules with visible light-sensitive moieties in the main chain as protection systems for a nontoxic perfume oil. They demonstrated how it is possible to control the remote release of the perfume with visible light irradiation. Using oil-in-water interfacial polymerization method, polyamide microcapsules with ortho-substituted azobenzene monomers were prepared. The microcapsule shells surface and cross-section morphology were observed by scanning electron microscopy (SEM). Triggered perfume release during light irradiation was monitored by means of gas-chromatography/mass spectroscopy analysis. The results of these studies confirm that the mechanism of perfume release from the microcapsule shell was linked to the decrease in modified azobenzene length during light illumination and, thus, the shrinking of the shell. The visible light-induced compression of microcapsules offers a new way to modulate the release of encapsulated materials. The same research group developed novel functional polymeric microcapsules, based on the same modified azobenzene moieties, used as building block for the polymeric shell. In this study, theoretical calculations were used to demonstrate that visible light can induce *trans-cis* isomerization and, thus, the release of encapsulated material, thanks to a "squeezing" mechanism. Interfacial polymerization leads to core-shell microcapsules; the photosensitive behavior was characterized by means of AFM, optical microscopy, SEM and light scattering. Optical micrographs of the obtained capsules, together with a simulation of the *trans* and *cis* structures of the azobenzene polymer used, are presented in Figure 2.15. These

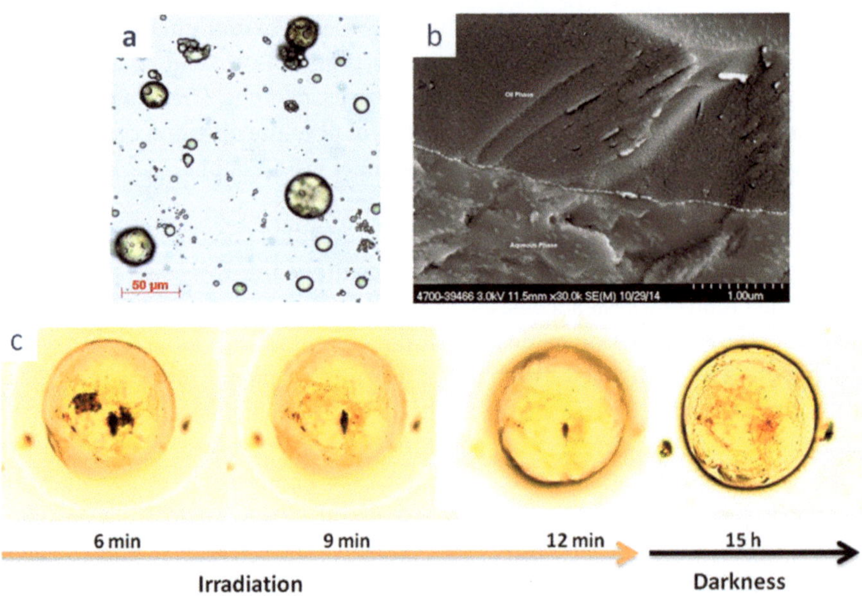

Fig. 2.15: (a) Optical micrograph of the capsules after preparation; (b) optical micrographs of microcapsule morphology changing during irradiation with visible light emitted from a microscopy bulb; (c) *trans* and *cis* structures of azobenzene polymer forming microcapsule shell.

analyses put into evidence that the microcapsules' size and surface morphology are strongly affected by irradiation under visible light and that these changes can be reverted by sample exposure to temperatures [30].

The same photosensitive microcapsule system was also studied by means of quantum chemical calculations. Computational investigation demonstrated that the polymeric main chain length forming the capsule shell decreases by approximately 28% in comparison to the initial state by *trans-cis* isomerization of azobenzene moieties. Finally, a mechanism of active release based on experimental and theoretical data was proposed [49].

2.6 Acknowledgements

Financial Support for Smartmem-Stimuli-responsive Membranes for consumer goods sustainability project under European Community's Horizon 2020 ITN Marie Curie grant agreement no. 675624 is gratefully acknowledged.

2.7 References

[1] Mitscherlich, E., Ueber das Stickstoffbenzid, Ann Pharm, **12** (1834) 311–314.
[2] Hartley, G. S., The cis-form of azobenzene, Nature, **140** (1937) 281–282.
[3] Zollinger, H., Color Chemistry: Syntheses, Properties, and Applications of Organic Dyes and Pigments. John Wiley & Sons, Hoboken, Nueva Jersey, USA 2003.
[4] Rabek, J. F., Scott, G. W., Photochemistry and Photophysics. CRC Press, Boca Ratón, Florida, USA 1989.
[5] Brown, E. V., Granneman, G. R., Cis-trans isomerism in the pyridyl analogs of azobenzene. Kinetic and molecular orbital analysis, May 1, 2002. Accessed April 12, 2019, at: https://pubs.acs.org/doi/abs/10.1021/ja00836a025.
[6] Dürr, H., Bouas-Laurent, H., Photochromism: Molecules and Systems. Elsevier, Amsterdam the Netherlands 2003.
[7] Morgenstern, K., Isomerization reactions on single adsorbed molecules, Acc Chem Res **42** (2009) 213–223.
[8] Koshima, H., Ojima, N., Uchimoto, H., Mechanical motion of azobenzene crystals upon photoirradiation, J Am Chem Soc **131** (2009) 6890–6891.
[9] Merino, E., Ribagorda, M., Control over molecular motion using the cis–trans photoisomerization of the azo group, Beilstein J Org Chem **8** (2012) 1071–1090.
[10] Repinec, S. T., Sension, R. J., Szarka, A. Z., Hochstrasser, R. M., Femtosecond laser studies of the cis-stilbene photoisomerization reactions: the cis-stilbene to dihydrophenanthrene reaction, May 1, 2002. Accessed April 12, 2019, at: https://pubs.acs.org/doi/abs/10.1021/j100178a026.
[11] Bédard, M., Skirtach, A. G., Sukhorukov, G. B., Optically driven encapsulation using novel polymeric hollow shells containing an azobenzene polymer, Macromol Rapid Commun **28** (2007) 1517–1521.
[12] Bandara, H. M. D., Burdette, S. C., Photoisomerization in different classes of azobenzene, Chem Soc Rev **41** (2012) 1809–1825.

[13] Ladányi, V., Dvořák, P., Anshori, J. A., Vetráková, Ľ., Wirz, J., Heger, D., Azobenzene photoisomerization quantum yields in methanol redetermined, Photochem Photobiol Sci **16** (2017) 1757–1761.
[14] Tamai, N., Miyasaka, H., Ultrafast dynamics of photochromic systems, Chem Rev **100** (2000) 1875–1890.
[15] Crecca, C. R., Roitberg, A. E., Theoretical study of the isomerization mechanism of azobenzene and disubstituted azobenzene derivatives, J Phys Chem A **110** (2006) 8188–8203.
[16] Lu, Y.-C., Diau, E. W.-G., Rau, H., Femtosecond fluorescence dynamics of rotation-restricted azobenzenophanes: new evidence on the mechanism of trans → cis photoisomerization of azobenzene, February 18, 2005. Accessed April 12, 2019, at: https://pubs.acs.org/doi/abs/10.1021/jp044934b.
[17] Titov, E., Granucci, G., Götze, J. P., Persico, M., Saalfrank, P., Dynamics of azobenzene dimer photoisomerization: electronic and steric effects, J Phys Chem Lett **7** (2016) 3591–3596.
[18] Mahimwalla, Z., Yager, K. G., Mamiya, J., Shishido, A., Priimagi, A., Barrett, C. J., Azobenzene photomechanics: prospects and potential applications, Polym Bull **69** (2012) 967–1006.
[19] Beharry, A. A., Sadovski, O., Woolley, G. A., Azobenzene photoswitching without ultraviolet light, J Am Chem Soc **133** (2011) 19684–19687.
[20] Wu, W., Yao, L., Yang, T., Yin, R., Li, F., Yu, Y., NIR-light-induced deformation of cross-linked liquid-crystal polymers using upconversion nanophosphors, J Am Chem Soc **133** (2011) 15810–15813.
[21] Mourot, A., *et al.*, Tuning photochromic ion channel blockers, ACS Chem Neurosci **2** (2011) pp. 536–543.
[22] Hamon, F., Djedaini-Pilard, F., Barbot, F., Len, C., Azobenzenes – synthesis and carbohydrate applications, Tetrahedron **65** (2009) pp. 10105–10123.
[23] Tylkowski, B., Giamberini, M., Underiner, T., Prieto, S. F., Smets, J., Photo-triggered microcapsules, Macromol Symp **360** (2016) pp. 192–198.
[24] Srinivasa, G. R., Abiraj, K., Gowda, D. C., The synthesis of azo compounds from nitro compounds using lead and triethylammonium formate, Tetrahedron Lett **44** (2003) pp. 5835–5837.
[25] Nishimura, N., Sueyoshi, T., Yamanaka, H., Imai, E., Yamamoto, S., Hasegawa, S., Thermal cis-to-trans isomerization of substituted azobenzenes II. Substituent and solvent effects, Bull Chem Soc Jpn **49** (1976) pp. 1381–1387.
[26] Sadovski, O., Beharry, A. A., Zhang, F., Woolley, G. A., Spectral tuning of azobenzene photoswitches for biological applications, Angew Chem Int Ed **48** (2009) pp. 1484–1486.
[27] Magennis, S. W., Mackay, F. S., Jones, A. C., Tait, K. M., Sadler, P. J., Two-photon-induced photoisomerization of an azo dye, Chem Mater, **17** (2005) pp. 2059–2062.
[28] Beltrame, P. L., Paglia, E. D., Castelli, A., Tantardini, G. F., Seves, A., Marcandalli, B., Thermal cis-trans isomerization of azo dyes in poly(methyl methacrylate) matrix: a kinetic study, J Appl Polym Sci **49** (1993) pp. 2235–2239.
[29] Jerca, V. V., Jerca, F. A., Rau, I., Manea, A. M., Vuluga, D. M., Kajzar, F., Advances in understanding the photoresponsive behavior of azobenzenes substituted with strong electron withdrawing groups, Opt Mater, **48** (2015) pp. 160–164.
[30] Del Pezzo, R., *et al.*, "Ortho-substituted azobenzene: shedding light on new benefits," Pure Appl Chem **0** 2018.
[31] Kano, K., Tanaka, Y., Ogawa, T., Shimomura, M., Okahata, Y., and Kunitake, T., Photoresponsive membranes. regulation of membrane properties by photoreversible cis-trans isomerization of azobenzenes, Chem Lett **9** (1980) pp. 421–424.
[32] Morgan, C. G., Thomas, E. W., Sandhu, S. S., Yianni, Y. P., Mitchell, A. C., Light-induced fusion of liposomes with release of trapped marker dye is sensitised by photochromic phospholipid, Biochim Biophys Acta BBA-Biomembr, **903** (1987) pp. 504–509.

[33] Bisby, R. H., Mead, C., Morgan, C. G., Photosensitive liposomes as 'cages' for laser-triggered solute delivery: the effect of bilayer cholesterol on kinetics of solute release, FEBS Lett **463** (1999) pp. 165–168.
[34] Bisby, R. H., Mead, C., Morgan, C. G., Wavelength-programmed solute release from photosensitive liposomes, Biochem Biophys Res Commun **276** (2000) pp. 169–173.
[35] Smith, A. M., Harris, J. J., Shelton, R. M., Perrie, Y., 3D culture of bone-derived cells immobilised in alginate following light-triggered gelation, J Control Release Off J Control Release Soc **119** (2007) pp. 94–101.
[36] Harada, A., Kataoka, K., Formation of polyion complex micelles in an aqueous milieu from a pair of oppositely-charged block copolymers with poly (ethylene glycol) segments, Macromolecules **28** (1995) pp. 5294–5299.
[37] Wang, Y., Han, P., Xu, H., Wang, Z., Zhang, X., Kabanov, A. V., Photocontrolled self-assembly and disassembly of block ionomer complex vesicles: a facile approach toward supramolecular polymer nanocontainers, Langmuir **26** (2010) pp. 709–715.
[38] Bronich, T. K., Kabanov, A. V., Kabanov, V. A., Yu, K., Eisenberg, A., Soluble complexes from poly (ethylene oxide)-block-polymethacrylate anions and N-alkylpyridinium cations, Macromolecules **30** (1997) pp. 3519–3525.
[39] Zhou, Y., Liu, B., Wang, X., Microspheres of polyurethanes functionalized with push-pull type azo chromophores and their photoinduced deformation behavior, Polymer **111** (2017) pp. 229–238.
[40] Yi, Q., Sukhorukov, G. B., UV light stimulated encapsulation and release by polyelectrolyte microcapsules, Adv Colloid Interface Sci **207** (2014) pp. 280–289.
[41] Tao, X., Li, J., Möhwald, H., Self-assembly, optical behavior, and permeability of a novel capsule based on an azo dye and polyelectrolytes, Chem – Eur J **10** (2004) pp. 3397–3403.
[42] Yi, Q., Sukhorukov, G. B., Externally triggered dual function of complex microcapsules, ACS Nano **7** (2013) pp. 8693–8705.
[43] Lin, H., Xiao, W., Qin, S.-Y., Cheng, S.-X., Zhang, X.-Z., Switch on/off microcapsules for controllable photosensitive drug release in a 'release-cease-recommence' mode, Polym Chem, **5** (2014) pp. 4437–4440.
[44] Xiao, W., Chen, W.-H., Zhang, J., Li, C., Zhuo, R.-X., Zhang, X.-Z., Design of a photoswitchable hollow microcapsular drug delivery system by using a supramolecular drug-loading approach, J Phys Chem B **115** (2011) pp. 13796–13802.
[45] Tylkowski, B., Pregowska, M., Jamowska, E., Garcia-Valls, R., Giamberini, M., Preparation of a new lightly cross-linked liquid crystalline polyamide by interfacial polymerization. Application to the obtainment of microcapsules with photo-triggered release, Eur Polym J **45** (2009) pp. 1420–1432.
[46] Tsuda, N., Ohtsubo, T., Fuji, M., Preparation of self-bursting microcapsules by interfacial polymerization, Adv Powder Technol **23** (2012) pp. 724–730.
[47] Salaün, F., Bedek, G., Devaux, E., Dupont, D., Gengembre, L., Microencapsulation of a cooling agent by interfacial polymerization: influence of the parameters of encapsulation on poly(urethane-urea) microparticles characteristics, J Membr Sci **370** (2011) pp. 23–33.
[48] Marturano, V., Ambrogi, V., Cerruti, P., Giamberini, M., Tylkowski, B., Photo-triggered release in polyamide nanosized capsules, AIP Conf Proc **1599** (2014) pp. 234–237.
[49] Trojanowska, A., *et al.*, Squeezing release mechanism of encapsulated compounds from photo-sensitive microcapsules, Appl Surf Sci **472** (2019) pp. 143–149.
[50] Garcia-Amorós, J., Martínez, M., Finkelmann, H., Velasco, D., Kinetico-mechanistic study of the thermal cis-to-trans isomerization of 4,4′-dialkoxyazoderivatives in nematic liquid crystals, J Phys Chem B **114** (2010) pp. 1287–1293.
[51] Baba, K., *et al.*, Kinetic study of thermal Z to e isomerization reactions of azobenzene and 4-dimethvlaniino-4′-nitroazobenzene in ionic liquids [1-R-3-methyliniidazolium bis(trifluoromethylsulfonyl)imide with R = Butyl, Pentyl, and Hexyl], Chem – Eur J **12** (2006) pp. 5328–5333.

[52] Bandara, H. M. D., Burdette, S. C., Photoisomerization in different classes of azobenzene, Chem Soc Rev **41** (2012) pp. 1809–1825.
[53] Jerca, F. A., Jerca, V. V., Anghelache, A. M., Vuluga, D. M., and Hoogenboom, R., Poly (2-isopropenyl-2-oxazoline) as a versatile platform towards thermoresponsive copolymers, Polym Chem **9** (2018) pp. 3473–3478.
[54] Jerca, F. A., Jerca, V. V., Kajzar, F., Manea, A. M., Rau, I., Vuluga, D. M., Simultaneous two and three photon resonant enhancement of third-order NLO susceptibility in an azo-dye functionalized polymer film, Phys Chem Chem Phys **15** (2013) pp. 7060–7063.
[55] Rodríguez, F. J., *et al.*, Optical anisotropy and non-linear optical properties of azobenzene methacrylic polymers, Polymer **45** (2004) pp. 2341–2348.
[56] Börger, V., Kuliskovska, O., Hubmann, K. G.-, Stumpe, J., Huber, M., Menzel, H., Novel polymers to study the influence of the azobenzene content on the photo-induced surface relief grating formation, Macromol Chem Phys **206** (2005) pp. 1488–1496.
[57] Merino, E., Ribagorda, M., Control over molecular motion using the cis-trans photoisomerization of the azo group, Beilstein J Org Chem **8** (2012) pp. 1071–1090.
[58] Möller, G., Harke, M., Motschmann, H., Prescher, D., Controlling microdroplet formation by light, Langmuir **14** (1988) pp. 4955–4957.
[59] El Halabieh, R. H., Mermut, O., Barrett, C. J., Using light to control physical properties of polymers and surfaces with azobenzene chromophores, Pure Appl Chem **76** (2004) pp. 1445–1465.
[60] Wang, X., Yang, Y., Liao, Y., Yang, Z., Jiang, M., Xie, X., Robust polyazobenzene microcapsules with photoresponsive pore channels and tunable release profiles, Eur Polym J **48** (2012) pp. 41–48.
[61] Kamiya, Y., Yamada, Y., Muro, T., Matsuura, K., Asanuma, H., DNA microcapsule for photo-triggered drug release systems, Chem Med Chem **12** (2017) pp. 2016–2021.

Bojana Boh Podgornik and Marica Starešinič

3 Microencapsulation technology and applications in added-value functional textiles

3.1 Introduction

3.1.1 Research and development trends

Microencapsulation is a knowledge-intensive and dynamic research field with an increasing growth of publications. Trends in patent vs. non-patent literature on microencapsulation illustrate the growth of basic research (scientific articles), as well as the fast growth of industrial research, represented in waves of patented inventions (Figure 3.1).

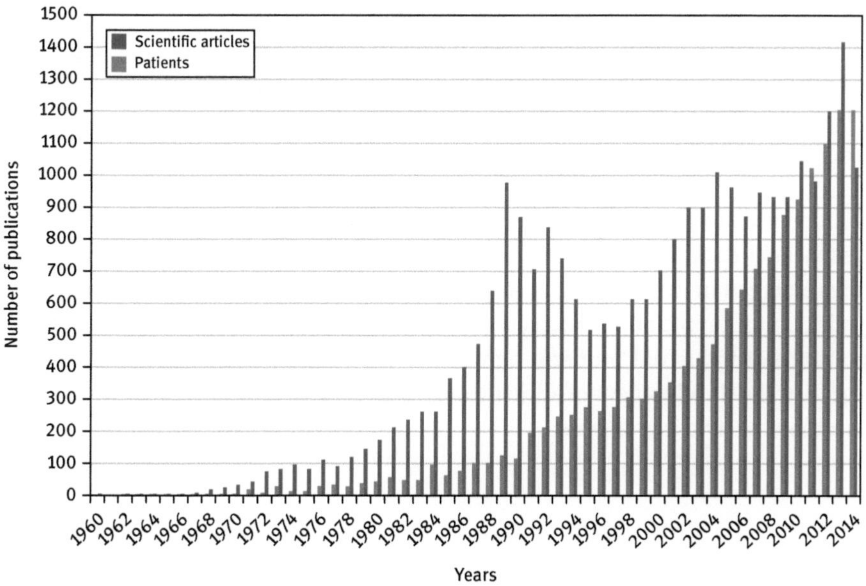

Fig. 3.1: Trends in scientific articles vs. patent documents on microencapsulation. Web of Science [1], advanced search: TS = (microcapsule* OR microencapsulat*) AND TS = (textile* OR cloth OR fabric OR garment*). Espacenet [2], advanced search: Title or abstract: (microcapsule* OR microencapsulat*) AND (textile* OR cloth OR fabric OR garment*).

Among numerous possible applications fields, microencapsulation offers many opportunities to improve the properties of textiles or to give them new functions.

https://doi.org/10.1515/9783110642070-003

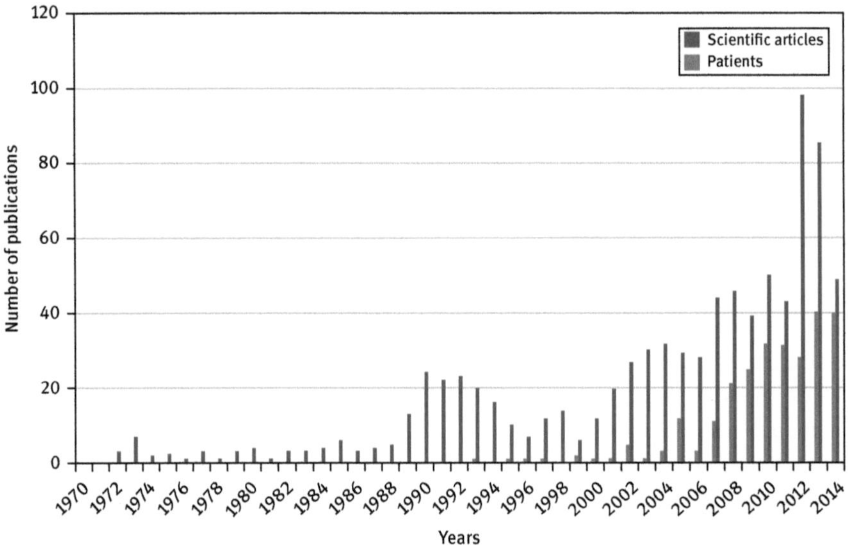

Fig. 3.2: Trends in scientific articles vs. patent documents on microencapsulation for textiles. Web of Science [1], advanced search: TS = (microcapsule* OR microencapsulat*) AND TS = (textile* OR cloth OR fabric OR garment*). Espacenet [2], advanced search: Title or abstract: (microcapsule* OR microencapsulat*) AND (textile* OR cloth OR fabric OR garment*).

A bibliometric analysis of scientific articles in the Web of Science [1], and patents in the Espacenet database [2] reveals that the first ideas of applying microcapsules in textiles emerged in the early 1970s, and that the majority of publications on microencapsulation for textile applications remain patents (Figure 3.2). This emphasises the importance of industrial property rights, and the strong participation of industrial research in the development of added-value functional textiles, invented with microencapsulated active ingredients.

3.2 Microencapsulation methods and processes of applying microcapsules to textiles

3.2.1 Microencapsulation methods

The selection of microencapsulation process for added-value textile applications depends on the desired characteristics and uses of the products. For example, microcapsule size, shape, wall material, active substance, release mechanism, method of application, and compatibility with other components of the formulation must be adapted to the requirements of textile processing methods, and uses of the final product.

Most often, microcapsules for textile applications have been prepared by one of the following technological possibilities:

Coacervation processes (e.g. gelatin-gum arabic microcapsule walls) taking place in colloid systems, where macromolecular colloid rich coacervate droplets surround dispersed microcapsule cores, and form a viscous microcapsule wall, which is solidified with crosslinking agents (Figure 3.3).

Polymerization methods, where monomers polymerize around droplets of an emulsion and form a solid polymeric wall. In *in situ* polymerization (e.g. aminoaldehyde resin walls), monomers or precondensates are added only to the aqueous phase of emulsion (Figure 3.4), while in interfacial polymerization (e.g. polyamide, polyester,

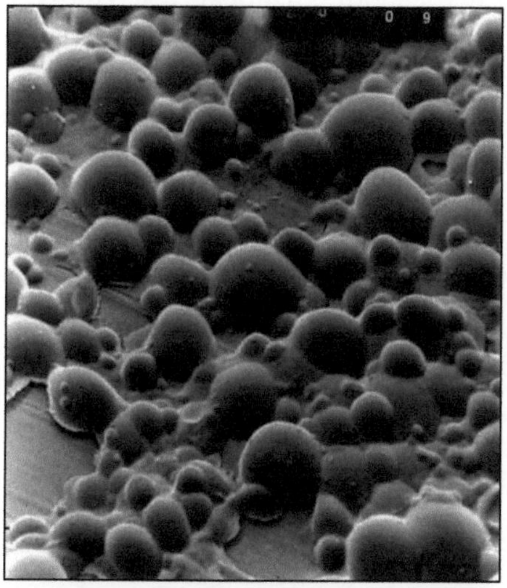

Fig. 3.3: Coating of microcapsules, produced by complex coacervation of gelatin and carboxymethyl cellulose (SEM, 630 ×) with softer, elastic microcapsule walls.

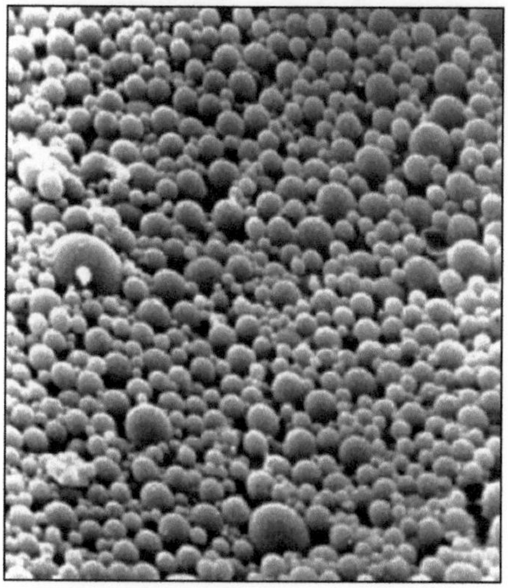

Fig. 3.4: Coating of microcapsules, produced by *in situ* polymerization of aminoaldehyde precondensates (SEM, 1900 ×) with impermeable, pressure-sensitive hard walls.

polyurethane walls), one of the monomers is dissolved in the aqueous phase and the other in a lipophylic solvent.

Physical/mechanical methods (e.g. spray-drying, fluidized bed coating, extrusion, deposition in vacuum, solvent evaporation from emulsions, ultrasonic liposome formation), where the microcapsule wall is mechanically applied, condensed or layered around the microcapsule core. Physical/mechanical microencapsulation methods are used to design microcapsules that release their content during textile dyeing, washing or drying; the walls are soluble or heat sensitive to dissolve or melt at a desired circumstance.

In situ polymerization is one of the chemical microencapsulation processes often used for technical applications, including textiles. The process takes place in oil-in-water emulsions; the result is nicely smooth, spherical, reservoir-type microcapsules with transparent polymeric pressure-sensitive microcapsule walls (Figures 3.5 and 3.6). Typical wall materials for *in situ* polymerization are aminoplast resins, such as melamine–formaldehyde, urea-formaldehyde, urea–melamine-formaldehyde or resorcinol-modified melamine–formaldehyde polymers. The

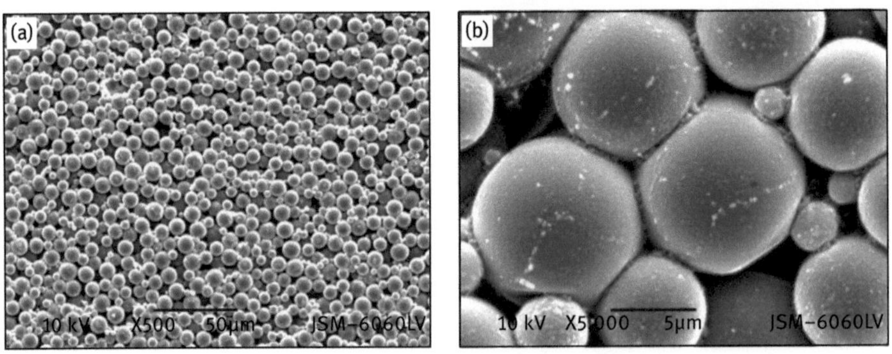

Fig. 3.5: Spherical, reservoir-type microcapsules, produced by *in situ* polymerization in oil-in-water emulsion (SEM, left 500 ×, right 5000 ×).

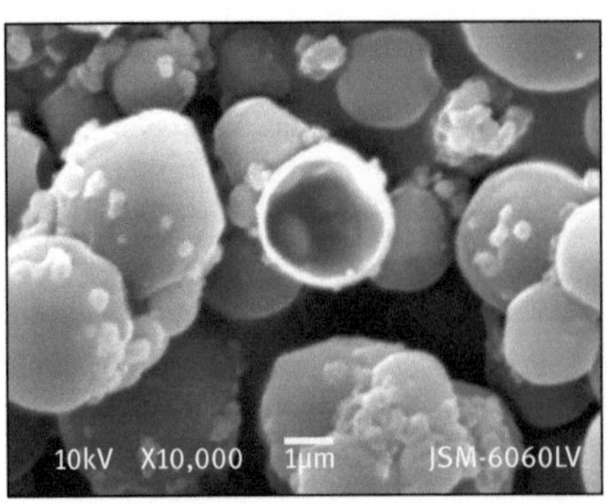

Fig. 3.6: Visualization of wall thickness in a container-type microcapsule, produced by *in situ* polymerization (SEM, 5000 ×).

in situ processes (Figure 3.7) can start either directly from amine and aldehyde monomers, or from the precondensates. Typically, all materials for the formation of microcapsule wall originate from the continuous aqueous phase of the oil-in-water emulsion system, and therefore have to be water-soluble. To achieve better process control and improved mechanical properties of microcapsules, modifying agents/ protective colloids are added, such as styrene-maleic acid anhydride copolymers, polyacrylic acid, or acrylamidopropylsulfonate and methacrylic acid/acrylic acid copolymers [3].

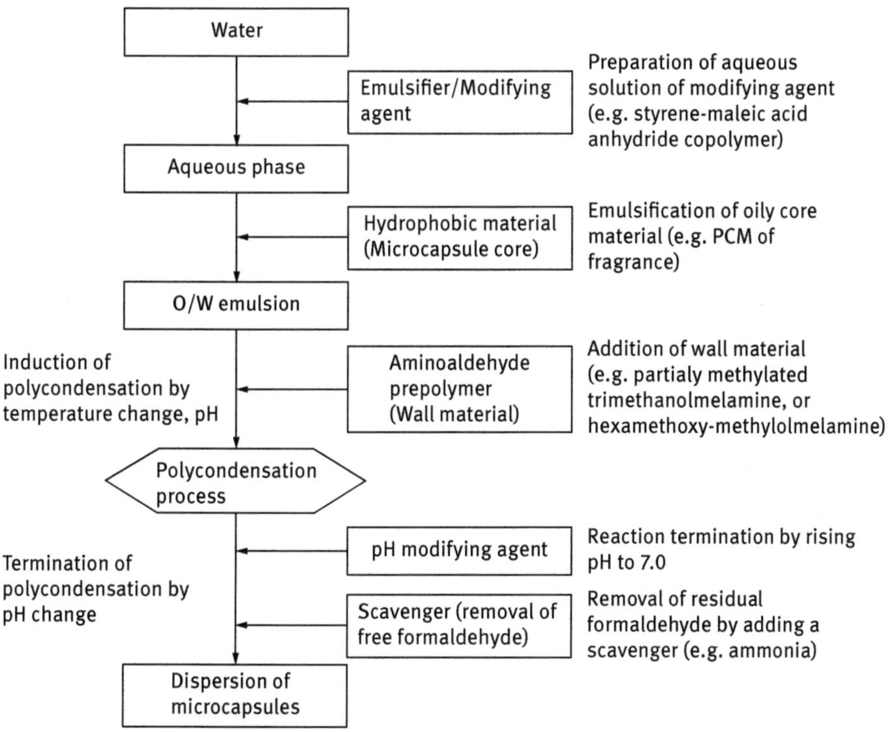

Fig. 3.7: Example of microcapsules synthesis by *in situ* polymerization process.

For some technical applications the *in situ* aminoaldehyde microcapsules remain irreplaceable, due to some superior characteristics, such as:

- the spherical reservoir-type shape with thin impermeable transparent walls (Figure 3.8);
- high chemical and thermal stability;
- high microcapsule resistance to harsh chemical environments (e.g. in detergents, softeners etc.);
- good storage stability;

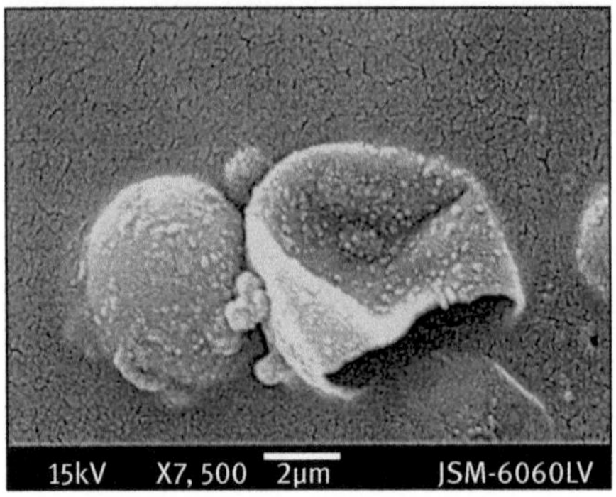

Fig. 3.8: Spherical type pressure-sensitive microcapsules, produced by *in situ* polymerization, after the release of encapsulated core material (SEM, 7500 ×).

- high microencapsulation yields (≥ 99%);
- effective microencapsulation process control;
- controllable microcapsule size and size distribution;
- good transferability of the *in situ* process to large-scale industrial production.

In addition, wall permeability and mechanical characteristics can be regulated and adapted, to obtain tailor-made pressure-sensitive or more elastic microcapsules with controled diffusion, to support different release mechanisms of the products [3, 4]. The main constraint of the *in situ* process is synthetic nature of aminoaldehyde microcapsule wall, and the residual formaldehyde in microcapsule suspension after the polycondensation process, which limits the *in situ* microcapsules to technical products. However, with the optimised selection of process parameters and application of formaldehyde scavengers, the concentration of free formaldehyde can be minimized to meet the technical standards for textiles [5–9].

3.2.2 Processes of applying microcapsules to textiles

Microcapsules have to be formulated for applications on woven or nonwoven textiles without substantially altering the feel or color of textile products. Formulation additives usually consist of binders, crosslinking agents, organic or inorganic pigments and fillers, antifoaming agents and/or other surfactants, and viscosity-controling agents/thickeners.

Binders play a crucial role in microcapsule formulations for textiles. To a large extent, they determine the quality, durability and washability of textile materials with microencapsulated ingredients. Typically, binders are selected from the groups of:

- water-soluble polymers, such as polyvinyl alcohol, carboxymethyl cellulose, starch and modified starches, xanthanes, alginates, and other natural gums;

- synthetic latexes, such as polyacrylate latexes, styrene-butadiene, polyvinyl-acetate, ethylene–vinyl acetate copolymers;
- synthetic resins, such as such as urea–and melamine–formaldehyde resins, dimethylol ethylene urea, dimethylol dihydroxy ethylene urea, dimethylol propylene urea, polyurethane and epoxy resins, vinyl acetate resins;
- synthetic rubbers, such as polyurethanes, nitrile and chloroprene rubbers;
- silicones.

Different techniques can be used for applications of microcapsules to textiles. Patents describe incorporation of microencapsulated compounds onto or into textiles by:
- coating with an air knife or rod coater;
- impregnation or immersion (Figure 3.9);
- printing techniques, such as screen-, photographic-, electrostatic-, pressure-transfer, thermal transfer and inkjet printing;
- spraying on the surface of textiles;
- inclusion of microcapsules into the textile fibers during the spinning process, such as polyester, nylon or modacryl fiber material;
- incorporation into polymer foams, coatings and multilayer composites that are placed or inserted into selected parts of textile clothing or footwear.

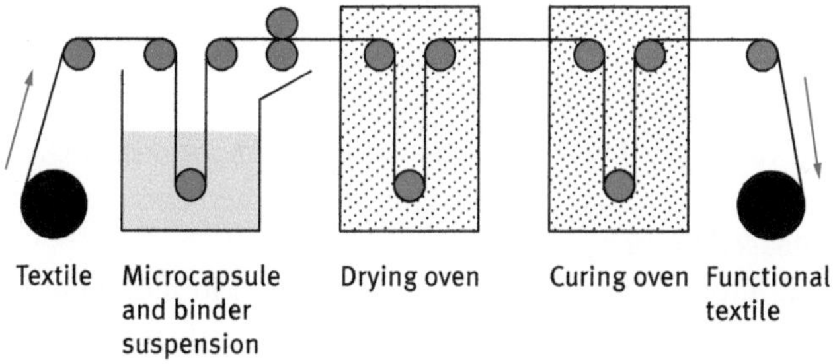

Fig. 3.9: An example of applying microcapsules to textile carriers by impregnation [10].

3.3 Purposes and release mechanisms of microcapsules in textile products

Mechanisms of releasing active ingredients from the microcapsule cores depend on the purpose of microencapsulation, on functions and desired effects of encapsulated components, and on the microcapsule wall characteristics, particularly on permeability. An overview of microencapsulation purposes and release mechanism in textile products is given below, with examples of patented inventions presented in Chapter 4.

In textile applications, microcapsules with permeable walls enable:

- Prolonged/sustained release of active components from the core. This principle is used in long-lasting perfumes and deodorants on textile carriers, in insect repellent textiles, and in sustained release cosmetic and medical textiles.
- Separation of low and high molecular weight molecules can be applied in microencapsulated enzymes in detergent compositions for machine washing of textiles.

Microcapsules with impermeable walls are used in formulations and products where temporary isolation and quick release of active components are necessary. Examples of useful functions and effects, achieved by applying impermeable microcapsules to textiles, include the following:

- Protection of substances against environmental effects: microcapsule walls protect unstable components against environmental influences, and release them only under the desired circumstances. For instance, microencapsulated vitamins, lipids and essential oils in cosmetic textiles are protected against oxidation; microencapsulated enzymes and oxidants are stabilized when added to laundry formulations for textiles.
- Separation of reactive components: this is used when leuco dyes are separated from color developers in thermochromic textiles, or to separate reactants in formulations of multicomponent adhesives and binders for textile bonding.
- Locally limited activity is applied to enable special color effects, such as reversible color changes, speckled patterns and glossy effects, or to reduce the migration of dyes in multicolor textile printing.
- Reduction/prevention of volatility: this ensures that volatile compounds, such as perfumes, fragrances and antimicrobial essential oils are retained in fragranced textiles until they are released in a target situation.
- Conversion from a liquid into a solid state: this is beneficial in formulations of powdered adhesives with microencapsulated solvents for textile-containing laminate bonding; liquid crystals are encapsulated and used in color changing textiles.

To release microencapsulated active components from microcapsule cores, numerous ways of release mechanisms have been invented and applied in added-value textile products (Figure 3.10), such as:

- The mechanism of external pressure, which breaks the microcapsule wall and releases the core, was the first developed and is still widely used, for instance in antimicrobial agents for socks and textile shoe inserts (mechanical pressure caused by walking), fragranced textiles, such as t-shirts, ties, handkerchiefs, pillows and linen (release by pressure and rubbing), and pressure-sensitive multicomponent adhesives for textile bonding (activation in a mechanical press).

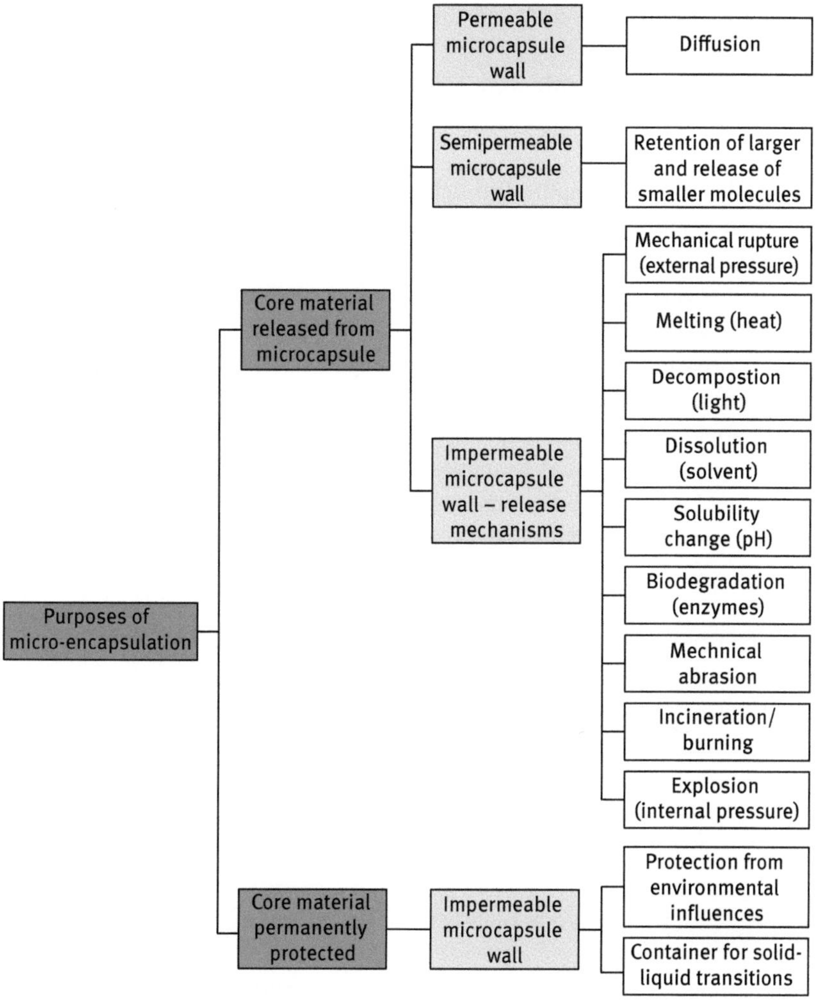

Fig. 3.10: Purposes of microencapsulation and release mechanisms of microencapsulated active ingredients in textile applications.

- In some applications, microcapsule wall breaks because of inner pressure. This happens if the core contains substances which, under special conditions (e.g. UV light), decompose into gaseous components. The effect is used in blowing agents in the production of light synthetic leather.
- The core substance can be released by abrasion of the microcapsule wall, e.g. in antistatics and fragrances in textile washing and drying.
- In many applications, core materials are released by heat that causes melting of microcapsule wall at a specifically designed temperature. Examples include components in cosmetic and medical textiles (release at body

temperature), and textile softeners and fragrances in formulations for dryers (release by heat).
- Microencapsulated fire retardants or extinguishers, released by burning, are used in fire-proof textile materials for carpets, curtains, fire-protecting clothes, and car interiors.
- Microcapsules in photographic and light-sensitive textile printing processes are decomposed or hardened by light.
- In textile washing/cleaning compositions, microcapsules with active ingredients dissolve in a specific solvent (most often water), sometimes only at a selected pH value of the washing cycle.
- In textile processing formulations, selected reagents may be released by enzymatic degradation of target microcapsules.
- In specific applications, permanent enclosure of the core material within the resistant microcapsules is essential. Examples include microencapsulated phase change materials (PCMs) for active thermal control, where microcapsules hold the PCM solid-liquid transitions, and for liquid crystals in reversible color changing textiles.

3.4 Applications of microcapsules in textile products

The possibilities for using microencapsulation technologies in textile products are numerous, and include coloring materials, enzymes, fire retardants, adhesives, fragrances, perfumes, insect repellents, disinfectants, cosmetic additives, decontaminants, PCMs, UV absorbers and self-healing agents (see Figure 3.11).

3.4.1 Microencapsulated dyes and pigments for textile dyeing and printing

Microencapsulation of dyes and pigments for dyeing and printing is one of the oldest microencapsulation applications in textile processing. The idea of including microencapsulated dyes and pigments found their place in different techniques, such as dyeing and printing by electrostatic fields, solvent dyeing, dot dyeing and multicolored speckled printing, pressure or thermal transfer printing, screen printing, photographic screen printing, and ink jet textile printing (Table 3.1).

3.4.2 Textiles with microencapsulated thermochromic materials

Thermochromism, the reversible dependence of color on temperature, utilizes temperature change to initiate color development or color fading. Thermochromic systems can involve inorganic compounds, such as transition metal and organometallic systems, or organic compounds, including liquid crystals, stereoisomerism and molecular rearrangement. Thermochromic systems based on liquid crystals and

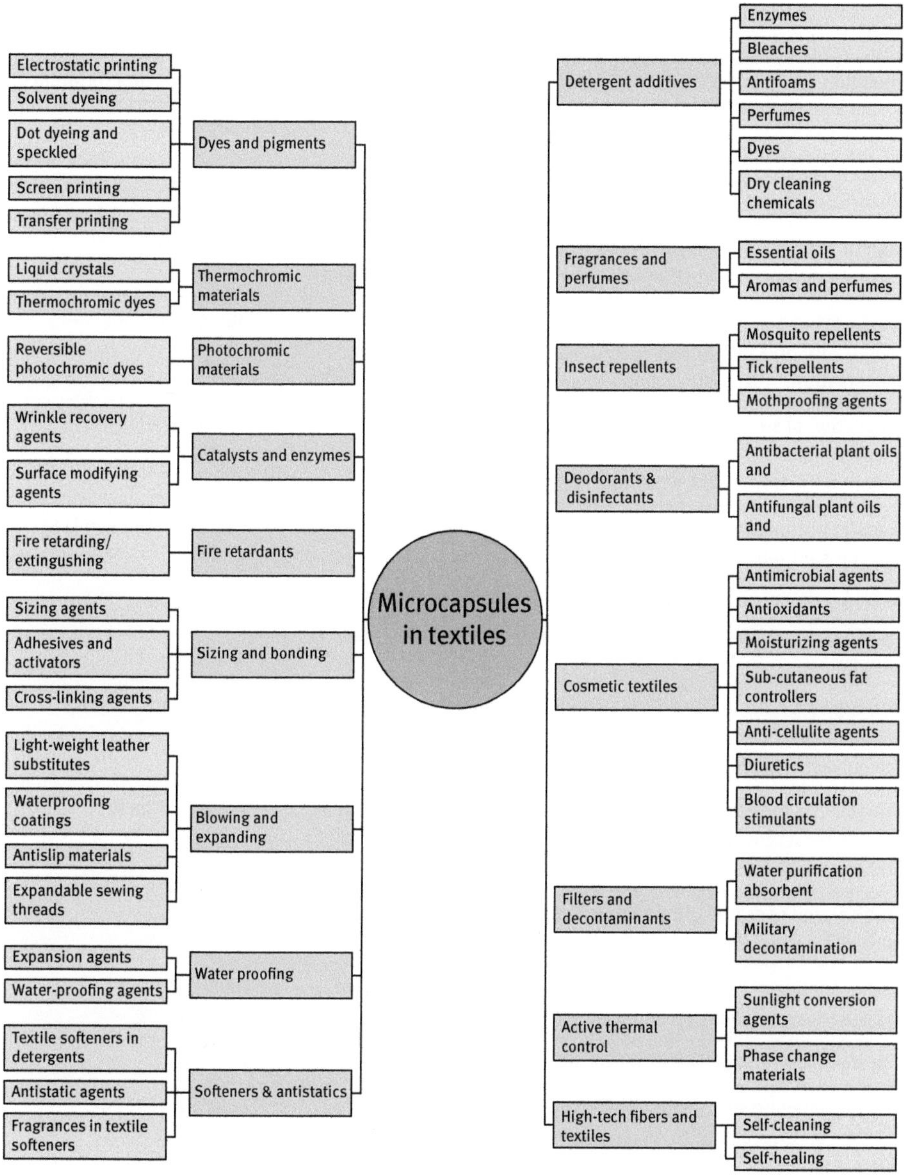

Fig. 3.11: Applications of microcapsules in added-value functional textile products.

Table 3.1: Examples of inventions applying microcapsules in textile dyeing and printing.

Invention	Patent applicant (reference)
Electrostatic textile printing with powders and microcapsules containing dyes in water or alcohol, to obtain sharper, clearer printings at lower electrostatic field strengths.	Sandoz [11]

Tab. 3.1 (*continued*)

Invention	Patent applicant (reference)
Solvent dyeing of polyester textiles with organic dyes, microencapsulated in polyethylene wax shells.	Sumitomo [12]
Microcapsules and dispersion pastes used for dot dyeing of textiles.	Totoki [13]
High contrast dot printing, achieved by gelatin-gum arabic coacervation microencapsulation of powder or liquid textile colorants.	Toa Gosei [14]
Woven nylon or silk fabric ribbon for mechanical transfer printing, coated with pressure-sensitive ink microcapsules, produced by coacervation.	National Cash Register [15]
Production of semi-permeable gelatin-gum arabic coacervate microcapsules with dyes for speckled screen printing on acrylic, cotton, polyester, nylon, wool, and rayon fabrics.	Sakai Textile [16, 17]
Transfer printing process with polyurethane microencapsulated dyes for synthetic textiles.	Dickinson Robinson [18]
Photographic screen printing on cotton, polyester, wool, or acrylic fabrics, with pastes containing microencapsulated yellow, red and blue dyes.	Nippon Kayaku [19]
Printing inks for transfer printing on polyester fabrics, containing microencapsulated trichlorobenzene or biphenyl solvents to swell the polyester fibers.	Seiren [20]
Printing inks for transfer printing on cotton fabrics with improved washfastness, containing microencapsulated reactive or acid dyes and microencapsulated dye fixing agents.	Fuji Photo Film [21]
Multicolored speckled printing on polyester, acrylic or wool, achieved by microencapsulated yellow, red and blue dyes.	Hayashi [22]
Textile printing process with microencapsulated dyes, resulting in reduced dye migration and improved pattern definition.	Milliken Research [23]
Transfer printing on textiles with microencapsulated indigo dyes.	Mihara [24]
Thermal imaging system for transferring photographic images to textiles, leather, ceramics, glass or plastics, with heat responsive dye-precursor microcapsules.	Foto-Wear [25, 26]
Technology for rapidly dyeing of polyester fiber cloth by dispersible dye microcapsules.	Jiangsu Shunyuan Textile [27]
Modified one-bath dyeing technology of polyester/rayon fabric, with dispersed microencapsulated active dye.	Shaoxing Dongshi Textile [28]
Method for dyeing natural protein textile fibres' with microcapsules containing polymeric pendants on the surface, to anchor to the fibres.	Ferrini [29]
Ink compositions for ink jet textile printing.	Seiko Epson [30]

molecular rearrangement have been applied successfully in textiles on a commercial scale [31]. Examples are presented in Table 3.2.

Table 3.2: Examples of microcapsule-based inventions for the production of thermochromic textiles.

Invention	Patent applicant (reference)
Textile material for clothing, comprising a fabric or leather support, coated with a composition of microencapsulated cholesteric liquid crystals.	Ruggeri [32]
Production of textile fibers exhibiting reversible color changes, based on microencapsulated color reversible thermochromic systems with leuco dyes (crystal violet lactone), color developers (benzyl-4-hydroxybenzoate) and stearyl phenoxyacetate.	Pilot Ink [33]
Thermochromic textiles with a broader color range for a given temperature, coated with binders and microencapsulated thermochromic pigments.	Pilot Ink [34]
Thermo- or photochromic cellulose fiber textiles, based on reversibly changeable microencapsulated thermochromic or photochromic materials. When worn, the resulting T-shirts exhibit color changes according to heat transmission from the body.	Matsui Shikiso [35, 36]
Color changing fabrics, based on synthetic fabrics, coated with microencapsulated thermochromic dyes. With the combination of four basic colors, each in two shades, a total of 56 fabric colors are achieved.	MATEO report [37]
Production of composite sensor fiber, comprising microencapsulated thermoresponsive materials, and their applications in fiber fabrics.	Commonwealth Scientific and Industrial Research Organisation [38]
Thermochromic pigments in microcapsules of very small sizes.	Chromatic Technologies [39]

3.4.3 Textiles with microencapsulated photochromic materials

Photochromic dyes absorb quanta in the visible or near-infrared light region. The excited state of a dye must last long enough to undergo a chemical reaction. Applications of photochromic dyes are known for invisible writing, erasable recording media, darkening of sunglasses, or darkening of textile products, such as curtains, t-shirts and sportswear [40]. Microencapsulation of photochromic dyes for textile applications is presented in Table 3.3.

3.4.4 Microencapsulated catalysts and enzymes for special textile effects

Patents on microencapsulated catalysts and enzymes in textile treatment describe methods for achieving special effects, such as wrinkle recovery, crease retention or

Table 3.3: Examples of inventions with microencapsulated photochromic materials.

Invention	Patent applicant (reference)
Light- and washfast reversible photochromic fabrics coated with spironaphthoxazine derivatives, microencapsulated in hollow porous inorganic microspheres, for applications in curtains.	Unitika [41]
Photochromic inks for textiles, containing reversible photochromic dyes, microencapsulated by a gelatin coacervation method, formulated with an acrylic binder, used for screen printing on cotton fabrics.	Japan Capsular Products & Mitsubishi [42, 43]
Reversible photochromic textiles, resistant to fastness due to abrasion or washing, containing microencapsulated photochromic substances and binders, used for the production of photochromic shirts.	Matsui Shikiso [44]
Textile printing pastes, comprising microencapsulated photochromic amine dyes, such as 6′-substituted spironaphthoxazines or 3-substituted naphthopyrans, and acrylic oligomers as binders.	Matsui Shikiso [45]
Production of microencapsulated photochromic substances for textile dyeing and printing.	Nippon Paint [46]
Photochromic double-shell microcapsules, prepared by interfacial and *in situ* polymerization, for textile printing applications.	China Tex. Acad. [47]

Table 3.4: Examples of microencapsulated catalysts and enzymes for textile treatment.

Invention	Patent applicant (reference)
Microencapsulated catalysts and crosslinking agents for improved wrinkle recovery and crease retention of durable press cotton, linen and regenerated cellulose fabrics.	Cluett, Peabody & Co. [48]
Sandoz's Sirrix Luna treatment, containing microencapsulated enzymes (e.g. hydrolases) to produce special effects on fabric surfaces – opalescence, moonlight effects, cat's eye reflection, and washed down appearances.	Sandoz [49]

biomechanical visual effects on fabric surfaces, resulting in opalescence, reflection, or matting (Table 3.4).

3.4.5 Textiles with microencapsulated fire retardants

One of the shortcomings of untreated textile materials used for decoration and construction purposes is their flammability. As a solution, flame retardant textiles have been developed with incorporated fire retardants. A review of microencapsulation of flame retardant formulations suitable for application in textiles was published by Salaün et al. [50]. Microencapsulation can be used to avoid reactions of fire retardants with textile polymers, prevent sublimation or exudation of fire retardants from the polymer, or to eliminate substance hydrophilicity. The idea of microencapsulated fire retardants for textiles was first launched by the industrial producers in the beginning of the 1970s. Textiles treated with microencapsulated fire retardants have been used for military and civilian clothing and tents, for carpets, furniture and car interiors (Table 3.5).

Table 3.5: Some inventions on fire-resistant textiles with microencapsulated fire retardants.

Invention	Patent applicant (reference)
Impregnation of textiles with microencapsulated active substances, including fire retardants (Unflame BP) in polyurethane, polyorganosiloxane, polyolefin, or epoxy resin walls.	Kanegafuchi Spinning [51]
Microencapsulated fire retardants, incorporated into synthetic fibers or applied on fabrics, released from microcapsules at the ignition temperature of textiles. Wall materials include polystyrene, polyvinyl alcohol, phenol–formaldehyde resins and urethane polymers.	Kanegafuchi Spinning [52]
Self-extinguishing and laundering resistant textiles, impregnated with a dispersion of microencapsulated fire retardants in polymeric walls, and acrylate binders.	Asahi Chemical [53]
Fire-resistant fibers, containing microencapsulated fire retardants and perfumes, formulated with acrylamide–methyl acrylate polymer binders.	Japan Exlan [54]
Flame resistant polyamide or polyester carpets, containing volatile fire retardants in heat sensitive urea–formaldehyde or melamine-formaldehyde polymeric wall microcapsules.	Champion International [55]
Washfast fire-resistant polyester fabrics, based on microencapsulated halogenated fireproofing agents and acrylate copolymer binders.	Matsumoto [56]
A fire retardant material, comprising both ammonium polyphosphate particles microencapsulated within a melamine or melamine-based resin, and melamine-based particles retained in a base material. The material can be applied to textiles.	Dartex Coatings [57]
Microcapsulation of flame retardants by *in situ* polymerization of aminoaldehyde resins, for applications in textiles and other technical products.	Aero [8]

3.4.6 Microencapsulated agents for textile sizing and adhesive bonding

Microencapsulated sizing agents, adhesives, adhesive activators and crosslinking agents have been used for textile sizing and bonding. The microcapsule core release mechanisms include pressure, heat or a combination of both (Table 3.6).

Table 3.6: Inventions on microencapsulated agents for textile sizing and bonding.

Invention	Patent applicant (reference)
Treatment of knitted fabrics with microencapsulated sizing agents, composed of vinyl adhesives, solvents, or other compounds susceptible to reactions with textiles.	Carlier [58]
Cold sealable textiles, based on microencapsulated adhesive components, applied in systems of cold adhesive bonding, reinforcing and stiffening, used in clothing, shoe, leather, and fur industries.	Hermann Windel Co. [59]
Textile adhesive composites with microencapsulated crosslinking agents, useful for reinforcement of textiles by hot pressing.	Lainiere de Picardie [60]

3.4.7 Microencapsulated blowing agents and expandable microcapsules for leather substitutes

Applications of expandable microcapsules in textile products include flexible light weight leather substitutes, waterproof coatings, anti-slip materials for carpets, and expandable sewing threads (Table 3.7).

Table 3.7: Examples of inventions on blowing agents and expandable microcapsules.

Invention	Patent applicant (reference)
Foamable microcapsules for the production of leather substitutes.	Achilles Corporation [61]
Composition of thermally expandable microcapsules in a polymeric resin, and production of light weight flexible leather substitute by hot pressing.	Bando Chemical [62]
Compositions of colored hollow silicate microspheres, coupling agents and organic polymer binders for light weight leather substitutes on textile fabric supports.	Meisei Rejinokara [63]
Anti-slip adhesive nonwoven textiles, containing polyester supports, laminated with a polyethylene film and microcapsules that expand by heating to produce an anti-slip foam.	Nippon Kako Seishi [64]

3.4.8 Microencapsulation for textile water proofing

Increased impermeability and water proofing of textile surfaces can be achieved by expanding a layer of microcapsules on a porous support into an impermeable layer, or by applying microencapsulated water proofing agents, and releasing them from microcapsule cores. In both cases, heat treatment plays a crucial role in microcapsule activation (Table 3.8).

Table 3.8: Examples of inventions using microcapsules for textile water proofing.

Invention	Patent applicant (reference)
Waterproof sewing threads, impregnated with heat expandable thermoplastic compositions containing microcapsules, used for sewing leather substitutes, textiles for raincoats, tents and shoes.	Nippon Rubber [65]
Textile coating compositions, containing waterproofing agents in inorganic porous microcapsules, give water resistance by heat treatment.	Toray Industries [66]
Thermally treated coating of synthetic resin microcapsules expands into an impermeable surface layer on porous polyurethane support.	Takashimaya Nippatsu [67]
Heat expandable microcapsules with a thermoplastic resin wall and a gas generating core, formulated with a binder, expand by heating and give flexible light weight waterproof woven, knit or nonwoven textiles.	Owari Seisen [68]

3.4.9 Microcapsules in textile softening and antistatic compositions

Fabric softeners and antistatics for textile washing and drying employ microcapsules to solve the incompatibility of antistatic compounds and anionic surfactants in detergents, to incorporate liquid ingredients into solid formulations, to add hydrophobic components into water-based formulations, and to achieve a prolonged release of fragrances (Table 3.9).

3.4.10 Microencapsulated ingredients in textile detergents

There are patents on microencapsulated components in detergent formulations for washing textile goods. The main applications include microencapsulated enzymes (Table 3.10); bleaching and whitening agents (Table 3.11); and perfumes and other additives, such as dry defoamers, dyes and cleaning chemicals (Tables 3.12 and 3.13).

Table 3.9: Examples of textile softening and antistatic compositions with microcapsules.

Invention	Patent applicant (reference)
Polyurethane foam sponge to be added to wet laundry in a rotating dryer, containing microencapsulated fabric softeners and perfumes in coacervation microcapsules.	Colgate Palmolive [69]
Pre-softener and washing composition with pressure-sensitive microcapsules, comprising a perfume core and urea resin wall.	Procter and Gamble [70]
Softening antistatic agents, microencapsulated by interfacial polymerization, released by heat and abrasion.	Procter and Gamble [71]
Fabric softening formulations comprising both microencapsulated and free fragrances; with microcapsule walls made of gelatin, dextrin, gum arabic, modified starch, urea-formaldehyde resin or other polymers.	International Flavors and Fragrances [72]
Aqueous textile softener compositions for the rinse stage of laundering, containing microencapsulated perfumes in a complex coacervate shells of gelatin and a polyanion.	Procter and Gamble [73]
Cationic polymer stabilized microcapsule compositions for fabric softeners.	Colgate Palmolive [74]
Friable perfume microcapsules for dryer-added fabric conditioning.	Procter & Gamble [75]
Use of a cross-linked cationic polymer to provide stability to microcapsules in a fabric softener composition.	Procter & Gamble [76]

Table 3.10: Examples of patents on microencapsulated enzymes for textile detergent formulations.

Invention	Patent applicant (reference)
Production of microencapsulated enzymes for detergents by spray-drying of compositions consisting of inorganic salts, water-soluble binders and enzymes.	Toyo Jozo & Fuji [77]
Spray cooled microencapsulated proteases with fatty acid or fatty alcohol walls, incorporated into sodium perborate bleach compositions.	Henkel [78]
Mixed granulate bleach and enzyme compositions, consisting of a dry peroxy acid bleach, and enzyme microcapsules.	Procter and Gamble [79]
Liquid detergent formulations with microencapsulated enzymes.	Showa Denko [80]
Microencapsulation of enzymes for detergents with a mixture of hard and soft waxes for wall materials.	Lever Brothers [81]

Tab. 3.10 (*continued*)

Invention	Patent applicant (reference)
Microencapsulation of proteases, lipases and amylases into dual-walled microcapsules, to achieve time release and prevent enzyme deactivation by halogen bleaches in mixed bleach/enzyme detergent compositions.	Olson [82]
Microencapsulation of enzymes with water-soluble alkali metal silicates and additives, to achieve prolonged enzyme storage stability in the presence of oxidant bleaches.	Clorox [83]
Liquid detergent concentrate with microencapsulated enzymes; microcapsule polymeric walls remain permeable for water and small molecules.	Novo Nordisk [84]
Liquid detergent composition, comprising microencapsulated enzymes; microcapsules are produced by crosslinking of a polybranched polyamine.	Novozymes [85]

3.4.10.1 Enzymes

In early patents, enzyme encapsulation improved detergent storage stability, reduced dusting and minimized health hazards in detergent factories and households. Subsequently microencapsulation was used to protect enzymes against the activity of aggressive additives, especially bleaches. Newer patents used the advantage of encapsulation to incorporate enzymes into liquid and gelled detergent formulations (Table 3.10).

3.4.10.2 Bleaching agents and whiteners

Microencapsulated bleaching agents in laundry formulations have the advantages of being separated from the oxidation sensitive components in the detergent compositions, to prevent reduction of their bleaching capacity, and to reduce the damage to fabrics (Table 3.11).

Table 3.11: Examples of patents on microencapsulated bleaching agents and whiteners in textile laundry formulations.

Invention	Patent applicant (reference)
Microencapsulated fluorescent whiteners, protected from the oxidative degradation by NaOCl bleach.	Prurex [86]
Spray-dried microcapsules of perborate activators for bleaching and washing liquors.	Henkel [87]
Microencapsulated fluorescent bis(triazinylamino) stilbenedisulfonate brighteners for white and colored fabrics.	Henkel [88]

Tab. 3.11 (*continued*)

Invention	Patent applicant (reference)
Fatty acid microencapsulated ethylenediamine tetraacetate for bleaches and detergents containing active oxygen.	Henkel [89]
Fatty acid microencapsulated tetraacetylglycoluril activator for sodium perborate in detergents.	Henkel [90]
Free flowing granular laundry detergents, containing chlorine donors and microencapsulated fluorescent whiteners in carboxymethyl cellulose microcapsules, produced by spray-drying.	Ciba-Geigy [91]
Molten fatty acids microencapsulation of peroxide bleaching agent activators, such as tetraacetylglycoluril and tetraacetilethylenediamine.	Nobel Hoechst Chimie [92]
Fluidized bed microencapsulation for the production of free flowing potassium dichloroisocyanourate bleach particles, encapsulated by an inner layer of fatty acid, and an outer layer of a water-soluble fatty acid salt.	Lever Brothers [93]
Fluidized bed technology, using poly(vinylpyrrolidone) wall material for microencapsulation of sodium perborate bleaching agents for detergent compositions.	Unilever [94]
Spray coating process of active chlorine bleach microgranules with two layer walls, utilizing a mixture of fatty acids and waxes.	Unilever [95]
Fluidized bed technology for the encapsulation of sodium percarbonate with a molten polyethylene wax.	Interox [96]
Microencapsulated bleaching agents consisting of an active halogen oxidizing material, and a fatty acid soap wall.	Lever Brothers [97]
Clear detergent gel compositions, comprising microencapsulated chlorine and oxygen bleaches, bleach precursors, enzymes, fabric softeners, surfactants and perfumes.	Lever Brothers [98]
Compositions of molecularly encapsulated preformed peroxyacids and bleach catalysts.	Procter & Gamble [99]

Table 3.12: Examples of microencapsulated antifoaming agents in textile detergents.

Invention	Patent applicant (reference)
Detergent composition for machine washing, containing antifoaming agents, microencapsulated by gelatin-gum arabic walls.	Unilever [100]
Low foaming detergent compositions, containing silicon defoamers, microencapsulated with methyl- or carboxymethyl cellulose.	Henkel [101]

Table 3.13: Examples of patented microencapsulated perfumes, dyes, softeners and other additives in laundry detergents.

Invention	Patent applicant (reference)
Microcapsule formulations for textile dry cleaning, containing microcapsules of foam forming liquid detergents or soaps, produced by coacervation.	Werner und Mertz [102]
Detergent compositions including a microencapsulated water-soluble dyes, bleaching agents, surfactants and a perfumes.	Dainichiseika Colour and Chemicals [103]
Laundry detergents containing perfumes in water-insoluble friable microcapsules. The microcapsules remain intact during laundering, and are fractured during handling of the laundered textiles, thus releasing the perfume.	Procter and Gamble [104]
Microencapsulation of cationic softeners with wall materials including waxes, fatty acids, fatty alcohols or fatty esters with the melting points above 50 °C; and their inclusion into powder detergents, to prevent undesired precipitation with anionic surfactants during the fabric laundering, but releasing the softener in a heated dryer.	Procter and Gamble [105]
Microencapsulated photoactivator dye compositions that are quickly soluble in water, for applications in detergents.	Procter & Gamble [106]
Synthesis of microcapsules containing fragrances or perfumes for laundry detergents or cleaning products.	BASF [107]
Structured liquid detergents formulations with incorporated microcapsules.	Unilever [108]
Storage stable microcapsules of scents in detergent compositions that are low in formaldehyde, produced by reacting aromatic alcohols or ethers and aldehydic components, and optionally a (meth)acrylate-polymers.	Henkel [109]
Laundry detergent compositions comprising microcapsules, pH tuneable di-amido gellants and surfactants.	Procter & Gamble [110]

3.4.11 Textiles with microencapsulated fragrances and perfumes

Fragranced textiles, containing microencapsulated essential oils, aromas and perfumes, have been developed to either slowly release their contents through permeable walls, or to have completely impermeable walls, and open only by application of mechanical pressure and rubbing whenever the wearer moves. A combination of both release mechanisms is also possible. After the problems of controling the release have been solved, and better washfast binders introduced, a new generation of aromatic textiles entered the market that remain fragrant over a prolonged period of time, resist dry cleaning, or keep the microcapsules over several washing cycles. Applications of microencapsulated fragrances, perfumes and antimicrobial essential oils in woven and nonwoven textiles range from perfumed curtains, bed linen, shirts, socks and hosiery to antimicrobial towels, shoe insoles, and textiles for seats in public transportation (Table 3.14). Figures 3.12–3.15 illustrate some examples of our own research [111].

Table 3.14: Examples of patents on fragrant microcapsules in textile products.

Invention	Patent applicant (reference)
Microcapsule-coating of fabrics for fragranced linings and ribbons, coated with gelatin-gum arabic coacervate microcapsules of lavender or pine oil.	Eurand [112]
Fragranced towels containing microencapsulated perfumes. Coating composition include acrylic polymer binders and pressure-sensitive microencapsulated perfumes.	Shibata Towel [113]
Fragrant textiles that retain fragrances after repeated washing. Coating compositions contain urea resin microencapsulated fragrances and silicone binders, such as epoxy modified dimethyl siloxane. Applications include dyed and softener treated fragrant silk scarves and handkerchiefs; washfast silk neckties with lasting fragrances; fragrant hand knitting and handicraft yarns; fragrant leather substitutes and cotton jersey shirts; fragrant cotton bedding with improved washfastness; waterproofed polyester textiles with jasmine microcapsules; and waterproofed woven fabrics, knits and yarns, treated with urea resin jasmine flower perfume microcapsules.	Kanebo [114–119]
Synthesis of essential oil microcapsules by *in situ* polymerization, and a technological process for preparing textile carriers saturated or coated with microencapsulated scents.	Aero [120]
Garments, interior materials, filters and automobile interiors, composed of nonwoven textiles, coated with essential oils in cyclodextrin and porous silica microspheres.	Osaka Juki [121]
Fragrant textiles with improved durability and slow release of fragrances, comprising hollow fibers with microencapsulated perfumes.	Kanebo [122]
Microencapsulated grass aroma in rattan-imitating mat surface materials in home textiles.	Shanghai Shuixing Home Textile [123]
Incorporation of a polyester microencapsulated fragrances into spinning materials to produce fragrant fibers.	Iangsu Zja New Material Co. [124]

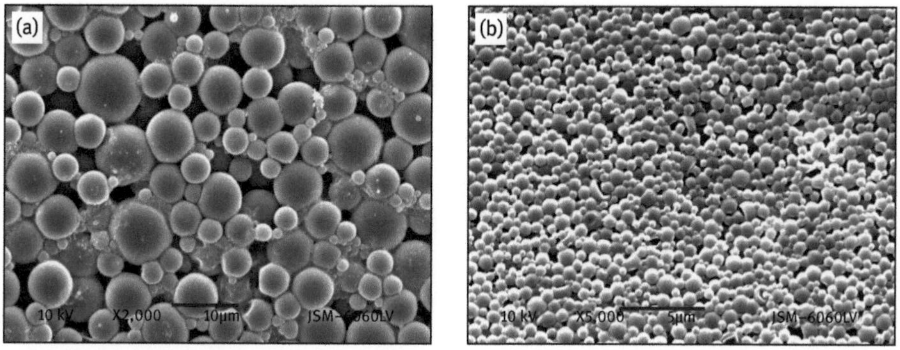

Fig. 3.12: Scanning electron microscope (SEM) photograph of microcapsules with a rose fragrance (left,), and eucalyptus essential oil (right), prepared by *in situ* polymerization, to be applied in fragranced textiles (SEM 2000 ×).

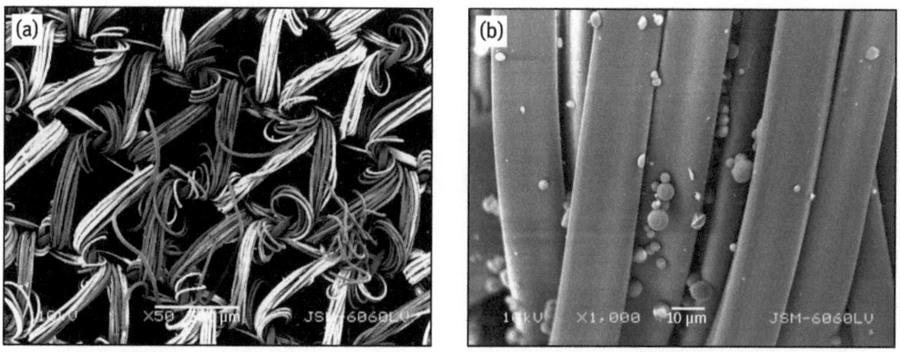

Fig. 3.13: Nylon pantyhose textile with microencapsulated rose oil in pressure-sensitive microcapsules, produced by *in situ* polymerization (SEM, left 50 ×, right 1000 ×).

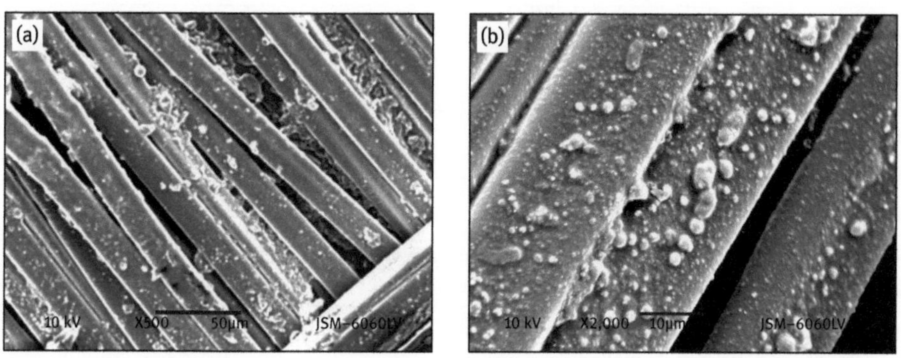

Fig. 3.14: Scanning electron microscope (SEM) of pressure-sensitive microencapsulated fragrances on a decorative wrapping ribbon; fragrances are released by mechanical pressure, applied by handling (SEM, left 500 ×, right 2000 ×).

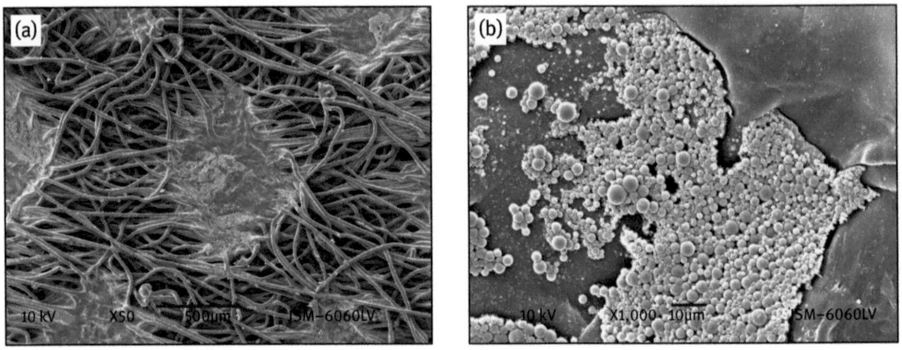

Fig. 3.15: Nonwoven textile handkerchief with microencapsulated decongestant eucalyptus oil (SEM, left 50 ×, right 1000).

3.4.12 Textiles with microencapsulated animal repellents

To obtain prolonged insecticidal and insect repellent effects of fibers and textiles, and to reduce the toxicity and volatility of active compounds, insect repellents and/or insecticides can be microencapsulated and applied to textiles (Table 3.15).

Table 3.15: Examples of patents on repellent and insecticide microcapsules in textile products.

Invention	Patent applicant (reference)
Insect repelling carpets, curtains and sheets, manufactured by the application of a sustained release microencapsulated diethyltoluamide insect repellent in a polyamine resin.	Hosokawa Textile [125]
Fabrics and panty hoses with durable insect repellence, finished with a composition containing aminoaldehyde polymer microcapsules of N,N-diethyl-m-toluamide repellent and acrylic polymer binder.	Toyobo [126]
Long-lasting mosquito repelling panty hoses, treated with microencapsulated repellents, such as Deet.	Toyobo [127]
Durable microencapsulated insect repellents as household sprays for textile products, e.g. for mothproofing.	Kanebo [128]
Improved fixing process of repellent microcapsules on textile fibers, contributing to better washfastness.	Hasokawa Textile [129]
Production of insecticidal microcapsules for fibers or textiles, and applications in insecticide-processed fibers or textiles.	Union Kagaku [130]
Insect repellent textiles comprising a natural or synthetic fabric, a microencapsulated insect repellent, and a binder.	Innovatec [131]
Microencapsulated biocide and repellent compositions with a double repellence action, used in textile garments.	Mateo Herrero María Pilar [132]

In addition to insect repellents, other animal repellents have been microencapsulated. For instance, prolonged release microencapsulated deer and rabbit repellents on nonwoven textiles were developed for horticultural and agricultural use [133].

3.4.13 Textiles with microencapsulated antimicrobial, disinfectant and deodorant components

Several essential oils and plant extracts have antimicrobial and deodorant properties. Because they are liquids, microencapsulation is required for the conversion into the solid state. At the same time, a prolonged activity of microencapsulated active substances can be achieved.

Inventions on microencapsulated antimicrobials for textile applications include various textile coating compositions with antimicrobial effects, as well as specific coating procedures and additives (Table 3.16).

Table 3.16: Examples of textile inventions with microencapsulated antimicrobial, disinfectant and deodorant components.

Invention	Patent applicant (reference)
Deodorant textiles, coated with porous microcapsules containing plant oils and extracts, such as wood oil and camellia leaf extract.	Toray [134]
Bactericidal printing compositions for garments, containing porous microcapsules with bactericides.	Tokyo Houlaisha [135]
Manufacturing of antibacterial garments by printing fabrics with a mixture of binders and porous bactericide microcapsules.	Tokyo Houlaisha [136]
Production of washfast antibacterial fabrics by immersing polyester-cotton knits in a dispersion of melamine resin wall microcapsules, containing N,N-diethyl-m-toluamide core, and a polyurethane binder.	Asahi [137]
Microencapsulated disinfectants, such as dimethyldidecylammonium chloride and glycerol, in ethylene-vinyl acetate copolymer walls, for applications in wound dressings, medical and surgical gloves, textiles, and paper products.	Flamel Technologies [138]
Textile fiber structures with adhered gelatin microcapsules, containing biocides, such as gamma-oryzanol in olive oil, and silicone binders.	Toray [139]
Preparation of textile carriers, coated or saturated with microencapsulated antimicrobial essential oils, produced by *in situ* polymerization of aminoaldehyde resins.	Aero [120]

Tab. 3.16 (*continued*)

Invention	Patent applicant (reference)
Industrial textiles with fungicidal and bactericidal activity, containing microencapsulated catechins and/or saponins.	Elb Company [140]
Anti-mite and antibacterial polyester staple fibers, prepared by spinning, containing micro anti-mite ceramic powders and silver-loaded nano titanium dioxide composites.	Shanghai Different Chemical Fibre Co. [141]

As an example of our work, we developed antimicrobial textile shoe insoles, based on nonwoven polyester textiles, impregnated with a mixture of microencapsulated essential oils of sage, lavender and rosemary (Figure 3.16). Pressure-sensitive aminoaldehyde resin microcapsules with partially permeable walls were prepared using a modified *in situ* polymerization method. For the impregnation of textiles, a technique for the transport of the textile carrier through the impregnation basin was used. Product testing proved the sustained release of essential oils from microcapsules in worn shoe insoles, and antimicrobial activity of the essential oil mixture against the microorganisms *Staphylococcus aureus, Candida albicans* and *Trichophyton mentagrophytes* [142, 143].

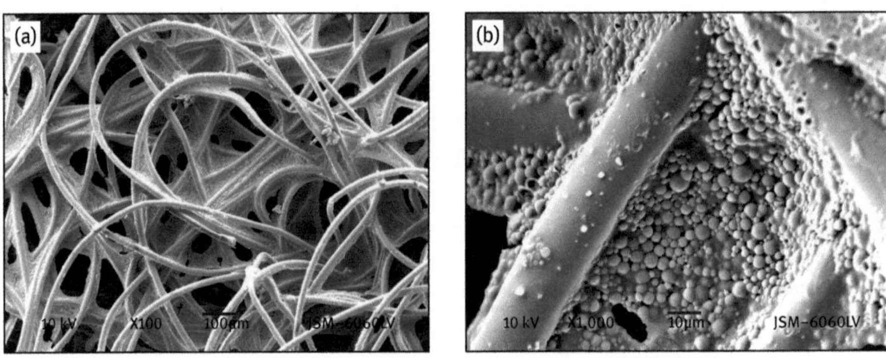

Fig. 3.16: Nonwoven textile for shoe insoles, impregnated with pressure-sensitive microcapsules, containing an antimicrobial composition. Essential oils are protected from oxidation until the microcapsules open by mechanical pressure during walking (SEM, left 50 ×, right 1000 ×).

3.4.14 Bioactive medical and cosmetic textiles with microencapsulated ingredients

In the 1990, the first inventions of medical and cosmetic textiles introduced added-value textile products with prolonged effects, such as antimicrobial effects, accelerating blood circulation, improving the physiological condition of skin, skin hydration, ageing prevention, or skin whitening. Soon other inventions followed, aiming at pain relief, itch suppression, accelerating the metabolism of water, reducing cellulite, and similar effects. The microcapsules are typically not broken when produced, processed, or laundered, but gradually burst open when the textiles are worn. The formulations are applied to the fabrics by soaking, coating or spraying; microcapsules can also be formulated as sprays, which tightly adhere the microcapsules to textile structures, such as hosiery, underwear, bedlinen, and bandages (Table 3.17).

Table 3.17: Examples of inventions including microcapsules in cosmetic and medical textiles.

Invention	Patent applicant (reference)
Pressure-sensitive microencapsulated physiologically active compounds, adhered to textile fibers by polymeric binders, with a prolonged antimicrobial and blood circulation accelerating effects.	Kanebo [144]
Textile structures with microencapsulated substances for improving the physiological conditions of human skin, such as functional vitamin C, vitamin E, seaweed extracts, antipruritic and analgesic agents, and/or aromatic agents – designed for medical auxiliary materials, bed clothes, stockings and underwear.	Kanebo [145]
Medical/cosmetic textiles with microencapsulated subcutaneous fat controllers – extracts of medicinal plants – to disperse fats, accelerate the metabolism of water, reduce stasis and cellulite.	Toray [146]
Bioactive moisturizing polyamide and silk protein elastic textiles with microencapsulated moisturizers, used in direct contact with skin as bandages, elastic supports, and hosiery.	Dim S. A [147]
Microencapsulation of bioactive body care substances by coacervation of gelatin-alginate complexes. Microcapsules are fixed on textile supports by crosslinking alginate or chitosan polysaccharides.	Ted Lapidus [148]
Pharmaceutical and medical functional textiles comprising a textile carrier and microencapsulated physiologically active substances.	Blücher GmbH [149]
Clothing for daily pharmacological treatment of fungal infections, consisting of microcapsules grafted on the textile material to manage moisture, and of microencapsulated antifungal agents.	InnovaTec [150]

3.4.15 Textile decontaminants, filters and odor absorbers

Some patents describe the incorporation of microcapsule bearing absorbents and decontaminants into textiles for special purposes, such as waste water purification, odor absorption, and military decontamination (Table 3.18).

Table 3.18: Examples of microencapsulated components in textile-based filters, odor absorbers and decontaminants.

Invention	Patent applicant
Microencapsulated conventional decontamination agents, effective for the deactivation of toxic mustard blistering agents or toxic nerve agents, applied to clothing fabrics in acrylic resinous binder finishes.	US Dept of the Army [151]
Coagulation filter textiles, coated with organic polymeric or inorganic coagulants, microencapsulated by microporous inorganic walls, for applications in waste water treatment.	Kanai Hiroyuki [152]
Textile odor-absorbing car interior linings with odor-absorbing microcapsules.	GM Global Technology Operations [153]
Functional textiles comprising a backing with microencapsulated active ingredients, used in protective clothing for civil and military use, and in filters for removing harmful materials, odors and poisons.	Blücher Gmbh [154]

3.4.16 Textiles for active thermal control

Textiles for active thermal control have been one of the fast growing product areas of microencapsulation technology applications (Table 3.19). In addition to attempts to convert sunlight energy into chemical and later thermal energy, a wave of inventions and practical applications utilized microencapsulated PCMs that absorb or emit heat at their phase change transition temperature (Figure 3.17). Typical examples of PCMs are strait chain paraffinic hydrocarbons with 13 to 28 carbon atoms, and the phase change temperatures ranging from −5.5 °C to +61 °C. As they are flammable and liquid above the phase transition temperature, microencapsulation is essential

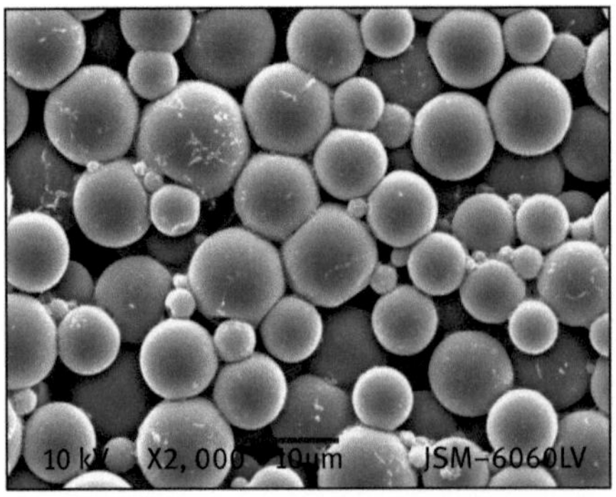

Fig. 3.17: Scanning electron microscope (SEM) photograph of microcapsules containing a paraffinic PCM, prepared by *in situ* polymerization, to be applied in functional textiles (SEM 2000 ×).

for their practical use in various thermal management applications. In functional textiles, microencapsulated PCMs function as heat absorbers or as barriers against cold, and are incorporated into products with enhanced thermal properties and active thermal control [155].

The choice of suitable PCMs depends on the latent heat of the phase change and the transition temperature. In general, the higher the PCM's latent heat of phase change, the more thermal energy a material can store. According to their phase change temperature ranges, the PCMs are categorized into three main groups – the heating, the cooling and the buffering PCMs [156]:

- The phase transition temperature of the heating PCMs is above the body's normal skin temperature. When a heating PCM is warmed above its transition temperature and placed in thermal contact with the skin, the temperature gradient flows from the PCM into the body.
- The cooling PCMs have a phase transition temperature below the body's normal skin temperature. When chilled below their transition temperature, the temperature gradient flows from the body into the PCM.
- The phase transition temperature of the buffering PCMs is slightly below the normal body temperature. These materials absorb or release heat depending on environmental and metabolic conditions.

To include PCM microcapsules into textile products, different systems have been developed, such as:

- the incorporation of PCM microcapsules into the textile fibers before or during the spinning process;
- the coating of fibers and fabrics with compositions of PCM microcapsules and binders;
- the insertion of polymer foams with microcapsules PCM into textile products;
- the preparation of complex composites with three or more layers.

Table 3.19: Examples of microcapsule involving inventions in functional textile products with heat storing and releasing properties.

Invention	Patent applicant (reference)
Incorporation of PCM microcapsules into textile fibers by adding microcapsules to the molten polymer or to the polymer solution before spinning.	Triangle R&D [157]
Textile materials with heat storing properties, capable of converting sunlight energy into chemical energy, and store it for later release, based on microencapsulated norbornadiene derivatives and catalysts, formulated in resin binders.	Kiyokawa [158]

Tab. 3.19 (*continued*)

Invention	Patent applicant (reference)
Coating compositions for textile fibers, containing PCM microcapsules, and binders as liquid polymers or polymer solutions (polyurethanes, nitrile and chloroprene rubbers, polyvinyl alcohol, silicones, ethylene/vinyl acetate copolymers, and acrylic polymers).	Triangle R&D [159]
Synthetic foam inserts with anisotropically distributed leak resistant dual-walled PCM microcapsules, used in textile products.	Bryant & Colvin [160, 161]
Applications of PCM microcapsule containing insulation materials in shoe insoles, ski boot liners, thermal socks, gloves and face masks for cold weather activities, diver's wet suits, heating or cooling blankets for treating hypothermia or fever patients, and therapeutic heating or cooling orthopaedic joint supports.	Buckley [156]
Applications of microencapsulated PCMs in thermal protection liners for diver's wetsuits, dry suits and hot water suits for extreme cold water diving, hot water diving, or as emergency backup heat sources.	USA Secretary of the Navy [162]
Inclusion of PCM microcapsules in composites with several layers, for improved thermal control and comfort of the wearer.	Buckley [163]
Three-layered thermal insulating fabric structure, containing microencapsulated paraffinic PCMs, providing a dynamic thermal response in clothing.	Outlast Technologies [164]
Metal oxide gel coated microcapsules, containing PCMs, with improved mechanical stress and flame resistance, to be incorporated into foams, fibers, slurries, coatings.	Frisby Technologies [165]
Nonwoven textiles with reversible enhanced thermal control, containing a web bonded by polymeric binder and a microencapsulated PCM.	Carl Freudenberg [166]
Nanostructured PCMs used for thermoregulatory coatings for use in a wide range of applications, including cooling textiles and wipes.	Bioastra Tech. [167]
Two-piece wearable absorbent articles, such as diapers, comprising microencapsulated PCMs and absorbent inserts.	Procter & Gamble [168]

3.4.17 Microcapsules in self-cleaning textiles and self-healing fibers

A new generation of high-tech functional textiles is emerging, known also as smart textiles; some of them contain various microencapsulated components to achieve self-refreshing, self-cleaning, abrasion-resistant, or self-healing properties (Table 3.20).

Table 3.20: Examples of inventions of self-cleaning surfaces and self-healing fibers, containing microencapsulated components.

Invention	Patent applicant (reference)
Microcapsules for self-refreshing textiles, containing microencapsulated polyols.	Despature et fils [169]
Silicone textile surface treatment (conditioning, hydrophobing, softening) with silicate wall microcapsules.	Dow Corning [170]
Particles with a structural surface, prepared by siloxane or silane emulsion polymerization, useful to produce abrasion resistant self-cleaning surfaces.	Wacker Chemie [171]
Hybrid high-strength carbon fiber/epoxy composites, reinforced with ultrathin toughening and self-healing core-shell fibers.	NDSU Res. Foundation [172]
Combination of a self-healing polymer matrix and carbon fiber reinforcement, designed to be used in space missions.	NASA [173]

3.5 Concluding remarks

The idea of using microencapsulation technology in added-value textile products was born soon after the introduction of the large-scale production of microcapsules for pressure-sensitive copying papers. Microencapsulation for textiles became a research and development area with a strong industrial intellectual property protection, as patent documents outnumbered scientific articles. In the past some reviews were prepared to summarize research and development achievements [111, 174–178]. The survey in this chapter, prepared by analysis of inventions from the beginning of the microencapsulation technology to the present day, reveals that the first burst of patents on microcapsules for textiles in the 1970s brought the following microencapsulated products: (i) dyes and pigments for special textile dyeing and printing techniques; (ii) catalysts, crosslinking agents and enzymes for textile treatment; (iii) reagents for textile sizing and bonding; (iv) fire retardants for fire-resistant textiles; (v) expandable microcapsules for the production of light weight leather substitutes and water proofing of porous textile surfaces; (vi) fragrant textiles with microencapsulated essential oils and aromas; (vii) ingredients in textile detergents and softeners, including enzymes, bleaches, softeners and antistatics for textile washing and drying compositions.

After a short stagnation of research in the beginning of the 1980s, there was a second wave of textile microcapsule patents, with new concepts of the following microecnapsulated products: (viii) thermochromic materials, which utilized temperature changes for color development and fading, and microencapsulated photochromic dyes – the results being thermochromic sports and leisure garments, and photochromic curtains, sportswear and shirts; (ix) blowing agents and expandable microcapsules for leather substitutes and textile water proofing; (x) components in textile filters, odor absorbers and decontaminants.

After 1990, the inventions were further extended and upgraded to: (xi) prolonged release bioactive medical and cosmetic textiles with microencapsulated bioactive/healing components; (xii) antimicrobial, disinfectant and deodorant textiles; (xiii) repellent and insecticidal textiles, (xiv) functional textiles with heat storing and releasing properties, based on microencapsulated PCMs, applied in sportswear and special technical apparel with active thermal control.

After the year 2000, new inventions appeared in almost all previously known application fields, particularly in the domains of microencapsulated thermochromic and photoschromic dyes for color changing fabrics and sensor fibers; new techniques and solutions in textile dyeing and printing, involving microcapsules; and microencapsulation of additives in sophisticated compositions of textile detergents and softeners.

Since 2010 a new generation of microcapsule-based inventions have been emerging, applying microencapsulated components to achieve (xv) self-cleaning and/or self-healing properties of high-tech smart textiles.

3.6 Acknowledgments

The research on microencapsulation was financially co-supported by: the Faculty of Natural Sciences and Engineering, University of Ljubljana; the Slovenian Research Agency (projects L2-5571, L1-6230 and L4-1562); and by the ERO, Chemical, Graphic and Paper Manufacturers, d.d. Celje, Slovenia. Samples of textiles with microcapsules for SEM imaging were kindly provided by Boštjan Šumiga, Ph.D., and Mr. Emil Knez from the AERO company.

3.7 References

[1] Web of Science. Thomson Reuters, 2015. Accessed 8 January 2015, at: http://home.izum.si/izum/ftbaze/wos.asp

[2] Espacenet, European Patent Office, 2015. Accessed 8 January 2015, from: http://worldwide.espacenet.com/advancedSearch?locale=enEP

[3] Boh, B., Šumiga, B., *In situ* polymerisation microcapsules, Bioencapsulation Innovations, 2013, 3–6.

[4] Knez, E. A., Method for preparing microcapsules, YU131984, Aero, 1986.

[5] Li, W., Wang, J., Wang, X., Wu, S., Zhang, X., Effects of ammonium chloride and heat treatment on residual formaldehyde contents of melamine-formaldehyde microcapsules, Colloid Polym Sci **285** (2007) 1691–1697.

[6] Wei, L., Zhang, X. X., Wang, X-C., Niu, J-J., Preparation and characterization of microencapsulated phase change material with low remnant formaldehyde content, Mater Chem Phy **106** (2007) 437–442.

[7] Berthier, D., Leon, G., Paret, N., Ouali, L., Stable formaldehyde-free microcapsules, WO 2011/161618, Firmenich, 2011.

[8] Knez, E., Vrtačnik, M., Ferk-Savec, V., Starešinič, M., Boh, B., Production of melamine-formaldehyde PCM microcapsules with ammonia scavenger used for residual formaldehyde reduction, Acta Chim Slov **58** (2011) 14–25.

[9] Golja, B., Boh, B., Šumiga, B., Forte-Taver, P., Printing of antimicrobial microcapsules on textiles, Color Technol **128** (2012) 95–102.

[10] Boh, B., Knez, E., Starešinič, M., Microcapsules in textile industry. In: Arshady, R., Boh, B., (eds.) Microcapsule patents and products, The MML series, Vol. 6. London: Citus, 2003. pp. 235–269.

[11] Goorhuis, H., Textile dyeing and printing with powders and microcapsules, DE2161381, Sandoz, 1972.

[12] Imada, K., Sueda, Y., Abeta, S., Yamada, E., Solvent dyeing of polyester with microcapsulated dyes, JP48092665, Sumitomo, 1973.

[13] Totoki, C., Dye capsules, JP49036985, Totoki, 1974.

[14] Ikeda, K., Sato, M., Kato, H., Colorants for textile printing, JP49128021, Toa Gosei, 1974.

[15] Moore, R. F., Schiller, F. C., Woven fabric printing ribbon having rupturable microcapsules bonded to its surface US3817773, National Cash Register, 1974.

[16] Oshiage, K., Takabayashi, T., Printing of textiles for speckled pattern, JP50065678, Sakai Textile, 1975.

[17] Oshiage, K., Takabayashi, T., Microencapsulation of dyes for printing of textiles, JP50067778, Sakai Textile, 1975.

[18] Bartlett, J. R., Shinner, C., Reprint strip material preprinted with dye and method of printing sheet material with it, DE2444140, Dickinson Robinson, 1975.

[19] Oshiage, K., Kubo, M., Photographic screen printing of textiles for sketchy pattern, JP51123379, Nippon Kayaku, 1976.

[20] Okuyama, H., Ishimaru, S., Asano, N., Transfer printing of polyester fabrics, JP51012755, Seiren, 1976.

[21] Kiritani, M., Transfer printing of cotton fabrics, JP51070379, Fuji Photo Film, 1976.

[22] Nakano, T., Sakaoka, K., Compositions for multicoloring fabrics with speckles, JP53042833, Hayashi Kagaku Kogyo, 1978.
[23] McBride, D. T., Godfrey, T. E., Process for improving pattern definition in dyeing of textiles, EP202856, Milliken Research, 1986.
[24] Mihara, K. G., Transfer printing of fabrics, JP 58030435, Mihara Kogaku Gijutsu, 1983.
[25] Hare, D. S., Williams, S. A., Imaging thermal transfer system and transferring a thermal recording image to a textile, leather, ceramic or wool, glass or plastic, WO9925917, Foto Wear, 1999.
[26] Hare, D. S., Williams, S. A., Imaging transfer system, WO9926111, Foto Wear, 1999.
[27] Zhang, X., Technology for rapidly dyeing polyester fiber cloth by using dispersible dye microcapsules, CN102454120, Jiangsu Shunyuan Textile Technology, 2012.
[28] Dong, J., Modified one-bath dyeing technology of polyester rayon fabric dispersed microcapsule active dye, CN102605657, Shaoxing Dongshi Textile Dyeing Printing Technology, 2012.
[29] Spogli, R., Method for colouring natural textile fibers, EP2628849, Ferrini - Societa 'a Responsabilita' Limitata, 2013.
[30] Saito, T., Ink composition for ink jet textile printing, EP2641943, Seiko Epson Corporation, 2013.
[31] Aitken, D., Burkinshaw, S. M., Towns, A. D., Textile applications of thermochromic systems, Rev Prog Color Relat Top **26** (1996) 1–9.
[32] Ruggeri, C., Textile material coated with liquid crystals, GB2116578, Ruggeri, 1983.
[33] Kito, T., Matsunami, N., Nakasuji, K., Shibahashi, Y., Fibers with color memory and exhibiting reversible color change from low temperature to high temperature, JP62156355 Pilot Ink, 1985.
[34] Shibahashi, Y., Nakasuji, N., Kataoka, T., Inagaki, H., Kito, T., Ozaki, M., et al., Thermochromic textiles, DE3602805, Pilot Ink, 1986.
[35] Kamata, K., Kitagawa, Y., Hoshikawa, R., Reversible photochromic colored textile good and its coloring method, JP6212579, Matsui Shikiso, 1994.
[36] Itigawa, Y., Hoshikawa, R., Reversibly color-variable coloring method by padding and colored textile product, JP7109681, Matsui Shikiso, 1995.
[37] MATEO Project. State of the art in Smart Textiles and Interactive Fabrics, EU Regional Framework Operation in the framework of the Interreg IIIC. 2013. Available at: www.mateo.ntc.zcu.cz/doc/State.doc (Accessed 24 February 2015).
[38] Cranston, R. W., Composite sensor fibres and applications therefor, WO2013131120, Commonwealth Scientific and Industrial Research Organisation, 2013.
[39] Clayton, T. S., Small scale microencapsulated pigments and uses thereof, US8883049, Chromatic Technologies, 2014.
[40] Sekar, N., Photochromic and thermochromic dyes and their applications, Colourage **45** (1998) 39–42.
[41] Furuta, T., Tanaka, M., Mitani, K., Light- and washfast photochromic fabrics coated with microencapsulated spironaphthoxazine derivatives, JP62289684, Unitika, 1987 and JP6051956, Unitika, 1994.
[42] Nakanishi, M., Iwasaki, T., Maeda, S., Microencapsulated photochromic materials for waterborne inks, WO8905335, Japan Capsular Products, Mitsubishi Kasei, 1989.
[43] Iwasaki, T., Maeda, S., Inks containing microcapsules, JP02110173, Japan Capsular Products, Mitsubishi Kasei, 1990.
[44] Kamata, K., Kitagawa, Y., Hoshikawa, R., Reversible photochromic colored textile good and its coloring method, JP6212579, Matsui Shikiso, 1994.
[45] Kitagawa, Y., Hoshikawa, R. A., Microencapsulated photochromic composition for textile printing paste and printed articles from, GB 2270321, Matsui Shikiso, 1994.
[46] Yabuchi, N., Mizuguchi, K., Ishii, K., Microencapsulated photochromic material, JP07159924, Nipppon Paint, 1995.

[47] Zhang, Y., Cui, G., Wang, Y., Zheng, X., Photochromic double-shell microcapsule and preparation method and application thereof, CN102886233, Jiangnan Branch of China Textile Academy, 2013.
[48] Pandell, N. W., Temin, S. C., Application of reactants and/or catalysts to textile fabrics in microencapsulated form, US 3632296, Cluett, Peabody & Co, 1972.
[49] Fornelli, S., Telluric treatment, Melliand Textil Int **2** (1996) 108–12.
[50] Salaün, F., Giraud, S., Vroman, I., Rault, F., A Review of Microencapsulation of Flame Retardant Formulations suitable for Application in Polypropylene Textile Substrates, Nova Science Publishers, 2015. Accessed 23 February 2015, at: https://www.novapublishers.com/catalog/productinfo.php?productsid=42209&osCsid=98acc41cb0e9616a0fd9d45ebd1ee111.
[51] Ida, S., Hosokawa, K., Impregnation of fibrous material with hydrophobic substances in microcapsules, DE2041899, Kanegafuchi Spinning, 1971.
[52] Nishijima, Y., Shimizu, K., Textile fire-retardant finishing, DE2164189, Kanegafuchi Spinning, 1972.
[53] Ikeda, S., Seya, T., Ueda, F., Matsunaga, M., Flame-proofing fibers, JP48012470, Ashai, 1973.
[54] Shimosaka, Y., Suzuki, H., Microcapsule. JP7448073, Japan Exlan, 1974.
[55] Vincent, D., Golden, R., Carpet with microcapsules containing volatile flame-retardant, US 3859151, Champion International, 1975.
[56] Nakayama, S., Flame retarder of fine particle. JP61042024, Matsumoto Yushi Seiyaku, 1980.
[57] Tilo Schwarzbach, T., Fire retardant compositions, EP1836342, Dartex Coatings Limited, 2007.
[58] Carlier, A. M., Treatment of knitted textiles with microencapsulated products. FR1582967, Carlier, 1969.
[59] Sroka, P., Cold-sealable article from textile flat structures DE 2460855, Hermann Windel, 1976.
[60] Groshens, P., Paire, C., Textile-adhesive composites containing microencapsulated crosslinking agents, FR2625745, Lainerie de Picardie, 1989.
[61] Sogabe, Y., Harada, K., Cellular polymer sheets, JP60149642, Achilles, 1985.
[62] Sato, H., Light-weight flexible leather substitutes, JP6335891, Bando Chemical, 1988.
[63] Kishi, S., Coloring of hollow microspheres, JP 57188435, Meisei Rejinokara, 1982.
[64] Hohara, K., Kamo, M., Shiraki, H., Nihei, N., Antislip adhesive nonwoven cloths, JP6273940, Nippon Kako Seishi, 1987.
[65] Otsubo, H., Yamagasi, S., Waterproof sewing threads, JP86296181, Nippon Rubber, 1986.
[66] Hatada, T., Mitsuyoshi, A., Masuda, S., Coated fabrics with improved functional properties. JP63012765, Toray Industries, 1988.
[67] Iwai, T., Manufacture of molded sheet integrated with skin, JP3045316, Takashimaya Nippatsu, 1991.
[68] Toyao, M., Ito, K., Waterproofing cloth, JP4316683, Owari Seisen, 1992.
[69] Grimm, J. E., Encapsulated fabric softener, US3896033, Colgate Palmolive, 1975.
[70] Schilling, K. J., Presoftener and washing composition mixture, DE2653259, Procter and Gamble, 1977.
[71] Pracht, H. J., Iding, S. H., Encapsulated liquid fabric conditioners, DE2632318, Procter and Gamble, 1977.
[72] Munteanu, M., Cseko, C., Oltarzewski, E. S., Lindauer, J. I., Withycombe, D. A., Liquid or solid fabric softener composition comprising microencapsulated fragrance suspension and process for preparing same, US4446032, International Flavors and Fragrances, 1984.
[73] Nimrick, T. L., Aqueous textile softener compositions for use in rinse stage of laundering, BR9004492, Procter and Gamble, 1991.
[74] Farooq, A., Jacques, A., Peeters, M., Eibel, M., Holmgren, M., Cationic Polymer Stabilized Microcapsule Composition, WO2008005693, Colgate Palmolive, 2008.
[75] Franklin, K. D., Patel Komal, G., Morgan, G., Jordan, G. T., Process of incorporating microcapsules into dryer-added fabric care articles, EP2027240, Procter & Gamble, 2009.

[76] Gizaw, Y., Bianchetti, G. O., Claeys, K. G., Bodet, F., Keijzer, O. P. D. T., Belanger, D. M., et al., Cationic polymer stabilized microcapsule composition, EP2674477, Procter & Gamble, 2013.
[77] Yamaguchi, T., Muroya, T., Kondo, T., Kitajima, M., Microcapsules containing an enzyme for a detergent, JP6968408, Toyo Jozo, Fuji Photo Film, 1969.
[78] Hachmann, K., Boeck, A., Jakobi, G., Jung, D., Storage-stable easily soluble detergent additive, DE2413561, Henkel, 1975.
[79] Herdeman, R. W., Dry bleach stable enzyme, US470287, Procter and Gamble, 1978.
[80] Onoda, T., Sugai, H., Sekiguchi, K., Enzyme-containing detergent composition, JP63105098, Showa Denko, 1988.
[81] Kamel, A., Hurckes, L. C., Morelli, M. M., Wax encapsulated actives and emulsion process for their production, US 4919841, Lever Brothers, 1990.
[82] Olson, K. E., Water insoluble encapsulated enzymes protected against deactivation by halogen bleaches, US4965012, Olson, 1990.
[83] Coyne, T. S., England, J. B., Haendler, B. L., Mitchell, F. E., Steichen, D. S., Johnson, C. L., Encapsulated enzyme in dry bleach composition, US5093021, Clorox, 1992.
[84] Lykke, M., Mistry, K. K., Simonsen, O., Symes, K. C., Enzyme-containing articles and liquid detergent concentrate, WO9724177, Novo Nordisk, 1997.
[85] Andersen, K. B., Foverskov, M,. Rasmussen, T., Simonsen, O., Jacobson, K., Noerby, M., et al., Microencapsulation of detergent enzymes, WO2014177709, Novozymes, 2014.
[86] Briggs, B., Encapsulated fluorescent whiteners stable in bleach, US3666680, Prurex, 1972.
[87] Weber, R., Opgenoorth, A., Solid powdered to granular agents for making cold-active bleaching and washing liquors, DE2048331, Henkel, 1972.
[88] Weber, R., Arends, D., Detergent compositions containing bis(triazinylamino) stilbenedisulfonate fluorescent brightener for white and coloured fabrics, DE2104975, Henkel, 1972.
[89] Kranz, H., Encapsulated ethylenediaminetetraacetate for use in bleaches and detergents containing active oxygen, DE2141280, Henkel, 1973.
[90] Hachmann, K., Saran, H., Sperling, G., Coated bleach activator, US 3925234, Henkel & Cie, 1975.
[91] Lohmann, F., Eckhard, C., Kleiber, K., Detergents containing chlorine donors and fluorescent whiteners, ZA7204248, Ciba Geigy, 1973.
[92] Christidis, Y., Diery, H., Coating and granulating tetraacetylglycoluril and tetraacetilethylenediamine, DE2535183, Nobel Hoechst, 1976.
[93] Alterman, D. S., Chun, K. W., Encapsulation of particles, US3983254, Lever Brothers, 1976.
[94] Alterman, D. S., Chun, K. W., Encapsulation process, GB1509797, Unilever, 1978.
[95] Mazzola, L. R., Encapsulated bleaches and methods for their preparation, US4078099, Lever Brothers, 1978.
[96] Brichard, J., Stabilizing particles containing peroxide compounds and bleaching compositions containing stabilized particles, EP030759, Interox, 1981.
[97] Chun, K. W., Lang, D. J., Santos, E., Encapsulated bleach particles coated with a mixture of C16-C18 and C12-C14 fatty acid soaps, US4655780, Lever Brothers, 1987.
[98] Corring, R., Gabriel, R., Clear detergent gel compositions having opaque particles dispersed therein, US 5141664, Lever Brothers, 1992.
[99] Appleby, D., Nelson, A., Brooker, A. T., A composition comprising a pre-formed peroxyacid and a bleach catalyst, EP1811014, Procter & Gamble, 2007.
[100] Best, P., Schweigl, O. F., Detergent compositions for machine washing, GB1207777, Unilever, 1970.
[101] Reuter, H., Saran, H., Witthaus, M., Low-foaming silicone-containing detergent, DE3128631, Henkel, 1983.
[102] Schneider, H., Cleaning textile surfaces with microcapsules of detergents, DE1287246, Werner und Mertz, 1969.

[103] Nakamura, M., Nakajima, K., Koimaru, I., Kito, S., Encapsulation of water-soluble dyes. JP7432922, Dainichiseika Color and Chemicals, 1974.
[104] Brain, D. K., Cummins, M. T., Washing composition, DE2653329, Procter and Gamble, 1977.
[105] Claus, A. D., Culver, G. E., Piatt, D. M., Wierenga, T. J., Detergent compatible, dryer released fabric softening/antistatic agents, EP0269179, Procter and Gamble, 1988.
[106] Thorengaard, B., York, D. W., Microencapsulated photoactivator dye compositions and detergents, NZ228315, Procter & Gamble, Danochemo, 1992.
[107] Boeckh, D., Jahns, E., Bertleff, W., Neumann, P., Microcapsule preparations and detergents and cleaning agents containing microcapsules, US6849591, BASF, 2005.
[108] Craven, R. M., Doyle, C. L., Hussey, I. J., Lavery, A. J., Philip, J., et al., Microcapsule incorporation in structured liquid detergents, WO2011120772, Unilever, 2011.
[109] Huchel, U., Bauer, A., Sunder, M., Microcapsule containing detergent or cleaning agent, US2013203642, Henkel, 2013.
[110] Smets, J., Prieto, S. F., Microcapsule compositions comprising pH tuneable di-amido gellants, WO2013039963, Procter & Gamble, 2013.
[111] Starešinič, M., Šumiga, B., Boh, B., Microencapsulation for textile applications and use of SEM image analysis for visualisation of microcapsules, Tekstil **54** (2011) 80–103.
[112] De Felice, I., Microcapsule-coating of fabrics, GB1401143, Eurand, 1975.
[113] Shibata Towel., Towels containing microencapsulated perfumes, JP58004886, Shibata Towel, 1983.
[114] Ono, H., Tokuoka, S., Fragrant fiber products, JP01266281, Kanebo, 1989.
[115] Ono, H., Tokuoka, S., Fragrant hand-knitting and handicraft yarns, JP1266282, Kanebo, 1989.
[116] Ono, H., Tokuoka, S., Neckties or ribbons with lasting fragrance, Japanese Patent JP3064504, Kanebo, 1991.
[117] Ono, H., Nunoo, T., Mudagami, S., Yamauchi, T., Omori, A., Fragrant curtains, JP 2055010, Kanebo, 1990.
[118] Ono, A., Fuse, T., Miyamoto, O., Makino, S., Yamato, Y., Kametani, H., et al., Fibrous structure having a durable fragrance and a process for preparing the same, US4917920, Kanebo, 1990.
[119] Nishigami, H., Manufacture of coated fabrics with lasting fragrance and bedding from them. JP3076875, Kanebo, 1991.
[120] Kukovic, M., Knez, E., Process for preparing carriers saturated or coated with microencapsulated scents, WO9609114, Aero, 1996.
[121] Okumura, K., Kamyama, S., Murata, S., Kawada, K., Kubota, M., Azuma, M., Imparting lasting fragrance to nonwoven fabrics, JP06116871, Osaka Juki, 1994.
[122] Sano, J., Une, T., Textile structure having fragrance, JP6228880, Kanebo, 1994.
[123] Shen, S. B., Preparation method of nano-negative ion and microcapsule grass aroma rattan-imitating mat surface material and application of mat surface material in home textile products, CN102991002, Shanghai Shuixing Home Textile, 2012.
[124] Zhou, H. M., Microencapsulated polyester fragrant masterbatch and preparation method thereof, CN103360731, Iangsu Zja New Material Co, 2013.
[125] Saito, Y., Carpets or other textiles containing diethyltoluamide insect repellent, JP3206003, Hosokawa Textile, 1991.
[126] Umibe, H., Hario, S., Inoue, S., Fabrics with durable insect repellence, JP03090682, Toyobo, 1991.
[127] Inoue, S., Hario, S., Umibe, H., Synergistic insect-repelling microcapsules containing toluamides and isobornyl compounds, for textiles, JP3148203, Toyobo, 1991.
[128] Ogawa, Y., Mutagami, S., Yamauchi, T., Spraying agent for mothproofing treatment, JP3127701, Kanebo, 1991.
[129] Saito, Y., Nakamura, M., Wash-fast insect repellent fabrics, JP3002101, Hosokawa Textile, 1991.
[130] Shirakawa, Y., Insecticidal processing agent for fibers or textiles and insecticide-processed fibers or textiles, JP10236903, Union Kagaku, 1998.
[131] Paya, J. G., Insect Repellent Textile, US20100183690, Innovatec, 2010.

[132] Pilar, M. H. M., Microencapsulated biocide repellent composition having a double repellency action, textile garment comprising same and use of said garment, WO2015011320, Mateo Herrero María Pilar, 2013.
[133] Boh, B., Kosir, I., Knez, E., Kukovic, M., Skerlavaj, V., Skvarc, A., Microencapsulation and testing of the agricultural animal repellent Daphne, J Microencapsul **16** (1999) 169–180.
[134] Mitsuyoshi, A., Miura, H., Hatada, T., Manufacture of deodorant cloths with washing durability, JP63256769, Toray, 1988.
[135] Katsuya, J., Azumaguchi, K., Garments printed with marking materials containing bactericides and their manufacture, JP2084501, Tokyo Houlaisha, 1990.
[136] Katsuya, J., Azumaguchi, K., Printing composition containing bactericides for marking garments, JP2084595, Tokyo Houlaisha, 1990.
[137] Kuramoto, N., Hiroshima, M., Washfast antibacterial fabrics, JP4100980, Ashai, 1992.
[138] Jorda, R., Autant, P., Rossin, R., Active principle-containing microcapsules, their applications, and their preparation, EP576377, Flamel Technologies, 1993.
[139] Takeda, K., Kawai, F., Amano, J., Modified fiber materials and manufacture thereof, JP06299466, Toray, 1994.
[140] Okamoto, H., Inoue, S., Miyamatsu, H., Yoshida, K., Functional microcapsules containing catechins and/or saponins and their composites, JP119106, Elb Company, 2000.
[141] Kong, C., Feng, Z., Liu, P., et al., Anti-mite antibacterial polyester staple fiber and preparation method thereof, CN103726125, Shanghai Different Chemical Fiber Co Ltd, 2014.
[142] Boh, B., Hodzar, D., Knez, E., Kukovic, M., Pipal, V., Voda, K., Development of microcapsules for textile finishing. Slovenski Kemijski dnevi, Proceedings (Glavic, P., Brodnjak Voncina, D., editors). Maribor, 1999, 762–767.
[143] Boh, B., Knez, E., Microencapsulated antimicrobials on non-woven textiles for shoe insoles. XVth International Workshop on Bioencapsulation, Vienna, 6–8 September, 2007, P4–01: 1–4.
[144] Murata, T., Microencapsulated alliins and allicins and fiber structures containing them, JP4108728, Kanebo, 1992.
[145] Toshikazu, F., Yoshikatsu, M., Kyoji, M., Microcapsules treating liquids containing the same, and textile structure having microcapsules adhering thereto, US5232769, Kanebo, 1993.
[146] Haruta, M., Takahashi, T., Saito, K., Textile, JP9296367, Toray, 1996.
[147] Dim, S. A., Bioactive textile comprising silk protein fibers and microencapsulated active agents, FR2780073, Dim, 1999.
[148] Lapidus, O., Brault, D., Lognone, V., Richard, J., Benoit, J. P., Morteau, S., Textile or clothing article, toiletries or body care product, bearing microcapsules, and methods for making same, WO005446, Ted Lapidus, 2000.
[149] Böhringer, B., Textile material, charged with microcapsules containing agents with physiological and/or technical effect and its use, EP1886714, Blücher GmbH, 2008.
[150] Paya, J. G., A method and composition to infuse an active ingredient into clothes and use of a binder agent for microcapsules of said composition, EP2682454, InnovaTec Sensorización y Communication S.L, 2014.
[151] Cowsar, D. R., Fabric containing microcapsules of chemical decontaminants encapsulated within semipermeable polymers, US4201822, US Dept of the Army, 1980.
[152] Ichinukizaka, I., Fabric filters for liquid separation, JP62106814 Kanai Hiroyuki, 1987.
[153] Pinder, P., Motor vehicle comprises a textile odor-absorbing interior lining equipped with odor-absorbing microcapsules, which are activated by friction and/or pressure and are designed for delivering odors and/or fragrances, DE102008027432, GM Global Technology Operations, Inc., Detroit, 2008.
[154] Bohringer, B., Functional textile material provided with microcapsules containing an active ingredient and use thereof, US7670968, Blucher Gmbh, 2010.

[155] Boh, B., Knez, E., Starešinič, M., Microencapsulation of higher hydrocarbon phase change materials by *in situ* polymerization, J Microencapsul **22** (2005) 715–735.
[156] Buckley, T., Phase change thermal materials, method and apparatus, US5722482. Buckley, 1998.
[157] Bryant. Y., Colvin, D., Fiber with reversible enhanced thermal storage properties and fabrics made therefrom, US4756958, Triangle Research and Development, 1988.
[158] Kiyokawa, H., Sunlight absorbing and thermal energy storage textile material and its production, JP5311579, Kiyokawa, 1992.
[159] Bryant, Y., Colvin, D., Fabric with reversible enhanced thermal properties, US5366801, Triangle Research and Development, 1994.
[160] Bryant, Y., Colvin, D., Moldable foam insole reversible enhanced thermal storage properties, US5499460, Bryant & Colvin, 1996.
[161] Bryant, Y., Colvin, D., Thermally enhanced foam insulation, US5637389, Bryant & Colvin, 1997.
[162] Nuckols, M., Hughes, R., Grupe, C., Fitzgibbon, S., Passive thermal capacitor for cold water diving garments, US6120530, US Secretary of Navy, 2000.
[163] Buckley, T., Flexible composite material with phase change thermal storage, US6004662, Buckley, 1999.
[164] Pause, B., Interactive thermal insulating system having a layer treated with coating of energy absorbing phase change material adjacent a layer of fibers containing energy absorbing phase change material, US6077597, Outlast Technologies, 2000.
[165] Holman, M., Gel-coated microcapsules, US6099894, Frisby Technologies, 2000.
[166] Grynaeus, P., Thermal control nonwoven material, US8449947, Carl Freudenberg, 2013.
[167] Rajagopalan, S., Nanostructured phase change materials for solid state thermal management, WO2014071528, Bioastra Technologies Inc., 2013.
[168] Roe, D. C., Wiggins, E. M., Norman, J. J., Insert with advantageous fastener configurations and end stiffness characteristics for two-piece wearable absorbent article, US20140046285, The Procter & Gamble Company, 2014.
[169] Olivier, M., Tillmann, B., Bedek, G., Salaun, F., Devaux, E., Dupont, D., et al., Microcapsules for self-refreshing textile, EP2218498, Despature et fils, 2010.
[170] Bekemeier, T., Deklippel, L., Dimitrova, T., Elms, R., Galeone, F., Lenoble, B., et al., Silicate Shell Microcapsules For Treating Textiles, EP2337839, Dow Corning, 2011.
[171] Sandmeyer, F., Particles with structured surface. US7972696, Wacker Chemie, 2013.
[172] Wu, X., Self-healing nanofibers, composites and methods for manufacturing, WO2014120321, NDSU Res, Foundation, 2014.
[173] Gordon, K. L., Siochi, E. J., Grimsley, B. W., Cano, R. J., Czaba, M. W., Puncture-healing thermoplastic resin carbon-fiber-reinforced composites, US20140066553, NASA, 2014.
[174] Nelson, G., Microencapsulation in textile finishing, Rev Prog Color Relat Top **31** (2001) 57–64.
[175] Nelson, G., Application of microencapsulation in textiles, Int J Pharm **242** (2002) 55–62.
[176] Boh, B., Knez, E., Microencapsulation of essential oils and phase change materials for applications in textile products, Indian J Fibre Tex **31** (2006) 72–82.
[177] Mondal, S., Phase change materials for smart textiles – An overview, Appl Therm Eng **28** (2008) 1536–1550.
[178] Sarier, N., Onder, E., Organic phase change materials and their textile applications: An overview, Thermochim Acta **540** (2012) 7–60.

Cinta Panisello Llatje, Tania Gumi and Ricard García Valls

4 Emerging application of vanillin microcapsules

4.1 Introduction

4.1.1 Vanillin

Vanillin (4-hydroxy-3-methoxybenzaldehyde) is one of the most important aromatic aldehydes and also one of the most-used flavoring agents worldwide. Its functional groups include aldehyde, hydroxyl, and ether. It is the primary component of the extract of the vanilla bean. However, synthetic vanillin is now used more often than natural vanilla extract. Figure 4.1 shows the chemical structure of vanillin [1] and Table 4.1 summarizes its main physicochemical properties [2].

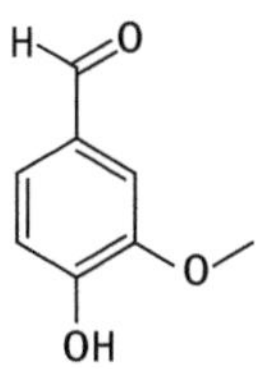

Fig. 4.1: Chemical structure of vanillin.

Table 4.1: Physicochemical properties of vanillin.

Vanillin physicochemical properties	
Boiling point	170 °C
Molar mass	152.15 g/mol
Water solubility	1 g/100 mL
Vapor pressure	0.01 mmHg at 25 °C
Melting point	81 °C
Flash point	147 °C
Odor threshold [3]	pp l09[a]

[a]Parts of compound in 10^9 parts of water (volume/volume).

Natural vanillin is extracted from vanilla pods, but this production cannot meet market demand alone [4]. Thus, vanillin is also produced industrially, through chemical synthesis, on a scale of more than 10,000 tonnes per year. Chemical synthesis is a well-established approach because it is economical. However, it has many drawbacks, such as the consequent environmental pollution and the lack of substrate

https://doi.org/10.1515/9783110642070-004

selectivity. These factors can reduce process efficiency and increase downstream processing cost [5]. Vanillin production is also not ideal through tissue culture techniques, because plants are slow growing and vanillin biosynthetic pathway is not very actively expressed [6]. Thus, alternative biotechnology-based approaches have been developed for the production of vanillin from lignin, phenolic stilbenes, isoeugenol, eugenol, ferulic acid, or aromatic amino acids and through biosynthesis by applying fungi, bacteria, plant cells, or genetically engineered microorganisms.

4.2 Properties and applications

Vanillin is considered to be one of the most appreciated flavor compounds, thus it is widely used for enhancing flavor in the food and beverage industry.

In addition, vanillin also has several bioactive properties [7] such as antimicrobial activities against yeasts, moulds [8] and bacteria [9, 10] and antioxidant properties [11, 12]. Because of this it is also used as a biopreservative.

Furthermore, it also possesses antimutagenic, anticlastogenic and antitumor activities, thus it is an important raw material in the pharmaceutical industries for production of drugs such as aldomet, dopamine, papaverine, and L-DOPA [13].

Finally, among other uses, vanillin is used to manufacture antifoaming agents for lubrication oils, and a great number of household products, deodorants, air fresheners, floor polishes, and herbicides [5, 14].

4.3 Microencapsulation of vanillin

4.3.1 Purpose of vanillin microencapsulation

In general, aromatic compounds are instable; they tend to degrade over time and their stability is mainly determined by chemical parameters. However, if they are protected from the environment then degradation could be stopped, or at least slowed down. Thus, the aim of microencapsulating a fragrance is to build a physical barrier between the aromatic molecule and its environment, in order to protect it or control its release. There are five principal reasons for encapsulating perfumes [1]:

- to protect perfume from oxygen in the air to avoid oxidation;
- to avoid hydrolysis;
- to avoid reactions of flavored ingredients with each other;
- to avoid evaporation;
- to convert liquids into powders, making them easier to handle.

In the particular case of vanillin, reported show that it is vulnerable to be degradation by sunlight in air, with a half-life of 4.7 h. Conversely, it is rather stable to hydrolysis in water. Finally, vanillin is readily biodegradable under aerobic conditions and it also degrades rapidly under anaerobic conditions [15].

Additionally, vanillin may be easily volatilized under some food processing conditions (baking, spray-drying, etc.) or – in the case of being added to detergents – it could be easily lost during the laundering operation.

Thus, vanillin microencapsulation could be mainly intended to increase its half-life in aggressive environments and/or to slow down its volatilization by controlling its release.

It is important to point out that vanillin, as with most flavors, is an expensive compound. Avoiding its degradation and controlling its release means that the same aromatic long-term effect could be obtained with the addition of less of the compound, which would lead to money savings for industry.

4.3.2 Materials and methods

Several materials and different methods for vanillin encapsulation have been reported.

Spray drying has been the most employed technique, possibly because it is the most commonly used microencapsulation method for food ingredients [16]; chitosan has been used as a wall material in order to encapsulate vanillin by this technique [17]. Mixtures of maltodextrin with gum arabic have been also employed in spray-drying processes [18].

However, carnauba wax has been employed as wall material for vanillin encapsulation by means of a melt dispersion technique [19].

Ethylcellulose microcapsules have been obtained by an oil-in-water solvent evaporation method. In addition, these were dip-coated by chitosan and the coating was crosslinked with nontoxic 1,2,3,4-butanetratecarboxylic acid. This modification was shown to improve the vanillin release [20].

Moreover, photosensitive microcapsules have been prepared by using a phase inversion technique. In this case the wall material was based on poly (α-methylstilbenesebacoate-co-α-methylstilbeneisophthalate), containing the photosensitive α-methylstilbene moiety [21].

Finally, this chapter is mainly focused on summarizing the work done within our research group on polysulfone/vanillin microcapsules prepared by a phase inversion precipitation technique [22].

4.4 Applications

Applications of vanillin microcapsules are found in different fields. Their main application is in the food industry, followed by cosmetics and household products.

Because it is a well known and widely used aromatic molecule, vanillin is also used in research as a reference compound for determining the encapsulation capacity of several preparations [21–23].

Finally, the encapsulation of vanillin and its controlled release offer potential applications for textile products. The antibacterial and aromatic properties of vanillin could be used in several fabric products [17, 24].

4.5 Polysulfone/vanillin microcapsules

4.5.1 Introduction

Polysulfone is one of the most employed polymers for the preparation of flat membranes and microcapsules, owing to its excellent chemical properties together with its thermal and mechanical resistance [25, 26]. In addition, it is a biocompatible polymer and thus can be used for medical applications [27–31].

Among applications of polysulfone capsules, special mention has to be given to their use for the elimination of toxic compounds from residual water [25], and the production of medical materials [32]. Nevertheless, the interest of this chapter is focused on the entrapment and controlled release of vanillin.

4.5.2 Preparation methods

The most-used methods for polysulfone microcapsule production are solvent evaporation [33] and phase inversion precipitation, with the latter method more common because capsule preparation is easy and fast [25, 34–38]. Phase inversion precipitation methods – immersion in liquid non-solvent or contact with vapor non-solvent – are described below.

4.5.3 Materials

For preparing microcapsules by phase inversion precipitation, at least three compounds are required: the polymer, the solvent and a non-solvent. As has been mentioned, polysulfone (Sigma Aldrich, St. Louis, MO, USA) was used as wall material because it is a polymer with excellent properties. Water has been the non-solvent used. As core material, vanillin was acquired from Sigma-Aldrich S.A., with purity > 99 %.

Solvent needs to be miscible with water and able to dissolve polysulfone. In that sense, dimethylformamide (DMF; scharlab > 99.8 %) was selected from a list of compatible solvents [39]. The reasons why more effective solvents were discarded lay in the fact that they were not completely miscible with water and/or they were flammable and thus, could not be used for safety reasons.

4.5.4 Methods

Phase inversion precipitation was used for the preparation of the capsules. This method is based on the interaction of at least three compounds – the polymer, a solvent in which the polymer is soluble and another substance, miscible with the solvent and in which the polymer is not soluble.

Mainly, it consists in contacting a polymeric solution (composed by the polymer, the solvent and the core material if required) either with a liquid or a vapor containing a non-solvent. This contact causes a change in the composition of the polymeric solution. By diffusion of the substances, the non-solvent concentration increases in the solution, together with the decrease of solvent concentration, and this leads the solution to a thermodynamically unstable state. When the solution is thermodynamically unstable, it spontaneously splits up into two different phases: in one of the phases polymer concentration is high (rich phase), whereas in the other one it is very low (poor phase). This process is known as liquid–liquid demixing. The rich in polymer phase precipitates, leading to the production of a solid membrane. The poor in polymer phase nuclei are responsible for pore formation in the membrane. After polymer precipitation, membrane morphology remains defined and invariable. Thus, membrane morphology is determined by the liquid–liquid demixing process. In the case of the ternary system used (PSf, DMF, water), when polymeric solution gets in contact with water, liquid–liquid demixing is fast and a solid structure is obtained in less than a second.

The first step in the production of microcapsules by the phase inversion precipitation is the preparation of a polymeric solution. It is obtained by solving the polymer in a suitable solvent, usually in a concentration range between 10 wt % and 30 wt %. This solution needs to be stirred for long enough to get the polymer completely dissolved in this case a 15 wt % polysulfone solution in DMF was used, which required stirring for 24 h. In addition, the bottles in which the solution is kept need to be hermetically closed, in order to avoid contact with atmospheric air, which contains moisture that would lead to premature precipitation of the polymer. In the case of the polysulfone/vanillin capsules herein described, the solution also contained a 10 wt % of vanillin, thus, bottles needed to be protected from light exposure and so amber bottles were preferred.

The second step is to disperse the solution into microdroplets, in order to consequently obtain microcapsules constituted by a spherical polysulfone membrane; atomization is used as the dispersion technique.

Two different atomization setup configurations have been reported for the preparation of polysulfone/vanillin microcapsules (see Figure 4.2). In Figure 4.2(a) precipitation is caused by the immersion in a precipitation bath containing a liquid non-solvent. The airbrush is positioned so that the outlet flow is perpendicular to the surface of the precipitation bath. When the polymeric solution microdroplets fall in the precipitation bath, microcapsules immediately floated on the surface and could be collected by filtration. In this chapter these capsules are referred to as IPS (immersion phase separation).

In Figure 4.2(b) precipitation is caused by the contact with vapor non-solvent inside a chamber. When water is the non-solvent the recommended relative humidity inside the chamber is 95 % at 20 °C. When the walls of the chamber are covered with microcapsules production should be stopped. The product is collected using a spatula. These capsules are identified in this chapter as VIPS (vapor induced phase separation).

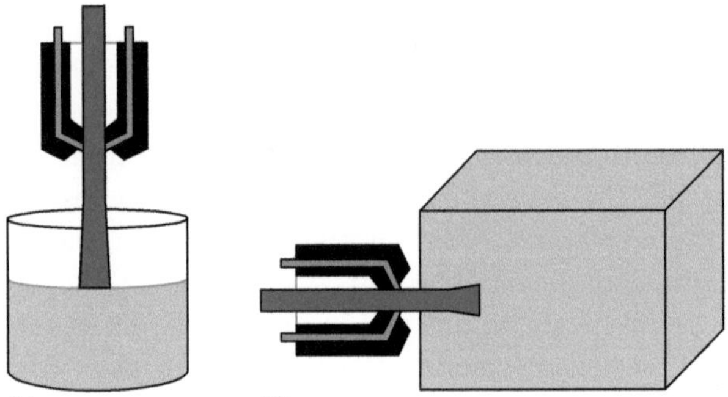

(a) (b)

Fig. 4.2: Scheme of atomization setup for capsules production by phase inversion: (a) by immersion in liquid non solvent (IPS), (b) by contact with non solvent vapor (VIPS).

4.6 Characterization

Polysulfone/vanillin microcapsules have been characterized using several techniques.

4.6.1 Scanning electron microscopy

Scanning electron microscopy (SEM), produces images of a sample by scanning it with a focused beam of electrons. The electrons interact with atoms in the sample, producing various signals that can be detected and that contain information about the sample's surface topography and composition.

In Figure 4.3 we can see surface images of microcapsules prepared by immersion precipitation or vapor non-solvent. In order to be observed by SEM, samples need to be electron conductive, thus the first step is to cover the samples with gold. The images on Figure 4.3 were obtained using a Jeol JSM-6400 working at 15–20 kV [40].

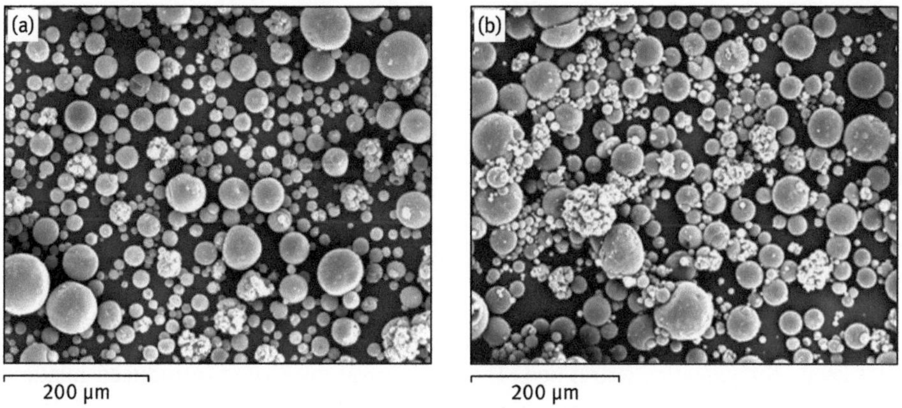

Fig. 4.3: Scanning electron microscope (SEM) micrographies of capsules obtained: (a) by IPS, (b) by VIPS.

Images like Figure 4.3 were analyzed at low magnifications using image-J® software in order to determine the mean size and size distribution of the samples. Table 4.2 shows the results obtained from measuring 500 capsules of each type. As can be observed, significant differences in size and size distribution between both preparations were not reported. This is explained by the fact that the size of the capsules is mainly determined not by the precipitation method but by the dispersion step [40].

Table 4.2: Mean size and size distribution of capsules prepared by IPS and by VIPS.

	Immersion precipitation	vapor-induced precipitation
Mean diameter (μm)	21.00	24.04
Standard deviation (μm)	12.62	16.10
Maximum diameter (μm)	83.32	95.23
Minimum diameter (μm)	6.17	4.71

In addition, if characterization of the capsules cross-section is required, capsules need to be cut. It has been found that cryogenic breaking is the best technique for cutting the capsules [41]. Figure 4.4 shows images of polysulfone/vanillin microcapsules obtained by both described methods. In order to obtain these images, capsules were mixed with a freezing medium (Jung Tisseu Freezing Medium, Leica Instrumental)

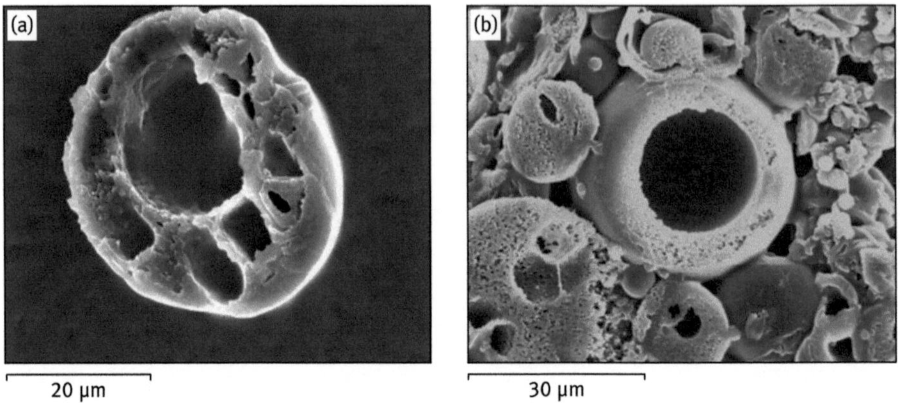

Fig. 4.4: Cross section images of capsules prepared by: (a) IPS, (b) VIPS.

and then immersed in liquid nitrogen. This media freezes fast and generates a matrix that keeps the capsules entrapped. Then frozen samples were introduced inside a cryochamber (Leica CM1850) at −22 °C and they were cut with a blade in slides of 15 μm thickness. Slides were place on a microscope glass and covered with gold for SEM observation [40, 41].

From Figure 4.4 it can be observed that microcapsules prepared by VIPS showed a uniform sponge-like structure, whereas on capsules prepared by VIPS very big pores (macrovoids) appeared. Differences on membrane cross-sections lay on the basis of the different precipitation techniques. More detailed information about liquid–liquid demixing processes in phase inversion can be found in the literature [40].

4.6.2 High performance liquid chromatography

Due to its versatility high performance liquid chromatography (HPLC) is one of the most employed analysis techniques based on separation, which allows its use in several fields.

The technique is based on the selective retention of the compounds in a sample when they flow through a column. Thus, components are solved in a suitable solvent (mobile phase) and they are forced (by applying high pressures) through the chromatographic column which contains a filling (stationary phase) that is able to selectively retain the different compounds in the sample. Because of this retention, different compounds flow out of the column separately and at different times, according to its retention time. In this way, a chromatogram is obtained, which allows the determination and quantification of the compounds in the sample.

In the work herein described, this technique was used to determine vanillin and DMF concentration in an aqueous media. The equipment used was an Agilent 1100 with photodiode array detector. The column used was a Supelcosil LC-8 (SUPELCO). The mobile phase was 80:20 water:acetonitrile. For all analyzes the flow rate was set at

1 mL/min, the column temperature was set at 40 °C, the analysis time was 8 min and the injection volume was 4 µL. Vanillin and DMF concentrations were determined at 229 nm, showing a typical retention time of 4.2–4.5 min and 2.0–2.2 min, respectively [42–44].

For the vanillin release experiments the medium was composed of 100 mL of distilled water in which 1 g of microcapsules was added. The preparation was stirred at 700 rpm for 72 h. Samples of 0.5 mL were taken from the release medium periodically and hermetically stored until analyzed by HPLC. Figure 4.5 shows the release results obtained for both types of microcapsules and the total mg of vanillin released to the medium in time, together with the percentage of vanillin released (related to the maximum amount released at the end of the experiment).

As can be observed, in both preparations the release was fast during the first 10 h of experiment but slowed down until it reached a plateau. This release tendency for IPS

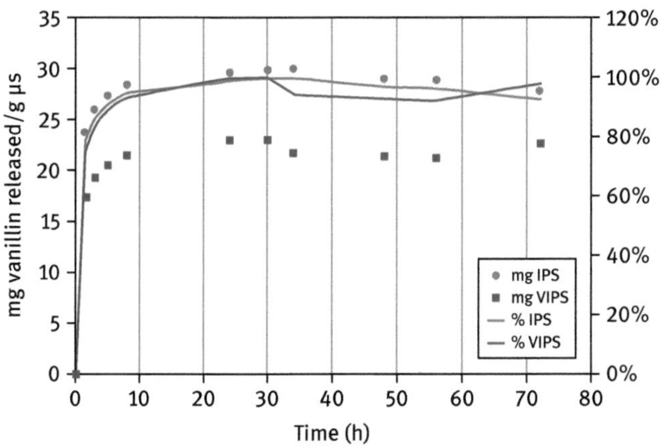

Fig. 4.5: Vanillin release from PSf/vanillin microcapsules.

capsules has been observed in several studies [22, 42]. These results showed that both preparations have similar permeability to vanillin (because the shape of the curve is the same). Conversely, although it may seem that microcapsules prepared by the VIPS methodology had encapsulated less vanillin, this is not necessarily true. It has been demonstrated in previous studies [40] that this difference was the result of higher amounts of DMF encapsulated in capsules prepared by VIPS, which affected their density.

4.6.3 Nitrogen adsorption/desorption analysis

Pore size distribution in the range of mesopores (from 2 to 50 nm) was determined by N_2 gas adsorption–desorption analysis. Prior to adsorption measurements, the

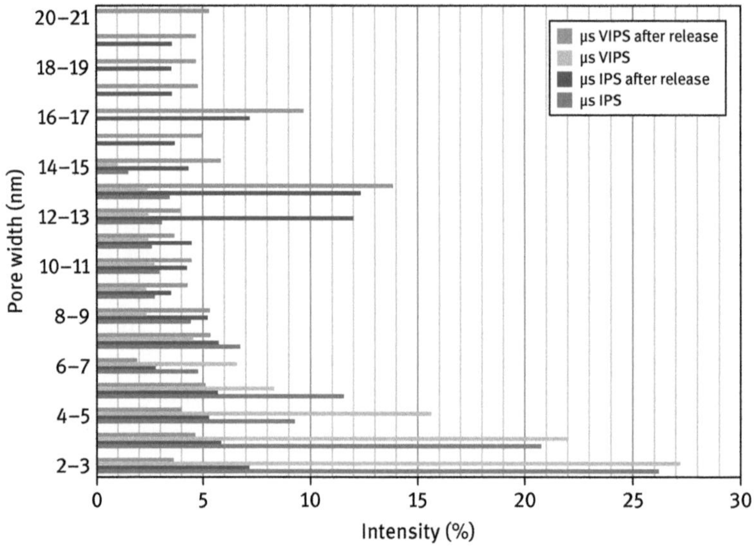

Fig. 4.6: Mesopore size distribution in microcapsules samples.

samples were degassed under vacuum for 12 h at 50 °C. Afterwards they were analyzed at 77 K with a Quadrasorb SI device [45]. Samples of microcapsules, taken before and after release experiments underwent N_2 adsorption/desorption analysis. Results are shown in Figure 4.6.

Significant differences between the two types of capsules were not detected. In both, before the release, the higher density was identified for pores in the range 2–6 nm. There were no pores over 15 nm. However, after the release, intensity of pores under 6 nm decreased, while larger pores were detected in the range between 12–20 nm. Thus, it is possible that many of the pores under 6 nm were, in fact, not small pores but larger ones partially blocked by vanillin, as determined in previous studies [45]. Thus, the precipitation technique is not affecting the mesopore size distribution.

4.6.4 Differential scanning calorimetry

In order to determine the effect of temperature on polysulfone/vanillin microcapsules, calorimetric curves were obtained by using a Mettler-Toledo 822 Differential Scanning Calorimetry (DSC) (Mettler-Toledo Inc., Schwerzenbach, Switzerland). DSC curves were obtained in a nitrogen atmosphere at 10 °C/min heating rate. The pan used was a 40 µL aluminium sealed crucible. Samples weight was approximately 8 mg. Capsules behavior in the range 20–100 °C was assessed. In addition, pure polysulfone and vanillin were analyzed.

As can be observed in Figure 4.7, the polymer did not suffer any changes in the range of temperatures assessed, because its glass transition temperature is found at 185 °C.

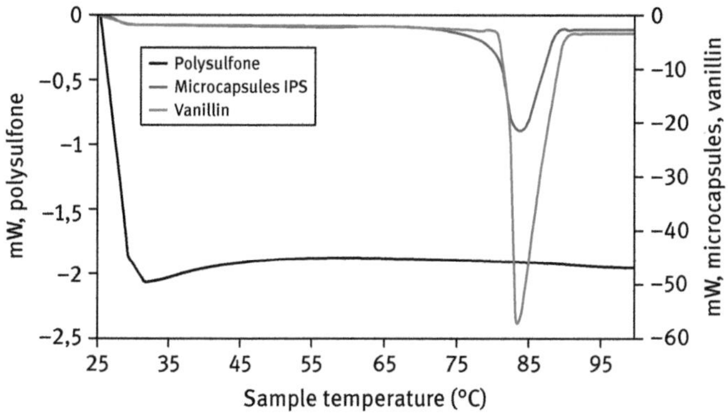

Fig. 4.7: DSC curves for vanillin, polysulfone and microcapsules.

This means that the wall of the capsules can resist temperatures considerably higher. Conversely, an endothermic peak was encountered for vanillin at 85 °C. This peak may corresponds to the melting of the compound, which occurs at 81.5 °C [2]. The same peak was appreciated in PSf/vanillin capsules, which again demonstrated the presence of vanillin in the capsules.

4.7 Antibacterial and aromatic finishing of fabrics

4.7.1 Introduction

As mentioned above, vanillin possesses several bioactive properties, such as antimicrobial activity. In particular, in this work the attention was focused on the inhibitory effect of vanillin against the growth of *Staphylococcus aureus* [13], which is one of the most common bacteria in postsurgical wound infections, causing several diseases ranging from mild to severe [46].

4.7.2 Antibacterial activity

Vanillin had previously shown significant inhibitory activities against *S. aureus* [13]. Thus, research was focused to determine if PSF/vanillin microcapsules could maintain this activity. A modified agar-well diffusion technique was employed [47] and petri dishes containing 15 mL of nutrient agar were filled with a mixture of 100 μL of standard 108 colony forming unit (CFU)/mL *S. aureus*. Different holes of 5 mm diameter were made, and they were filled with:

- Different solutions of vanillin in ethanol at 1.25 wt %, 2.5 wt %, 5 wt %, 10 wt %.
- Solid vanillin, 0.1 g.

- PSf/vanillin microcapsules prepared by IPS, 0.1 g.
- Ultra pure water and ethanol as control solutions.

The inhibitory activity of vanillin against *S. aureus* was tested by incubating the plates for 1 week at 37 °C. A clear zone around the holes indicated the inhibitory effect. Figure 4.8 shows images obtained one day after the beginning of the experiment.

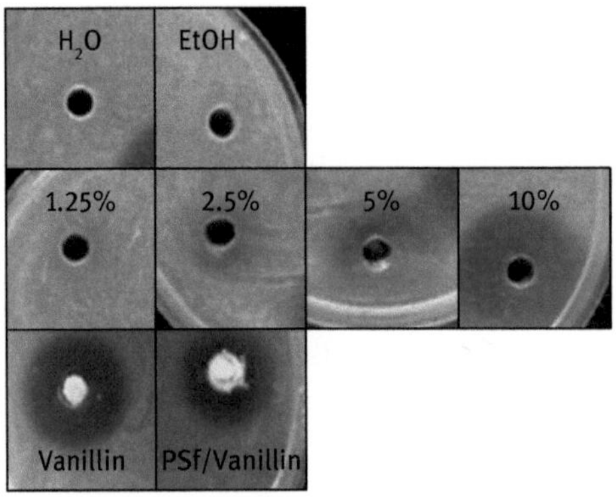

Fig. 4.8: Inhibitory activity of vanillin against *S. aureus*.

Significant inhibitory effects could be observed in the most concentrated vanillin/ethanol solutions. Similar results had been previously reported [13]. This is logical because dissolved vanillin is able to diffuse through the cultivation medium. However, the most interesting results were that significant inhibitions were also observed when vanillin was added in a solid state, or even in the encapsulated product. Although inhibition diameters in the solid products were shorter than in the ethanol solutions, this could be explained by the fact that when vanillin was in solid state its diffusion to the medium was hindered. According to the published results [24], the inhibition diameter remain constant for at least 1 week, which means that PSf/vanillin microcapsules inhibited *S. aureus* for at least 1 week, showing promising results for the inhibition of the bacteria growth.

4.7.3 Microcapsules adhesion to fabrics

Microcapsules prepared by IPS, which had shown to have an antibacterial effect, were incorporated in cotton fabric samples. For this purpose, the first step was to coat the fabric samples with Seitex 100 using a casting knife, which provides coatings of uniform thickness (50 μm) pushed by an applicator working at constant velocity (K-Pain applicator, UK).

The second step was to place a layer of capsules on a tray.

Finally, fabric samples were deposited over the tray, allowing their coated side to be in contact with the capsules while exerting a slight pressure in order to facilitate the adhesion of the capsules to the coating. This procedure ensured fixing the maximum number of capsules per area.

The durability of the adhesion was tested by exposing the fabrics to several washing cycles. The experiments were conducted in a commercial washing machine (Bosch Maxx WFO-2063). According to ASTM D2960-05 (standard test method of controlled laundering test using naturally soiled fabrics and household appliances), the load weight was fixed to 3 kg. The laundry program was intended to be as close as possible to the standard normal home laundry test conditions fixed by the American Association of Textile Chemists and Colorists. Thus, washing time was 15 min, followed by rinsing for 10 min and finally spinning for 10 min. The pieces of fabrics were weighted and observed by SEM before and after each washing in order to determine the weight loss and the number of capsules per area.

Figure 4.9 shows images of the fabric with capsules added, before the laundry and after the first, second, third, fourth and fifth washing cycles. It can be observed that the density of capsules suffered an important decrease after the second washing cycle.

Fig. 4.9: Scanning electron microscope (SEM) images from (a) microcapsules, (b) fabrics, (c) fabrics with microcapsules before washing, (d) fabrics with microcapsules after first washing, (e) after second washing, (f) after third washing, (g) after fourth washing, (h) after fifth washing.

Although several rounded particles were observed (Figure 4.9 f, h, g) they were not PSf/vanillin microcapsules but were zeolites from the detergent, as was demonstrated by elemental analysis and justified in an already published study [24].

As can be observed, the number of capsules decreased every washing cycle (see Table 4.3, extracted from a previous study [24]). However, an interesting fact was that smaller capsules seemed to resist more washing cycles.

Table 4.3: Amount of capsules and diameter range after each washing cycle.

Microcapsules/cm^2	Diameter range (µm)	Washing cycles
110000 ± 18000	2.5–40	0
79000 ± 18000	2.5–28	1
14000 ± 1100	2.5–24	2
12000 ± 4900	2.5–28	3
2900 ± 460	2.5–17	4
1900 ± 460	2.5–13	5

This observation makes us think that maybe durability of the adhesions could be improved by adjusting the diameter of the microcapsules to 10 µm. However, more precise nozzles were required and this was out of the scope of the work.

The relation between the amount of capsules encountered and weight loss is shown in Figure 4.10.

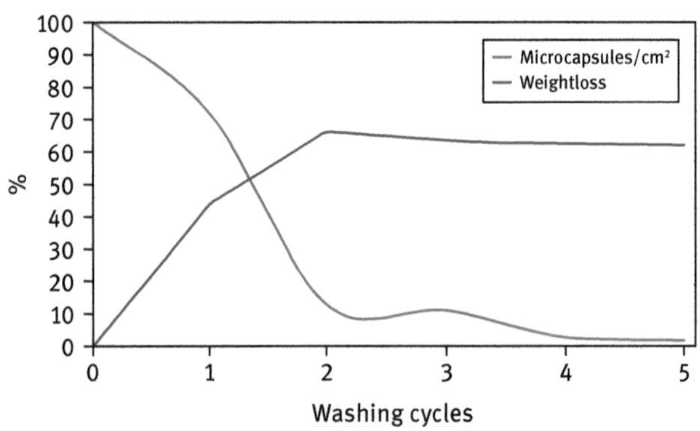

Fig. 4.10: Relation between microcapsules/cm^2 encountered and the weight loss measured after every washing cycle.

The most loss of weigh happened during the second washing cycle. After that the weight lost reached a plateau. However, in the case of capsules, even more of them were lost during the first and second washing cycles, and they were still being lost along all the experimentation. It is comprehensible that the loss of capsules did not affect significantly the weight of the samples, due to its low density (144 kg/m^3). In addition, zeolites were being attached to the fabrics. However, it was interpreted that the main weight changes were due to the binder, Seitex 100, which was used to attach the capsules. According to that, most of the binder would be lost in the first and second washing cycles, but that a thin layer (in direct contact with the textile) probably remained, maintaining some capsules stuck to the fabrics [24].

4.7.4 Aroma durability

A small survey was conducted in order to assess the perception of vanillin aroma and its durability in the fabrics coated with vanillin microcapsules. A population sample of three volunteers were asked to rate the aroma intensity for each piece of fabric. Thus, reproducibility and inclusion of different smelling sensitivities was assured. The volunteers smelled and rated their perception of vanillin aroma before and after each washing cycle.

The experimental design included performing five washing cycles. From every cycle, five different pieces of fabrics were assessed. Thus, variables in the study were: the observer (persons 1–3), the aroma intensity (not detecting aroma = 0, detecting = 1, strong detection = 2), the washing cycle (1–5) and the aroma detection (yes = 1, no = 0).

The two main hypotheses of the survey were:

- All the observers perceive the same aroma intensity.
- Washing cycles influence aroma intensity.

These hypotheses validated their statistical significance for a confidence interval of 95% against their null hypotheses. The statistical analysis and hypothesis contrast were performed by using the statistical software JMP® Pro 9.0.3 from SAS Institute Inc. (Cary, NC, USA).

Finally, a model was constructed that was able to predict the probability of smelling vanillin in fabrics after being subjected to different number of washings.

Before the first washing cycle, all the observers agreed that all the samples released a strong vanillin aroma, thus scoring a value of 2 in terms of aroma intensity. More details about the statistical analysis can be found in the published paper [24]. However, its main conclusions were:

- There existed differences in the aroma perception among the three observers. Thus, survey data was recorded from people with different olfactory sensibilities, which gives robustness to the modeling done.
- There was a statistically significant correlation between the perfume release of a fabric and the number of washing cycles. Validation of this hypothesis allowed building a model in order to quantify this relationship.
- A model was built using the number of washing cycles as the factor, and the aroma detection as the response variable.
- The model was validated against the experimental data and – as can be observed in Figure 4.11 – it fitted well with the sample data. A black line plots the model fit equation that represents the whole population, while a gray line plots the data obtained from the survey sample.

This model was useful because it allowed the determination of the probability to maintain fabrics perfumed for any washing time. Finally, according to the validated model, we could predict that the probability of maintaining the aroma after two

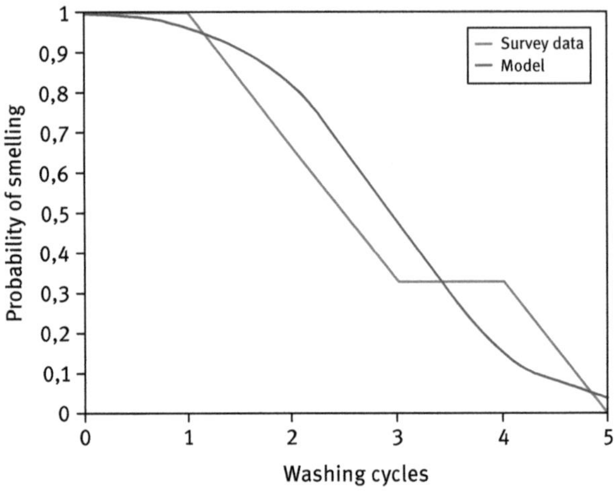

Fig. 4.11: Validation of the model against survey data.

washing cycles was 82%, while after the third and the fourth washing cycles the probability was only 48% and 15%, respectively.

4.8 Conclusions

In conclusion, polysulfone/vanillin microcapsules can be effectively produced by phase inversion precipitation. Two different approaches – IPS and VIPS – have shown to be successful for this purpose when using DMF as solvent and water as non-solvent. Vanillin encapsulation was a challenge because it is a polar compound, which could easily leak to the water phase. However, the fast precipitation of the polymer succeeded in entrapping the flavor. In addition, the performance of both products, in terms of vanillin release, was similar. Presence of vanillin in the capsules has been demonstrated by DSC and HPLC analysis.

Conversely, SEM analysis showed that the products had a different cross-section structure. Macrovoids appeared in microcapsules produced by IPS, whereas these large pores were not found in capsules prepared by VIPS.

After having successfully prepared and characterized polysulfone/vanillin microcapsules, the objective of the research was to use them for providing fabrics of antimicrobial and aromatic properties through a microcapsules coating. Thus it was neceaasry to design and assess a suitable method.

First of all, antimicrobial activity of the capsules against *S. aureus* was confirmed. Capsules were shown to inhibit the growth of the bacteria for at least 1 week.

Afterwards, microcapsules were incorporated to 100% cotton fabric samples, which underwent several washing cycles in a conventional washing machine. Resistance of the adhesion and durability of the aroma were investigated.

Over 50% of the capsules were lost during the first and second washing cycles; however, smaller capsules (around 10 μm) resisted more washing cycles. In fact, they were still being encountered after the fifth washing.

Aroma durability was determined by a perception survey, which concluded that there was a correspondence between durability of the aroma and the washing cycle. A model was built and validated in order to predict the probability of maintaining the aroma after different washing cycles.

The studies in the literature that have been described in this chapter set the basis for further development of fabrics with antimicrobial activity and pleasant aroma finishing based on polysulfone/vanillin capsules. Further work in this area should be focused on narrowing the size distribution of the capsules and improving their adhesion to the fabrics.

4.9 References

[1] Rowe, D., (editor). Chemistry and Technology of Flavors and Fragrances, Blackwell Publishing, Oxford, 2005.

[2] Green, D. W., Perry's Chemical Engineers' Handbook, McGraw-Hill, New York, 2008.

[3] Buttery, R. G., Ling, L. C., Volatile flavor components of corn tortillas and related products, J Agric Food Chem (1995) 1878–1882.

[4] Krings, U., Berger, R. G., Biotechnological production of flavours and fragrances, Appl Microbiol Biotechnol **49** (1998) 1–8.

[5] Kaur, B., Chakraborty, D., Biotechnological and molecular approaches for vanillin production: a review, Appl Biochem Biotechnol **169** (2013) 1353–1372.

[6] Ramachandra Rao, S., Ravishankar, G., Vanilla flavour: production by conventional and biotechnological routes, J Sci Food Agric **80** (2000) 289–304.

[7] Sinha, A., Sharma, U., Sharma, N., A comprehensive review on vanilla flavor: extraction, isolation and quantification of vanillin and others constituents, Int J Food Sci Nutr **59** (2008) 299–326.

[8] Cerrutti, P., Alzamora, S. M., Inhibitory effects of vanillin on some food spoilage yeasts in laboratory media and fruit purées, Int J Food Microbiol **29** (1996) 379–386.

[9] Rupasinghe, H. P. V., Boulter Bitzer, J., Ahn, T., Odumeru, J. A., Vanillin inhibits pathogenic and spoilage microorganisms in vitro and aerobic microbial growth in fresh-cut apples, Food Res Int **39** (2006) 575–580.

[10] Delaquis, P., Stanich, K., Toivonen, P., Stanich, K., Toivonen, P., Effect of pH on the inhibition of Listeria spp. by vanillin and vanillic acid, J Food Prot **68** (2005) 1472–1476.

[11] Burri, J., Graf, M., Lambelet, P., Loliger, J., Vanillin: more than a flavoring agent — a potent antioxidant, J Sci Food Agric **48** (1989) 49–56.

[12] Mourtzinos, I., Konteles, S., Kalogeropoulos, N., Karathanos, V., Thermal oxidation of vanillin affects its antioxidant and antimicrobial properties, Food Chem **114** (2009) 791–797.

[13] Rakchoy, S., Suppakul, P., Jinkarn, T., Antimicrobial effects of vanillin coated solution for coating paperboard intended for packaging bakery products, Asian J Food Agro-Industry **2** (2009) 137–184.

[14] Walton, N. J., Mayer, M. J., Narbad, A., Vanillin, Phytochemistry **63** (2003) 505–515.

[15] OECD. SIDS – Vanillin, Available from: http://www.inchem.org/documents/sids/sids/121335.pdf

[16] Gharsallaoui, A., Roudaut, G., Chambin, O., Voilley, A., Saurel, R., Applications of spray-drying in microencapsulation of food ingredients: An overview, Food Res Int **40** (2007) 1107–1121.

[17] Yang, Z., Zeng, Z., Xiao, Z., Ji, H., Preparation and controllable release of chitosan/vanillin microcapsules and their application to cotton fabric, Flavour Fragr J **29** (2014) 114–120.

[18] Janiszewska, E., Arciszewska, M., Witrowa-Rajchert, D., Correlation between efficiency of vanillin aroma micro-encapsulation and physical properties of powders obtained, Zywn Technol Jakosc **20** (2013) 174–186.

[19] Milanovic, J., Manojlovic, V., Levic, S., Rajic, N., Nedovic, V., Bugarski, B., Microencapsulation of flavors in carnauba wax, Sensors **10** (2010) 901–912.

[20] Feczko, T., Kokol, V., Voncina, B., Preparation and characterization of Ethylcellulose-based microcapsules for sustaining release of a model fragrance, Macromol Res **18** (2010) 636–640.

[21] Bogdanowicz, K. A., Tylkowski, B., Giamberini, M., Preparation and characterization of light-sensitive microcapsules based on a liquid crystalline polyester, Langmuir **29** (2013) 1601–1608.

[22] Peña, B., Panisello, C., Aresté, G., Garcia-Valls, R., Gumí, T., Preparation and characterization of polysulfone microcapsules for perfume release, Chem Eng J **179** (2012) 394–403.

[23] Bruno, P., Malucelli, G., Tylkowski, B., Ferré, J., Giamberini, M., Acrylic microspheres as drug-delivery systems: synthesis through in situ microemulsion photoinduced polymerization and characterization, Polym Int **62** (2013) 304–309.

[24] Panisello, C., Peña, B., Gilabert Oriol, G., Constantí, M., Gumí, T., Garcia-Valls, R., Polysulfone/vanillin microcapsules for antibacterial and aromatic finishing of fabrics, Ind Eng Chem Res **52** (2013) 9995–10003.

[25] Peña, B., Gumí, T., State of the art of polysulfone microcapsules, Curr Org Chem Special Is: **17** (2013) 22–29.

[26] Guillen, G., Pan, Y., Li, M., Preparation and characterization of membranes formed by nonsolvent induced phase separation: a review, Ind Eng Chem Res **50** (2011) 3798–3817.

[27] Wenz, L. M., Merritt, K., Brown, S. A., Moet, A., Steffee, A. D., Invitro biocompatibility of polyetheretherketone and polysulfone composites, J Biomed Mater Res **24** (1990) 207–215.

[28] Sivaraman, K., Kellenberger, C., Pane, S., Ergeneman, O., Luehmann, T., Porous polysulfone coatings for enhanced drug delivery, Biomed Microdevices **14** (2012) 603–612.

[29] Rahimy, M. H., Peyman, G. A., Chin, S. Y., Golshani, R., Aras, C., Borhani, H., et al., Polysulfone capillary fiber for intraocular drug-delivery – in-vitro and in-vivo evaluations, J Drug Target **2** (1994) 289–298.

[30] Gastaldello, K., Melot, C., Kahn, R. J., Vanherweghem, J. L., Vincent, J. L., Tielemans, C., Comparison of cellulose diacetate and polysulfone membranes in the outcome of acute renal failure. A prospective randomized study, Nephrol Dial Transplant **15** (2000) 224–230.

[31] Chlopek, J., Rosol, P., Morawska-Chochol, A., Durability of polymer-ceramics composite implants determined in creep tests, Compos Sci Technol **66** (2006) 1615–1622.

[32] Zhao, C., Liu, X. D., Nomizu, M., Nishi, N., Preparation of polysulfone hollow microspheres encapsulating DNA and their functional utilization, J Microencapsul **21** (2004) 283–291.

[33] Ozcan, S., Tor, A., Aydin, M., Removal of Cr(VI) from aqueous solution by polysulfone microcapsules containing Cyanex 923 as extraction reagent, Desalination **259** (2010) 179–186.

[34] Yin, J., Chen, R., Ji, Y., Zhao, C., Zhao, G., Adsorption of phenols by magnetic polysulfone microcapsules containing tributyl phosphate, Chem Eng J **157** (2010) 466–474.

[35] Gong, X., Lu, Y., Xiang, Z., Luo, G., Preparation of polysulfone microcapsules containing 1-octanol for the recovery of caprolactam, J Microencapsul **26** (2009) 104–110.

[36] Van den Berg, C., Preparation and analysis of high capacity polysulfone capsules, React Funct Polym **69** (2009) 766–770.

[37] Gong, X. C., Luo, G. S., Yang, W. W., Wu, F. Y., Separation of organic acids by newly developed polysulfone microcapsules containing triotylamine, Sep Purif Technol **48** (2006) 235–243.
[38] Yang, W. W., Lu, Y. C., Xiang, Z. Y., Luo, G. S., Monodispersed microcapsules enclosing ionic liquid of 1-butyl-3-methylimidazolium hexafluorophosphate, React Funct Polym **67** (2007) 81–86.
[39] Olabisi, O., Volatility of solvents from polysulfone melt, J Appl Polym Sci **22** 1021–1028.
[40] Panisello, C., Garcia-Valls, R., Polysulfone/vanillin microcapsules production based on vapor induced phase inversion precipitation, Ind Eng Chem Res **51** (2012) 15509–15516.
[41] Torras, C., Pitol, L., Garcia Valls, R., Two methods for morphological characterization of internal microcapsule structures, J Memb Sci **305** (2007) 1–4.
[42] Peña, B., Casals, M., Torras, C., Gumi, T., Garcia-Valls, R., Vanillin release from polysulfone macrocapsules, Ind Eng Chem Res **48** (2009) 1562–1565.
[43] Gumi, T., Gascon, S., Torras, C., Garcia-Valls, R., Vanillin release from macrocapsules, Desalination **245** (2009) 769–775.
[44] Waliszewski, K., Pardio, V., Ovando, S., A simple and rapid HPLC technique for vanillin determination in alcohol extract, Food Chem **101** (2007) 1059–1062.
[45] Peña, B., de Menorval, L-C., Garcia Valls, R., Gumi, T., Characterization of polysulfone and polysulfone/vanillin microcapsules by 1H NMR spectroscopy, solid-state (13)C CP/MAS-NMR spectroscopy, and N(2) adsorption desorption analyses, Appl Mater Interfaces **3** (2011) 4420–4430.
[46] Brook, I., Microbiology and management of post-surgical wounds infection in children, Pediatr Rehabil **5** (2002) 171–176.
[47] Chung, K. T., Wu Yuan, C. D., Thomasson, W. R., Growth inhibition of selected food-borne bacteria, particularly Listeria monocytogenes, by plant extracts, J Appl Bacteriol **69** (1990) 498–503.

Monika Haponska, Marcin Luczak, Patryk Nowak, Anna Bajek, Bartosz Tylkowski and Irene Tsibranska

5 Polyphenol encapsulation – application of innovative technologies to improve stability of natural products

5.1 Microencapsulation in food industry

Microencapsulation is considered as one of the major interdisciplinary, knowledge-intensive, and dynamic industrialized technologies of the last decade. Developed approximately 65 years ago, microencapsulation is defined as a technology of packaging solids, liquids, or gaseous materials in miniature, sealed capsules that can release their contents at controlled rates under specific conditions [1]. The main objective of encapsulation is to protect the core material from adverse environmental conditions, such as undesirable effects of light, moisture, and oxygen, thereby contributing to an increase in the shelf life of the product and promoting a controlled liberation of the encapsulate [2–4]. This method allows the use of wide range of reagents and substances, including those sensitive to temperature or pH that could not be enclosed using different techniques, and can be used to entrap different types of materials (solids, liquids, drugs, proteins, bacterial cells, etc.) in polymeric/inorganic structures [5]. The final products of microencapsulation are called microparticles. They can be divided due to internal structure and morphology into two main groups: microspheres and microcapsules. Within those microparticles, we can distinguish polycore capsules, continuous shells/core capsules, continuous core capsules with multiple layers of wall material, or the matrix capsules, also known as spheres, in which the encapsulated active compound is incorporated in the shell material [6].

In the food industry, the microencapsulation process can be applied for a variety of reasons, which have been summarized by Desai and Park [1] as follows:
- protection of the core material from degradation by reducing its reactivity to its outside environment;
- reduction of the evaporation or transfer rate of the core material to the outside environment;
- modification of the physical characteristics of the original material to allow easier handling;
- tailoring the release of the core material slowly over time, or at a particular time;
- to mask an unwanted flavor or taste of the core material;

https://doi.org/10.1515/9783110642070-005

- dilution of the core material when only small amounts are required, while achieving uniform dispersion in the host material; and
- to help separate the components of the mixture that would otherwise react with one another.

Food ingredients of acidulates, flavoring agents, sweeteners, colorants, lipids, vitamins and minerals, enzymes, and microorganisms, are encapsulated using different technologies [7].

5.2 Polyphenols

Polyphenols are important fruit, vegetable, and grain components that not only contribute to aroma, color, and taste but also protect the plant from ultraviolet (UV) radiation and pathogens attack. The results of studies outlined in several reviews provide a current understanding on the biological effects of polyphenols and their relevance to human health. From the medical point of view, these compounds show a wide spectrum of biological properties such as antioxidant, anti-inflammatory, antibacterial, and antiviral activities. High redox potential, together with the ability to act as reducing agents, hydrogen donors, and singlet oxygen quenchers, makes polyphenols effective antioxidants for the treatment of diseases associated with free radicals. Therefore, polyphenols become potential therapeutic agents against cancer, diabetes, and cardiovascular disorders, acting against reactive oxygen species generated by exogenous chemicals or endogenous metabolism and preventing cell damages caused by oxidative stress. Moreover, they offer great hope for the prevention of chronic human diseases [8–12].

Polyphenols are a large family of substances, ranging from simple molecules to complex structures. Several thousand molecules having a polyphenol structure (i.e. several hydroxyl groups on aromatic rings) have been identified in higher plants, and several hundred are found in edible plants. These molecules are secondary metabolites of plants and are generally involved in defense against UV radiation or aggression by pathogens. These compounds may be classified into different groups as a function of the number of phenol rings that they contain and of the structural elements that bind these rings to one another. Distinctions are thus made between the phenolic acids, flavonoids, stilbenes, and lignans (Figure 5.1). The flavonoids, which share a common structure consisting of two aromatic rings (A and B) that are bound together by three carbon atoms that form an oxygenated heterocycle (ring C), may themselves be divided into four subclasses as a function of the type of heterocycle involved: flavonols, flavones, isoflavones, and flavanones (Figure 5.2). In addition to this diversity, polyphenols may be associated with various carbohydrates and organic acids and with one another.

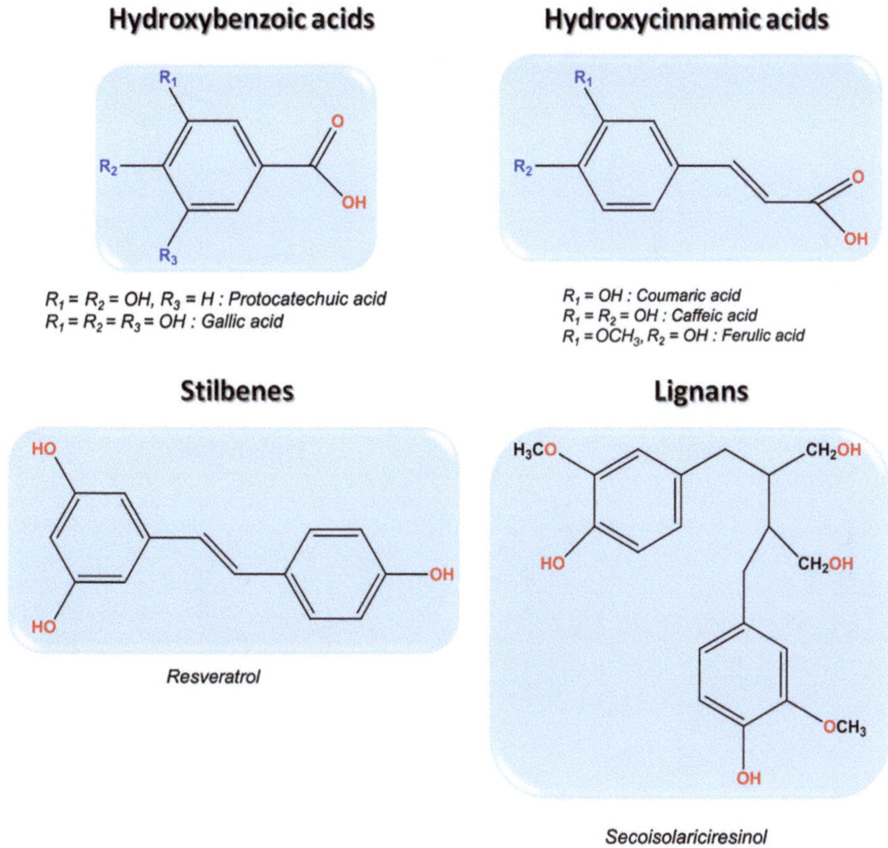

Fig. 5.1: Structures of phenolic acids, flavonoids, stilbenes, and lignans.

5.2.1 Phenolic acids

Two classes of phenolic acids present in plant materials at different levels can be distinguished basing on C1C6 and C3C6 backbones: derivatives of benzoic acid and derivatives of cinnamic acid. The hydroxybenzoic acid content of edible plants is generally very low, with the exception of certain red fruits, black radish, and onions, which can have concentrations of several tens of milligrams per kilogram fresh weight [13, 14]. Tea is an important source of gallic acid: tea leaves may contain up to 4.5 g/kg fresh wt. [15–17]. Furthermore, hydroxybenzoic acids are components of complex structures such as hydrolyzable tannins (gallotannins in mangoes and ellagitannins in red fruits such as strawberries, raspberries, and blackberries). Because these hydroxybenzoic acids, both free and esterified, are found in only a few plants

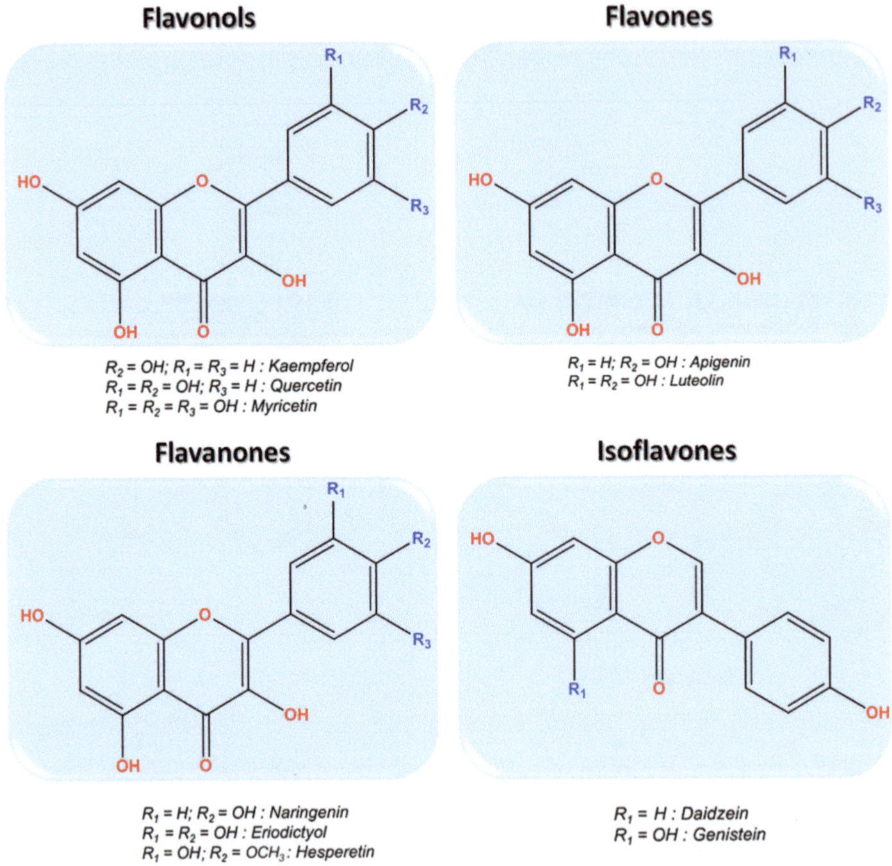

Fig. 5.2: Structures of flavonols, flavones, isoflavones, and flavanones.

eaten by humans, they have not been extensively studied and are not currently considered to be of great nutritional interest [18].

The hydroxycinnamic acids are more common than are the hydroxybenzoic acids and consist mainly of p-coumaric, caffeic, ferulic, and sinapic acids. These acids are rarely found in the free form, except in processed food that has undergone freezing, sterilization, or fermentation. The bound forms are glycosylated derivatives or esters of quinic acid, shikimic acid, and tartaric acid. Caffeic and quinic acid combine to form chlorogenic acid, which is found in many types of fruit and in high concentrations in coffee: a single cup may contain 70–350 mg chlorogenic acid [19]. The types of fruit having the highest content (blueberries, kiwis, plums, cherries, apples) contain 0.5–2 g hydroxycinnamic acids/kg fresh wt.

Caffeic acid, also known as 3,4-dihydroxycinnamic acid, both free and esterified, is generally the most abundant phenolic acid and represents between 75% and 100%

of the total hydroxycinnamic acid content of most fruit. Caffeic acid can be found in apples, blueberries, coffee, cider, and propolis. Hydroxycinnamic acids are found in all parts of the fruit, although the highest concentrations are seen in the outer parts of ripe fruit. Concentrations generally decrease during the course of ripening, but total quantities increase as the fruit increases in size [20, 21].

Ferulic acid is the most abundant phenolic acid found in cereal grains, which constitute its main dietary source. The ferulic acid content of wheat grain is ≈0.8–2 g/kg dry wt., which may represent up to 90% of total polyphenols [22]. Ferulic acid is found chiefly in the outer parts of the grain. The aleurone layer and the pericarp of wheat grain contain 98% of the total ferulic acid. The ferulic acid content of different wheat flours is thus directly related to levels of sieving, and bran is the main source of polyphenols. Rice and oat flours contain approximately the same quantity of phenolic acids as wheat flour (63 mg/kg), although the content in maize flour is about 3 times as high. The *trans* form of ferulic acid is esterified to arabinoxylans and hemicelluloses in the aleurone and pericarp. Only 10% of ferulic acid is found in soluble free form in wheat bran. Monomers and dimers are covalently conjugated through ester linkage with polysaccharides, glycoproteins, polyamines, lignin, and hydroxyl fatty acids. Those dimers can be found in cereals and form bridge structures between chains of hemicellulose [23–25].

5.2.2 Flavonoids

Flavonols are the most common flavonoids in foods. The best known, typical, and most abundant flavonol example is quercetin. Other common flavonol compounds are kaempferol and myricetin, which are present in food; tamarixetin and isorhamnetin, which can be found in plasma or tissues after quercetin consumption; and fisetin and morin [26, 27]. Flavonols are generally present at relatively low concentrations of ≈15–30 mg/kg fresh wt. The richest sources are onions (up to 1.2 g/kg fresh wt.), apples, curly kale, cider, leeks, broccoli, grapes, and blueberries. Red wine and tea also contain up to 45 mg flavonols/L. These compounds are present in glycosylated forms. The associated sugar moiety is very often glucose or rhamnose, but other sugars may also be involved (e.g. galactose, arabinose, xylose, glucuronic acid). Fruit often contains between 5 and 10 different flavonol glycosides. These flavonols accumulate in the outer and aerial tissues (skin and leaves) because their biosynthesis is stimulated by light. Marked differences in concentration exist between pieces of fruit on the same tree and even between different sides of a single piece of fruit, depending on exposure to sunlight [28]. Similarly, in leafy vegetables such as lettuce and cabbage, the glycoside concentration is ≥10 times as high in the green outer leaves as in the inner light-colored leaves [29]. This phenomenon also accounts for the higher flavonol content of cherry tomatoes than of standard tomatoes, because they have different proportions of skin to whole fruit.

Flavones are much less common than flavonols in fruit and vegetables. Flavones consist chiefly of glycosides of luteolin and apigenin. The only important edible sources of flavones identified to date are parsley and celery. Cereals such as millet and wheat contain C-glycosides of flavones [30]. The skin of citrus fruit contains large quantities of polymethoxylated flavones: tangeretin, nobiletin, and sinensetin (up to 6.5 g/L of essential oil of mandarin). These polymethoxylated flavones are the most hydrophobic flavonoids [31].

In human foods, flavanones are found in tomatoes and certain aromatic plants such as mint, but they are present in high concentrations only in citrus fruit. The main aglycones are naringenin in grapefruit, hesperetin in oranges, and eriodictyol in lemons. Flavanones are generally glycosylated by a disaccharide at position 7: either a neohesperidose, which imparts a bitter taste (such as to naringin in grapefruit), or a rutinose, which is flavorless. Orange juice contains between 200 and 600 mg hesperidin/L and 15–85 mg narirutin/L, and a single glass of orange juice may contain between 40 and 140 mg flavanone glycosides. Because the solid parts of citrus fruit, particularly the albedo (the white spongy portion) and the membranes separating the segments, have a very high flavanone content, the whole fruit may contain up to 5 times as much as a glass of orange juice [32, 33].

Isoflavones such as genistein and daidzein are commonly regarded to be phytoestrogens because of their estrogenic activity in certain animal models. A major dietary source of isoflavonoids is soy products. There are at least 12 known isoflavone compounds in soybeans (3 aglycones, 3 glucosides, 3 acetyl-ester glucosides, and 3 malonyl-ester glucosides). Significant amounts of the isoflavone genistein as its glucosyl glucoside have also been reported in the tubers of the American groundnut (*Apios americana*) [34]. Mazur et al. [35] estimated the isoflavone concentrations in 68 cultivars of 19 common leguminous food species and four forage legumes. The highest total isoflavone concentration was found in kudzu root (*Pueraria lobata*) (>2 mg/g dry weight). Puerarin has been reported to be the major isoflavonoid in kudzu dietary supplements [36].

5.2.3 Lignans

Lignans are naturally occurring plant phenols that are derived biosynthetically from phenylpropanoids. Most lignans occur freely in plants, but a small proportion of them coexist with sugars to form glycosides in wood and resin of plants. Lignans are commonly referred to as dimers, with complex skeletons and characteristic chemical functions, but a few are trimers or tetramers. The widely distributed lignans are important components of food and medicine that are derived from plants, and they have been target compounds for organic synthesis and biological-function research because of the many types of bonding of the C6-C3 units and oxidation of the structures. Lignans, which occur almost in all morphological parts of the plants including xylem, roots, leaves, flowers, fruits, rhizomes, stems, and seeds, are secondary metabolites with low molecular weight. Lignans can be classified into three main types according to their structures: lignans, neolignans, and hybrid lignans, which can be further divided into eight categories (furofuran, furan, dibenzylbutane, dibenzylbutyrolactone, aryltetralin, arylnaphthalene, dibenzocyclooctadiene, and dibenzylbutyrolactol) [37]. Various types of lignans have attracted considerable attention because of their numerous pharmacological features such as the antitumor, hepatoprotective, platelet activating factor antagonistic, insecticidal and estrogenic, antifungal, antihypertensive, sedative, and antioxidant activities. Plants with high lignan contents have been used as folk medicine in China, Japan, and the Eastern World since ca. 1,000 years. Nowadays, their extensive use in traditional medicine makes lignans an important family of lead compounds for the development of new therapeutic agents based on structural modifications [38, 39].

5.2.4 Stilbenes

Stilbenes, synthesized through the common phenylpropanoid way, which also leads to flavonoids and lignin formation in plants, are found in very low quantities in the human diet [40, 41]. One of these, resveratrol, for which anticarcinogenic effects have been shown during screening of medicinal plants and which has been extensively studied, is found in low quantities in wine (0.3–7 mg aglycones/L and 15 mg glycosides/L in red wine). However, because resveratrol is found in such small quantities in the diet, any protective effect of this molecule is unlikely at normal nutritional intakes [39, 42].

5.3 Encapsulation of polyphenols

The functional properties of a microcapsule are size, morphology, encapsulation efficiency, stability, and release [43]. In case of natural polyphenols, the most important property is their biological activity, related to scavenging free radicals and interaction with proteins, which makes them potentially interesting for a variety of applications, whose realization is limited by the inherent instability of these phytochemicals [44, 45]. Instability is observed as degradation during processing and storage (temperature, oxygen, light) or within *in vivo* administration (pH, enzymes in the gastrointestinal tract). Encapsulating polyphenols (usually plant extracts) is aimed to preserve the biological activity and to improve the stability of the active compounds, as well as to ensure controlled release of the latter. Encapsulation has the advantage of being a nonthermal stabilization approach, suitable for temperature sensitive natural biologically active compounds, as the ones extracted from different plants with medical application. The wall material usually improves the stability of the active compounds by protecting them from direct exposure to air and light. In this way, the polyphenols' inherent antioxidant activity is preserved and even improved [46]. Encapsulation and coencapsulation of different biologically active substances are investigated on micro- and nano-scale levels [10, 46–50], the former being in the focus of the present review. It is a promising approach to improve the performance of medicines, as the capsules exhibit a controlled-release profile for the contained polyphenols, leading to better bioavailability. Encapsulation allows us to solve two main reasons for the low bioavailability of polyphenols as free compounds and the need to administer higher concentrations of molecules of interest: the impossibility to maintain the active molecular form until the time of consumption and the insufficient gastric residence time, which usually leads to concentrations of an order of magnitude lower than the effective ones, determined from *in vitro* assays [51, 52]. The investigations concern mainly *in vitro* studies, many of them under conditions, which best simulate the expected *in vivo* environment (see Table 5.1), but the number of *in vivo* studies reporting improved bioavailability within oral administration of bioactive compounds and extracts is increasing [48, 52–54]. Resveratrol, which exhibits antioxidant properties and cancer preventive activities, and provides cardiovascular protection, is at the same time an oxygen- and photo-sensitive compound. In recent studies, after encapsulation, the permeability and photostability of resveratrol on cells were enhanced [55, 56]. Another important polyphenol, curcumin, having antioxidant, antiseptic, anti-inflammatory, and analgesic properties, exhibits low oral bioavailability due to physicochemical instability and water insolubility. The bioavailability of curcumin was increased by coating with peptide [57]. By nanoencapsulation, the nanoparticles can also be used for intravenous injection (when their average diameter is less than 200 nm). In general, regulating the particle size enables targeted delivery to different organs [10].

5.3.1 Improved stability

The improved stability of the active substances after encapsulation is the focus of a number of studies concerning polyphenols encapsulation [58–63]. These results are important for enlarging commercial applications, requiring longer storage with preserved biological activities. Usually, encapsulation in solid microcapsules is applied, but there are also attempts to encapsulate bioactive flavonoids (e.g. rutin) and anthocianins in multiple emulsions using a spinning disc reactor [64].

The effect of the encapsulation method, the choice of the carrier, particle size or size distribution, shape and smoothness of the outer surface, as well as swelling behavior are important factors affecting many properties of the microcapsules, among which are storage stability, core material retention (encapsulation efficiency), and controlled release. Spray drying [50, 58, 65–67], phase separation [68], emulsification/internal gelation [59, 69], supercritical antisolvent: precipitation and coprecipitation [70, 71], rapid expansion of supercritical solution [59], and electrostatic extrusion in microbeads [60] are some of the methods used for polyphenol encapsulation; comparative reviews on the subject being given in [51, 72].

The possibility to obtain higher retention of phenolics within the capsules is discussed in view of the encapsulating material, as well as the solvent used for extraction (aqueous or ethanolic) [65]. The reported stability data concern light, temperature, oxygen, and moisture exposure. Cocrystallization with sucrose is investigated in order to improve the properties of antioxidant powders obtained from plant extracts with better physicochemical stability during storage [73]. The small size of the particles (3–5 mm) with a narrow particle size distribution usually leads to a high degree of agglomeration [71], which is to be avoided by encapsulation. By nanoscale applications (e.g. lipid-core nanocapsules), the formation of very stable colloidal suspensions is reported [48], with unchanged particle size characteristics, polydispersity index, zeta potential, and drug loading.

Extensive reviews on recent advances in encapsulation of polyphenols are found in [51, 52], where the whole spectrum of biological activities is discussed, as well as accent given on the methods of encapsulation, coating materials and achieved encapsulating efficiencies. The encapsulation techniques are discussed in relation to the degree of stabilization achieved [51, 72].

5.4 Controlled release

Nowadays, the food industry focuses not only on the improvement of the stability and bioavailability of bioactive compounds but also on their controlled release [74]. Controlled release means not constant, but rather slowed-down release (compared to the direct application of polyphenols containing extracts), which can be predicted adequately and used in the appropriate way for the human body. The slowed down release occurs because instead of a pure dissolution process, diffusion through the capsules wall is taking place, whose rate coefficient (the internal diffusion coefficient through the pores of the encapsulating material) is essentially lower than the value of the molecular diffusion coefficient. Furthermore, the release can be controlled by the characteristic diffusion time, which includes the square of the characteristic particle size and the rate coefficient (the effective diffusion coefficient in the solid capsule). Diffusion can be coupled with erosion and partial dissolution of the encapsulating material, resulting in a transport rate between that of pore diffusion in the solid and molecular diffusion in the free volume of the surrounding liquid. Swelling or shrinking of the encapsulating material contributes to the complexity of the mass transfer mechanism, especially for polymeric encapsulating materials like hydrogels.

Crude extracts are multicomponent in nature, and their encapsulation is limited by the differences in solubility, distribution, and potential interaction among the contained biologically active compounds. The present review deals with *in vitro* investigations of the release kinetics of encapsulated natural extracts of polyphenols and flavonoids. They are illustrated by a typical kinetic curve, giving the increase of the concentration of the target compounds in the liquid around the microcapsules versus time. These curves, in general, start with a steep initial slope (burst effect), followed by a much lower and eventually constant rate of increasing concentrations (sustained release) and ending by a plateau, which corresponds to the equilibrium for a given set of experimental conditions (temperature, particle size, and solid-to-liquid ratio). The experimental kinetic curves are either reported in terms of needed contact time to reach the plateau or the final value of the released compound's concentration or are further treated mathematically by simple semiempirical models to define the value of the rate coefficients.

Table 5.1 gives a summary of recent data on the release of encapsulated polyphenols from natural extracts in view of the experimental conditions employed. Factors controlling the release kinetics are the amount of incorporated polyphenols and the composition of the coating material [75], the size of the microcapsules, and the temperature and pH of the release media (being especially important for pH- or temperature-sensitive hydrogels). Surface diffusion and the so-called burst effect are observed especially with smaller sizes (micro- or nano-scale) of the encapsulated particles [10].

The various techniques used for encapsulation concern also the diversity of microcapsule morphologies, which is important for the adequate mathematical modeling of the release process through them [51].

5.4.1 Mechanism of mass transfer and modeling

The increasing diversity of controlled delivery systems requires an adequate mathematical modeling of the release process to help reveal the mechanism(s) of bioactive release and to facilitate the optimization of the carrier systems by avoiding excessive experimentation [76]. Two major groups of models are observed:

- Semiempirical, geometry independent, based on simple mathematical functions (power or exponential), and accounting for one or two (additive) controlling release mechanisms;
- Detailed models with larger physical basis, based on analytical or rather numerical solutions of a set of differential equations for different geometry and different boundary conditions. Here, two- and three-dimensional solutions are also of interest.

The mathematical apparatus developed for controlled drug delivery is also used in the particular case of polyphenols release. Furthermore, the same approach for understanding the release kinetics and its mathematical modeling is applied to nanoencapsulation of polyphenols [49, 76].

Table 5.1: Investigations dealing with polyphenol release kinetics from microcapsules.

Coating material	Method of encapsulation	Source Material/ extraction method	Ref.	Release data			
				experimental			Modeling
				Release time	Particle size distribution	Release media	
Ethyl cellulose	Emulsion phase separation	Bayberry, microwave extraction with 80% ethanol	[68]	40 min	17 µm to 93 µm	Simulated-intestinal fluid	–
Whey protein	Spray drying	Blueberry pomace extract with 80% ethanol	[76]		48.5 µm volumetric mean diameter (D4,3)		
Sodium alginate	Emulsification/internal gelation	Cocoa extract	[69]	1,500 min	39 to 321 µm	Polyoxyethylene sorbitan monolaurate (Tween_20)	Peppas-Sahlin eq.
Whey protein, carboxymethyl cellulose, and pectin	Film formation (according to simplex centroid experimental design)	Roselle extract (aqueous extract of Roselle calyx)	[75]	500 min (300 min to reach the plateau)	1 cm^2 pieces with an approximate weight of 0.1 g	NaCl aqueous solution	Peppas eq.
Sucrose matrix calcium alginate hydrogel calcium alginate-chitosan	Entrapment by cocrystallization Encapsulation by ionic gelation	Aqueous extracts of yerba mate	[79] [77]	45 s dissolute. 200 min release (40 min to reach the plateau) 800 min	0.2 and 2 mm	Water gastric and intestinal simulated fluids.	Peppas eq. Kopcha, Lordi eq.

Tab. 5.1 *(continued)*

Coating material	**Method of encapsulation**	**Source Material/ extraction method**	**Ref.**	**Release data**			
				experimental			**Modeling**
				Release time	**Particle size distribution**	**Release media**	
Cellulose-lignin hydrogels	Immersion of dried hydrogel in the extract and swelling	Extract of grapes seeds from Chambourcin with ethanol	[73]	550 min	Pore size given: 169 and 431 μm	Water-ethanol	Peppas (Korsmeyer-Peppas)
Poloxamers	Coprecipitation by supercritical antisolvent process	Ethanolic extract of rosemary leaves	[70]	1 h	<1 mm agglomerates 5 to 20 μm	Aqueous	–
Poly-ε-caprolactone	High-pressure antisolvent coprecipitation	Green tea extract in acetone	[71]	90 h (20 h to reach the plateau)	3–5 mm High degree of agglomeration	Phosphate buffer	–
Polyethyleneglycol	Ionic gelification	Ethanolic extract of anthocyanins from jabuticaba skins (*Myrciaria cauliflora*)	[59]	30 min	2.8–3.2 mm	Hydrochloric acid/ potassium chloride buffer	First-order kinetics
Alginate-chitosan microbeads	Electrostatic extrusion	Raspberry leaf, hawthorn, nettle, yarrow, ground ivy, olive leaf	[60]	30 min	781–1785 μm	Water	–
Whey protein gels		Bilberry extract from *Vaccinium myrtillus*	[61]	300 min	5 mm	Simulated gastric fluid	Peppas (Ritger Peppas)

The majority of investigations concerning encapsulated polyphenols release report diffusional mechanism – Fick's type or case 2 anomalous diffusion according to the semi-empirical power law equation, known as Peppas equation [77]:

$$\frac{M_t}{M_\infty} = kt^n \tag{5.1}$$

Here, M_t and M_∞ denote the cumulative amount released at time t and infinite time, respectively.

The mechanism of transfer is revealed in the value of the exponent n, obtained by regression analysis on the experimental data: $n = 0.5$ indicates Fickian diffusion mechanism (Higuchi model [78]):

$$\frac{M_t}{M_\infty} = k_H \sqrt{t} \tag{5.2}$$

while a value $0.5 < n < 1$ indicates an anomalous or non-Fickian behavior, as observed for gallic acid encapsulation [62]. When $n = 1$, a case II transport mechanism is involved, where the rate is largely determined by swelling and relaxation of the polymer chains, while $n > 1$ is known as special (or super) case II transport. The latter is related to the plasticization of the polymer system and increased mobility, facilitating the active compound release. Super case II transport was observed during natural extract release from chitosan microcapsules [79]. Many authors obtained values of the exponent lower than 0.5 (or 0.45, which is theoretically expected for sphere geometry). Some of them comment $n < 0.45$ as diffusional mechanism [61, 80], but attention should be paid also to the relation between n and the rate constants, obtained by regression. Keeping the same rate constants, the authors [61] have recalculated the value of n much closer to 0.45. Others [69] attribute this fact to the influence of a relaxation/dissolution mechanism, which suggests that relaxation has an influence on the release process (example with polyphenols release from alginate microspheres in an aqueous medium). Such observations were made also by numerical simulation of the release kinetics from particles with pronounced deformation due to swelling or shrinking. The power-law equation was used to study the mechanism of diffusion of quercetin or rutin into the skin following release from liposome/hydrogel complex system [62]. Numerous applications of Eq. (5.1) are found in studies concerning polyphenol release from encapsulated natural extracts (Roselle, grapes seeds, bilberry) [61, 75, 76].

When the remaining of actives in the solid amount is considered, the respective form of the power-law dependence (1) becomes:

$$(M_0 - M_t)/M_0 = kt^a \tag{5.3}$$

where M_0 and M_t correspond to solid phase content at initial "0" and at time "t." Considering the mechanism of transfer, this relation is used for diffusion-controlled release of antioxidants (carnosine and gallic acid) together with the parabolic diffusion model [81]:

$$(1 - M_t/M)/t = kt^{-0.5} + b \tag{5.4}$$

Both expressions are single kinetics models supposing one rate-determining mechanism of release. They are used to treat the release data of antioxidants (e.g. gallic acid) controlled by intraparticle or surface diffusion [78].

Empirical kinetic models for more than one controlling release mechanisms are illustrated by the next examples.

A) An intercept like the term "b" in Eq. (5.4) can be included in the semiempirical power-law equation (Eq. [5.1]) to account for the release due to the initial burst effect, considered as time independent. This approach is employed in both diffusion-controlled and swelling-controlled release systems, including antioxidants from encapsulated natural extracts (yerba mate) in food additives [81, 82].

An alternative to Eq. (5.1) simple mathematical description of the release kinetics is the monoexponential relation. In the form of Eq. (5.5), it was used to calculate the concentration of a bioactive compound entrapped within a nanocarrier [80]:

$$C = C_0 e^{-kt} \tag{5.5}$$

as well as its two-step version – the biexponential equation:

$$C = ae^{-k_1 t} + be^{-k_2 t} \tag{5.6}$$

where the rate constants k_1 and k_2 are related to burst and sustained release, respectively.

B) In the case of hydrogels, an anomalous transport mechanism is preferred, coupling Fickian diffusion with relaxation of the hydrogel network. The Peppas and Sahlin equation supposes diffusional mechanism in the first term, whereas the second one stands for the "Case-II transport" contribution [53]:

$$\frac{M_t}{M_\infty} = k_1 t^m + k_2 t^{2m} \tag{5.7}$$

This semiempirical expression is supposed to have better chances of treating experimental data with contributions of Fickian diffusion and matrix relaxation/dissolution. Examples are found in release data of polyphenols from natural extracts (yerba mate, cocoa extract) encapsulated in microspheres [69, 83].

C) In the case of yerba mate extract encapsulated in calcium alginate beads, a release mechanism combining erosion and diffusion was observed [80] and is described by the semiempirical model of Kopcha and Lordi [83]:

$$M = At^{1/2} + Bt \tag{5.8}$$

where M is the percentage of polyphenols released at time t, while A and B are the diffusion and erosion terms, respectively. This simple expression combines two additive time-dependent power terms – a diffusion one (like in the Higuchi equation $n = 0.5$) and an erosion one ($n = 1$).

Analytical and numerical solutions, based on Fickian diffusion for different particle geometries [84], available in the field of controlled drug delivery, are rarely used for polyphenol release from micro- and nano-capsules. Such models are successfully developed as two-dimensional ones [82, 85] complicated by moving boundary conditions reflecting swelling/shrinking [82, 85] and/or variable diffusion coefficients to describe more sophisticated internal diffusional transport [82].

5.5 Conclusions

Activity preservation and controlled release are the keywords for the importance of the encapsulation method. Polyphenols represent a wide area of application because of their diversity and wealth of bioactive properties, as well as the inherent limited stability and/or solubility, which results in restricted commercial application. Microencapsulation is an innovative approach allowing protection against oxidation and thermal degradation and improving their bioavailability *in vivo* and *in vitro*. Numerous investigations of the latter are reported, but also studies on release *in vivo* conditions are constantly increasing, answering the challenge of testing the release in a specific surrounding environment. Nanoencapsulation of polyphenols with therapeutic importance offers a way to obtain a minimally invasive delivery of sustained concentrations to a specific site. The future of encapsulated polyphenols and site-specific carrier targeting is promising in order to improve the performance of medicines and functional foods and use to the best the health benefits of these bioactives. Natural extracts, with their unique multicomponent composition, as well as individual polyphenols, have been an object of encapsulation and extensive study concerning the material of the carrier and the potential of the different encapsulation methods to obtain high amounts of entrapped polyphenols and a desired kinetic of their release. Coencapsulation methodologies, where two or more bioactive ingredients can be combined to have a synergistic effect, are also gaining interest. Providing controlled amount and rate of active compound release, the encapsulating of naturally derived polyphenolics is of primary interest for the food industry, as well as for pharmaceutical and cosmetic purposes, thus answering the demand for healthy products and foods containing naturally derived preservation ingredients.

Mathematical modeling has a significant role in understanding the release mechanisms and to quantify the rate parameters of the process. Models of different complexities require reliable experimental validation in order to reveal the detailed phenomena and to possess a predictive capability, which can be particularly fruitful if integrated in the process of product development.

5.6 Acknowledgements

We acknowledge the Bulgarian National Science Fund – Ministry of Education and Science under Contract DN 07/11-15.12.2016. We also acknowledge the following: the EURECAT Internal Research Program: Plan of Research and Innovation 2019 (PRI 2019) and ACCIÓ – Regional Agency for the Business Competitiveness of the Generalitat de Catalunya for financial support for the "HealthCap Development of Biodegradables Capsules to Prevent/Manage Side Effects of Skin Cancer Treatment" project and the Ministerio de Ciencia, Innovación y Universidades Spain for financial support for "MICROESENCI-Desarrollo y validación de nuevas MICROcápsulas con aceites ESENCIales para productos con efecto antimicrobiano y desinfectante" (grant RTC-2017-6352-1).

5.7 References

[1] Desai, K. G. H., Jin Park, H., Recent developments in microencapsulation of food ingredients, Dry Technol **23** (2005) 1361–1394. doi:10.1081/DRT-200063478.

[2] Shahidi, F., Han, X., Encapsulation of food ingredients, Crit Rev Food Sci Nutr **33** (1993) 501–547. doi:10.1080/10408399309527645.

[3] Paulo, F., Santos, L., Design of experiments for microencapsulation applications: a review, Mater Sci Eng C **77** (2017) 1327–1340. doi:https://doi.org/10.1016/j.msec.2017.03.219.

[4] Meirowitz, R., Smart Textile Coatings and Laminates, In: William C Smith (ed.), 5 – Microencapsulation technology for coating and laminating. Elsevier, Woodhead Publishing: Amsterdam, The Netherlands, 2019. pp. 117–154. doi:https://doi.org/10.1016/B978-0-08-102428-7.00005-5.

[5] Gonçalves, A., Estevinho, B. N., Rocha, F., Microencapsulation of vitamin A: a review, Trends Food Sci Technol **51** (2016) 76–87. doi:https://doi.org/10.1016/j.tifs.2016.03.001.

[6] Herrero-Vanrell, R., Bravo-Osuna, I., Andrés-Guerrero, V., Vicario-de-la-Torre, M., Molina-Martínez, I. T., The potential of using biodegradable microspheres in retinal diseases and other intraocular pathologies, Prog Retin Eye Res **42** (2014) 27–43. doi:https://doi.org/10.1016/j.preteyeres.2014.04.002.

[7] Ranadheera, C. S., Evans, C. A., Adams, M. C., Baines, S. K., Microencapsulation of Lactobacillus acidophilus LA-5, Bifidobacterium animalis subsp. lactis BB-12 and Propionibacterium jensenii 702 by spray drying in goat's milk, Small Rumin Res **123** (2015) 155–159. doi:https://doi.org/10.1016/j.smallrumres.2014.10.012.

[8] Aliakbarian, B., Casazza, A. A., Perego, P., Valorization of olive oil solid waste using high pressure–high temperature reactor, Food Chem **128** (2011) 704–710. doi:https://doi.org/10.1016/j.foodchem.2011.03.092.

[9] Kelly, N. P., Kelly, A. L., O'Mahony, J. A., Strategies for enrichment and purification of polyphenols from fruit-based materials, Trends Food Sci Technol **83** (2019) 248–258. doi:https://doi.org/10.1016/j.tifs.2018.11.010.

[10] Kumari, A., Yadav, S., Pakade, Y., Kumar, V., Singh, B., Chaudhary, A., Yadav, S., Nano encapsulation and characterization of Albizia chinensis isolated antioxidant quercitrin on PLA nanoparticles, Colloids Surf B Biointerfaces **82** (2011) 224–232. doi:10.1016/j.colsurfb.2010.08.046.

[11] Georgetti, S. R., Casagrande, R., Souza, C. R. F., Oliveira, W. P., Fonseca, M. J. V., Spray drying of the soybean extract: effects on chemical properties and antioxidant activity, LWT – Food Sci Technol **41** (2008) 1521–1527. doi:https://doi.org/10.1016/j.lwt.2007.09.001.

[12] Moure, A., Cruz, J. M., Franco, D., Domínguez, J. M., Sineiro, J., Domínguez, H., José Núñez, M., Parajó, J. C., Natural antioxidants from residual sources, Food Chem **72** (2001) 145–171. doi:https://doi.org/10.1016/S0308-8146(00)00223-5.

[13] Herrmann, K., Nagel, C. W., Occurrence and content of hydroxycinnamic and hydroxybenzoic acid compounds in foods, Crit Rev Food Sci Nutr **28** (1989) 315–347. doi:10.1080/10408398909527504.

[14] Chandrasekara, A., Encyclopedia of Food Chemistry, In: Varelis, P., Melton, L., Shahidi, F. (eds.), Phenolic acids. Elsevier, Oxford, 2019. pp. 535–545. doi:https://doi.org/10.1016/B978-0-08-100596-5.22395-0.

[15] Tomás-Barberán, F. A., Clifford, M. N., Dietary hydroxybenzoic acid derivatives – nature, occurrence and dietary burden, J Sci Food Agric **80** (2000) 1024–1032. doi:10.1002/(SICI)1097-0010(20000515)80:7<1024::AID-JSFA567>3.0.CO;2-S.

[16] Tsibranska, I. H., Tylkowski, B., Concentration of ethanolic extracts from Sideritis ssp. L. by nanofiltration: comparison of dead-end and cross-flow modes, Food Bioprod Process. **91** (2013) 169–174. doi:https://doi.org/10.1016/j.fbp.2012.09.004.

[17] Li, Y., Jiang, X., Hao, J., Zhang, Y., Huang, R., Tea polyphenols: the application in oral microorganism infectious diseases control, Arch Oral Biol **102** (2019) 74–82. doi:https://doi.org/10.1016/j.archoralbio.2019.03.027.

[18] Chagne, D., Krieger, C., Rassam, M., Sullivan, M., Fraser, J., Andre, C., Pindo, M., Troggio, M., Gardiner, S., Henry-Kirk, R., Allan, A. C., Mcghie, T., Laing, W. A., QTL and candidate gene mapping for polyphenolic composition in apple fruit, BMC Plant Biol **12** (2012). doi:10.1186/1471-2229-12-12.

[19] Clifford, M., Chlorogenic acids and other cinnamates – nature, occurrence, dietary burden, absorption and metabolism, J Food Sci Agric **80** (2000) 1033–1043. doi:10.1002/(SICI)1097-0010(20000515)80:73.0.CO;2-T.

[20] Rodríguez, H., Landete, J. M., de las Rivas, B., Muñoz, R., Metabolism of food phenolic acids by Lactobacillus plantarum CECT 748T, Food Chem. **107** (2008) 1393–1398. doi:https://doi.org/10.1016/j.foodchem.2007.09.067.

[21] Spagnol, C. M., Assis, R. P., Brunetti, I. L., Isaac, V. L. B., Salgado, H. R. N., Corrêa, M. A., In vitro methods to determine the antioxidant activity of caffeic acid, Spectrochim. Acta Part A Mol. Biomol. Spectrosc. **219** (2019) 358–366. doi:https://doi.org/10.1016/j.saa.2019.04.025.

[22] Lempereur, I., Rouau, X., Abecassis, J., Genetic and agronomic variation in arabinoxylan and Ferulic acid contents of durum wheat (Triticum durum L.) grain and its milling fractions, J. Cereal Sci. **25** (1997) 103–110. doi:https://doi.org/10.1006/jcrs.1996.0090.

[23] Scalbert, A., Rémésy, C., Morand, C., Jiménez, L., Manach, C., Polyphenols: food sources and bioavailability, Am. J. Clin. Nutr. **79** (2004) 727–747. doi:10.1093/ajcn/79.5.727.

[24] Manach, C., Williamson, G., Morand, C., Scalbert, A., Rémésy, C., Bioavailability and bioefficacy of polyphenols in human. I. Review of 97 bioavailability studies, Am. J. Clin. Nutr. **81** (2005) 230S–242S. doi:10.1093/ajcn/81.1.230S.

[25] Granata, G., Consoli, G. M. L., Lo Nigro, R., Geraci, C., Hydroxycinnamic acids loaded in lipid-core nanocapsules, Food Chem. **245** (2018) 551–556. doi:https://doi.org/10.1016/j.foodchem.2017.10.106.

[26] Perez-Vizcaino, F., Duarte, J., Flavonols and cardiovascular disease, Mol. Aspects Med. **31** (2010) 478–494. doi:https://doi.org/10.1016/j.mam.2010.09.002.

[27] D'Andrea, G., Quercetin: a flavonol with multifaceted therapeutic applications? Fitoterapia. **106** (2015) 256–271. doi:https://doi.org/10.1016/j.fitote.2015.09.018.

[28] Price, S. F., Breen, P. J., Valladao, M., Watson, B. T., Cluster sun exposure and quercetin in pinot Noir grapes and wine, Am. J. Enol.Viticult. **46** (1995) 187–194.
[29] HERRMANN, K., Flavonols and flavones in food plants: a review, Int. J. Food Sci. Technol. **11** (1976) 433–448. doi:10.1111/j.1365-2621.1976.tb00743.x.
[30] Sartelet, H., Serghat, S., Lobstein, A., Ingenbleek, Y., Anton, R., Petitfrère, E., Aguie-Aguie, G., Martiny, L., Haye, B., Flavonoids extracted from fonio millet (Digitaria exilis) reveal potent antithyroid properties, Nutrition. **12** (1996) 100–106. doi:https://doi.org/10.1016/0899-9007(96)90707-8.
[31] Drewnowski, A., Gomez-Carneros, C., Bitter taste, phytonutrients, and the consumer: a review, Am. J. Clin. Nutr. **72** (2000) 1424–1435. doi:10.1093/ajcn/72.6.1424.
[32] Gil, E. S., Couto, R. O., Flavonoid electrochemistry: a review on the electroanalytical applications, Rev. Bras. Farmacogn. **23** (2013) 542–558. doi:https://doi.org/10.1590/S0102-695X2013005000031.
[33] Crozier, A., Jaganath, I. B., Clifford, M. N., Dietary phenolics: chemistry, bioavailability and effects on health, Nat. Prod. Rep. **26** (2009) 1001–1043. doi:10.1039/B802662A.
[34] Barnes, S., Wang, C. C., Kirk, M., Johnson, M., Coward, L., Barnes, N. C., Vance, G., Boersma-Maland, B., HPLC-mass spectrometry of isoflavonoids in soy and the American groundnut, Apios Americana, 2002. doi:10.1007/978-1-4757-5235-9_7.
[35] Mazur, W. M., Duke, J. A., Wähälä, K., Rasku, S., Adlercreutz, H., Isoflavonoids and lignans in legumes: nutritional and health aspects in humans 11 The method development and synthesis of the standards and deuterium-labelled compounds was supported by National Institutes of Health Grants No. 1 R01 CA56289-01 and No. 2 R01 CA56289-04, and analytical work by the EU research contract FAIR-CT95-0894, J. Nutr. Biochem. **9** (1998) 193–200. doi:https://doi.org/10.1016/S0955-2863(97)00184-8.
[36] Prasain, J., Jones, K., Kirk, M., Wilson, L., Johnson, M., Weaver, C., Barnes, S., Profiling and Quantification of isoflavonoids in kudzu dietary supplements by high-performance liquid chromatography and electrospray ionization tandem mass spectrometry, J. Agric. Food Chem. **51** (2003) 4213–4218. doi:10.1021/jf030174a.
[37] Kiyama, R., Biological effects induced by estrogenic activity of lignans, Trends Food Sci. Technol. **54** (2016) 186–196. doi:https://doi.org/10.1016/j.tifs.2016.06.007.
[38] Zhang, J., Chen, J., Liang, Z., Zhao, C., New lignans and their biological activities, Chem. Biodivers. **11** (2014) 1–54. doi:10.1002/cbdv.201100433.
[39] Ignat, I., Volf, I., Popa, V. I., A critical review of methods for characterisation of polyphenolic compounds in fruits and vegetables, Food Chem. **126** (2011) 1821–1835. doi:https://doi.org/10.1016/j.foodchem.2010.12.026.
[40] Fernández-Marín, M. I., Guerrero, R. F., García-Parrilla, M. C., Puertas, B., Richard, T., Rodriguez-Werner, M. A., Winterhalter, P., Monti, J. -P., Cantos-Villar, E., Isorhapontigenin: a novel bioactive stilbene from wine grapes, Food Chem. **135** (2012) 1353–1359. doi:https://doi.org/10.1016/j.foodchem.2012.05.086.
[41] Jeandet, P., Sobarzo-Sánchez, E., Clément, C., Nabavi, S. F., Habtemariam, S., Nabavi, S. M., Cordelier, S., Engineering stilbene metabolic pathways in microbial cells, Biotechnol. Adv. **36** (2018) 2264–2283. doi:https://doi.org/10.1016/j.biotechadv.2018.11.002.
[42] Soleas, G. J., Diamandis, E. P., Goldberg, D. M., Resveratrol: a molecule whose time has come? And gone? Clin. Biochem. **30** (1997) 91–113. doi:https://doi.org/10.1016/S0009-9120(96)00155-5.
[43] Ye, Q., Georges, N., Selomulya, C., Microencapsulation of active ingredients in functional foods: from research stage to commercial food products, Trends Food Sci. Technol. **78** (2018) 167–179. doi:https://doi.org/10.1016/j.tifs.2018.05.025.
[44] Aguiar, J., Estevinho, B. N., Santos, L., Microencapsulation of natural antioxidants for food application – the specific case of coffee antioxidants – a review, Trends Food Sci. Technol. **58** (2016) 21–39. doi:https://doi.org/10.1016/j.tifs.2016.10.012.

[45] Ribeiro, A. M., Estevinho, B. N., Rocha, F., Microencapsulation of polyphenols – the specific case of the microencapsulation of Sambucus Nigra L. extracts – a review, Trends Food Sci. Technol. (2019) article in press. doi:https://doi.org/10.1016/j.tifs.2019.03.011.
[46] Coradini, K., Lima, F. O., Oliveira, C. M., Chaves, P. S., Athayde, M. L., Carvalho, L. M., Beck, R. C. R., Co-encapsulation of resveratrol and curcumin in lipid-core nanocapsules improves their in vitro antioxidant effects, Eur. J. Pharm. Biopharm. **88** (2014) 178–185. doi:https://doi.org/10.1016/j.ejpb.2014.04.009.
[47] Asanuma, H., Liang, X., Nishioka, H., Matsunaga, D., Liu, M., Komiyama, M., Synthesis of azobenzene-tethered DNA for reversible photo-regulation of DNA functions: hybridization and transcription, Nat. Protoc. **2** (2007) 203. https://doi.org/10.1038/nprot.2006.465.
[48] Zanotto-Filho, A., Coradini, K., Braganhol, E., Schröder, R., Melo de Oliveira, C., Simões-Pires, A., Battastini, A., Pohlmann, A., Stanisçuaski Guterres, S., Forcelini, C., Beck, R., Moreira, J. C., Curcumin-loaded lipid-core nanocapsules as a strategy to improve pharmacological efficacy of curcumin in glioma treatment, Eur. J. Pharm. Biopharm. **83** (2013) 156–67. doi:10.1016/j.ejpb.2012.10.019.
[49] Wu, T. -H., Yen, F. -L., Lin, L. -T., Tsai, T. -R., Lin, C. -C., Cham, T. -M., Preparation, physicochemical characterization, and antioxidant effects of quercetin nanoparticles, Int. J. Pharm. **346** (2008) 160–168. doi:https://doi.org/10.1016/j.ijpharm.2007.06.036.
[50] Choi, K. -O., Kim, D., Lim, J. D., Ko, S., Hong, G. -P., Lee, S., Functional enhancement of ultrafine Angelica gigas powder by spray-drying microencapsulation, LWT. **101** (2019) 161–166. doi:https://doi.org/10.1016/j.lwt.2018.10.097.
[51] Fang, Z., Bhandari, B., Encapsulation of polyphenols – a review, Trends Food Sci. Technol. **21** (2010) 510–523. doi:https://doi.org/10.1016/j.tifs.2010.08.003.
[52] Ting, Y., Jiang, Y., Ho, C.- T., Huang, Q., Common delivery systems for enhancing in vivo bioavailability and biological efficacy of nutraceuticals, 2014. doi:10.1016/j.jff.2013.12.010.
[53] Arifin, D. Y., Lee, L. Y., Wang, C.-H., Mathematical modeling and simulation of drug release from microspheres: implications to drug delivery systems, Adv. Drug Deliv. Rev. **58** (2006) 1274–1325. doi:https://doi.org/10.1016/j.addr.2006.09.007.
[54] Grassi, M., Lamberti, G., Cascone, S., Grassi, G., Mathematical modeling of simultaneous drug release and in vivo absorption, Int. J. Pharm. **418** (2011) 130–141. doi:https://doi.org/10.1016/j.ijpharm.2010.12.044.
[55] Davidov-Pardo, G., McClements, D. J., Nutraceutical delivery systems: Resveratrol encapsulation in grape seed oil nanoemulsions formed by spontaneous emulsification, Food Chem. **167** (2015) 205–212. doi:https://doi.org/10.1016/j.foodchem.2014.06.082.
[56] Sessa, M., Balestrieri, M. L., Ferrari, G., Servillo, L., Castaldo, D., D'Onofrio, N., Donsì, F., Tsao, R., Bioavailability of encapsulated resveratrol into nanoemulsion-based delivery systems, Food Chem. **147** (2014) 42–50. doi:https://doi.org/10.1016/j.foodchem.2013.09.088.
[57] Altunbas, A., Lee, S. J., Rajasekaran, S. A., Schneider, J. P., Pochan, D. J., Encapsulation of curcumin in self-assembling peptide hydrogels as injectable drug delivery vehicles, Biomaterials. **32** (2011) 5906–5914. doi:https://doi.org/10.1016/j.biomaterials.2011.04.069.
[58] Mahdavi, S. A., Jafari, S. M., Ghorbani, M., Assadpoor, E., Spray-drying microencapsulation of anthocyanins by natural biopolymers: a review, Dry. Technol. **32** (2014) 509–518. doi:10.1080/07373937.2013.839562.
[59] Santos, D. T., Albarelli, J. Q., Beppu, M. M., Meireles, M. A. A., Stabilization of anthocyanin extract from jabuticaba skins by encapsulation using supercritical CO2 as solvent, Food Res. Int. **50** (2013) 617–624. doi:https://doi.org/10.1016/j.foodres.2011.04.019.
[60] Belščak-Cvitanović, A., Stojanović, R., Manojlović, V., Komes, D., Cindrić, I. J., Nedović, V., Bugarski, B., Encapsulation of polyphenolic antioxidants from medicinal plant extracts in alginate-hitosan system enhanced with ascorbic acid by electrostatic extrusion, Food Res. Int. **44** (2011) 1094–1101. doi:https://doi.org/10.1016/j.foodres.2011.03.030.

[61] Betz, M., Kulozik, U., Whey protein gels for the entrapment of bioactive anthocyanins from bilberry extract, Int. Dairy J. **21** (2011) 703–710. doi:https://doi.org/10.1016/j.idairyj.2011.04.003.
[62] Park, S. N., Lee, M. H., Kim, S. J., Yu, E. R., Preparation of quercetin and rutin-loaded ceramide liposomes and drug-releasing effect in liposome-in-hydrogel complex system, Biochem. Biophys. Res. Commun. **435** (2013) 361–366. doi:https://doi.org/10.1016/j.bbrc.2013.04.093.
[63] Sun, X., Cameron, R. G., Bai, J., Microencapsulation and antimicrobial activity of carvacrol in a pectin-alginate matrix, Food Hydrocoll. **92** (2019) 69–73. doi:https://doi.org/10.1016/j.foodhyd.2019.01.006.
[64] Akhtar, M., Murray, B. S., Afeisume, E. I., Khew, S. H., Encapsulation of flavonoid in multiple emulsion using spinning disc reactor technology, Food Hydrocoll. **34** (2014) 62–67. doi:https://doi.org/10.1016/j.foodhyd.2012.12.025.
[65] Flores, F. P., Singh, R. K., Kong, F., Physical and storage properties of spray-dried blueberry pomace extract with whey protein isolate as wall material, J. Food Eng. **137** (2014) 1–6. doi:https://doi.org/10.1016/j.jfoodeng.2014.03.034.
[66] Hoyos-Leyva, J. D., Bello-Perez, L. A., Agama-Acevedo, J. E., Alvarez-Ramirez, J., Jaramillo-Echeverry, L. M., Characterization of spray drying microencapsulation of almond oil into taro starch spherical aggregates, LWT. **101** (2019) 526–533. doi:https://doi.org/10.1016/j.lwt.2018.11.079.
[67] Li, K., Woo, M. W., Patel, H., Selomulya, C., Enhancing the stability of protein-polysaccharides emulsions via Maillard reaction for better oil encapsulation in spray-dried powders by pH adjustment, Food Hydrocoll. **69** (2017) 121–131. doi:https://doi.org/10.1016/j.foodhyd.2017.01.031.
[68] Zheng, L., Ding, Z., Zhang, M., Sun, J., Microencapsulation of bayberry polyphenols by ethyl cellulose: preparation and characterization, J. Food Eng. **104** (2011) 89–95. doi:https://doi.org/10.1016/j.jfoodeng.2010.11.031.
[69] Lupo, B., Maestro, A., Porras, M., Gutiérrez, J. M., González, C., Preparation of alginate microspheres by emulsification/internal gelation to encapsulate cocoa polyphenols, Food Hydrocoll. **38** (2014) 56–65. doi:https://doi.org/10.1016/j.foodhyd.2013.11.003.
[70] Visentin, A., Rodríguez-Rojo, S., Navarrete, A., Maestri, D., Cocero, M. J., Precipitation and encapsulation of rosemary antioxidants by supercritical antisolvent process, J. Food Eng. **109** (2012) 9–15. doi:https://doi.org/10.1016/j.jfoodeng.2011.10.015.
[71] Sosa, M. V., Rodríguez-Rojo, S., Mattea, F., Cismondi, M., Cocero, M. J., Green tea encapsulation by means of high pressure antisolvent coprecipitation, J. Supercrit. Fluids. **56** (2011) 304–311. doi:https://doi.org/10.1016/j.supflu.2010.10.038.
[72] Cavalcanti, R. N., Santos, D. T., Meireles, M. A. A., Non-thermal stabilization mechanisms of anthocyanins in model and food systems – an overview, Food Res. Int. **44** (2011) 499–509. doi:https://doi.org/10.1016/j.foodres.2010.12.007.
[73] López-Córdoba, A., Deladino, L., Agudelo-Mesa, L., Martino, M., Yerba mate antioxidant powders obtained by co-crystallization: stability during storage, J. Food Eng. **124** (2014) 158–165. doi:https://doi.org/10.1016/j.jfoodeng.2013.10.010.
[74] Pulicharla, R., Marques, C., Das, R. K., Rouissi, T., Brar, S. K., Encapsulation and release studies of strawberry polyphenols in biodegradable chitosan nanoformulation, Int. J. Biol. Macromol. **88** (2016) 171–178. doi:https://doi.org/10.1016/j.ijbiomac.2016.03.036.
[75] Ciolacu, D., Oprea, A. M., Anghel, N., Cazacu, G., Cazacu, M., New cellulose-lignin hydrogels and their application in controlled release of polyphenols, Mater. Sci. Eng. C. **32** (2012) 452–463. doi:https://doi.org/10.1016/j.msec.2011.11.018.
[76] Flores, F. P., Singh, R. K., Kerr, W. L., Pegg, R. B., Kong, F., Total phenolics content and antioxidant capacities of microencapsulated blueberry anthocyanins during in vitro digestion, Food Chem. **153** (2014) 272–278. doi:https://doi.org/10.1016/j.foodchem.2013.12.063.

[77] Kopcha, M., Lordi, N. G., Tojo, K. J., Evaluation of release from selected thermosoftening vehicles, 1991. doi:10.1111/j.2042-7158.1991.tb03493.x.
[78] Fathi, M., Mozafari, M. R., Mohebbi, M., Nanoencapsulation of food ingredients using lipid based delivery systems, Trends Food Sci. Technol. **23** (2012) 13–27. doi:https://doi.org/10.1016/j.tifs.2011.08.003.
[79] Deladino, L., Anbinder, P. S., Navarro, A. S., Martino, M. N., Encapsulation of natural antioxidants extracted from Ilex paraguariensis, Carbohydr. Polym. **71** (2008) 126–134. doi:https://doi.org/10.1016/j.carbpol.2007.05.030.
[80] Ritger, P. L., Peppas, N. A., A simple equation for description of solute release I. Fickian and non-fickian release from non-swellable devices in the form of slabs, spheres, cylinders or discs, J. Control. Release. **5** (1987) 23–36. doi:https://doi.org/10.1016/0168-3659(87)90034-4.
[81] Kong, X., Jin, L., Wei, M., Duan, X., Antioxidant drugs intercalated into layered double hydroxide: structure and in vitro release, Appl. Clay Sci. **49** (2010) 324–329. doi:https://doi.org/10.1016/j.clay.2010.06.017.
[82] López Córdoba, A., Deladino, L., Martino, M., Effect of starch filler on calcium-alginate hydrogels loaded with yerba mate antioxidants, Carbohydr. Polym. **95** (2013) 315–323. doi:https://doi.org/10.1016/j.carbpol.2013.03.019.
[83] Naddaf, A. A., Tsibranska, I., Bart, H. -J., Kinetics of BSA release from poly(N-isopropylacrylamide) hydrogels, Chem. Eng. Process. Process Intensif. **49** (2010) 581–588. doi:https://doi.org/10.1016/j.cep.2010.05.008.
[84] Deladino, L., Navarro, A. S., Martino, M. N., Carrier systems for yerba mate extract (Ilex paraguariensis) to enrich instant soups. Release mechanisms under different pH conditions, LWT – Food Sci. Technol. **53** (2013) 163–169. doi:https://doi.org/10.1016/j.lwt.2013.01.030.
[85] Siepmann, J., Podual, K., Sriwongjanya, M., Peppas, N. A., Bodmeier, R., A new model describing the swelling and drug release kinetics from hydroxypropyl methylcellulose tablets, J. Pharm. Sci. **88** (1999) 65–72. doi:10.1021/js9802291.

Justyna Kozlowska, Anna Bajek, Natalia Stachowiak,
Weronika Prus-Walendziak and Bartosz Tylkowski

6 Application of microencapsulation in medical and pharmaceutical industry

6.1 Microencapsulation of antibiotics

The last decade has been focused on the development of novel and efficient drug delivery systems to various applications, especially in the medical and pharmaceutical industry. Polymeric microparticles have shown some advantages as drug carriers, such as high stability, good biocompatibility, and multifunctionality. This potential has been examined for encapsulating antibiotics in microparticles, which allows controlled or sustained release of the drug and improves the pharmacokinetics profile of antibiotics [1–3]. Hence, drug-loaded microparticles offer reaching the therapeutic level of drug concentration without exceeding the maximum tolerable dose, as well as maintaining the concentrations for extended periods of time till the desired therapeutic effect is reached [4, 5]

Many recent studies have described the incorporation of antibiotics into microparticles. This aspect was approached by Durán et al. in an interesting work about the preparation of microparticles based on rifampicin (RIF) and biodegradable polymer poly(3-hydroxybutyrate-co-3-hydroxyvalerate) (PHBV) for oral administration. RIF, belonging to the chemical class of compounds termed ansamycins, is an antituberculosis antibiotic with excellent sterilizing activity (Figure 6.1). The results indicated that the encapsulation efficiency of drug-loaded microparticles was about 14%, and almost 90% of the antibiotic was released after 24 h. It was demonstrated that the microencapsulated RIF exhibited a similar inhibition value as free

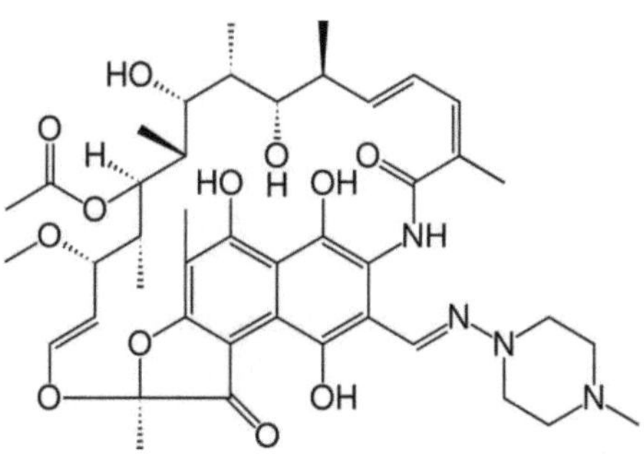

Fig. 6.1: Chemical structure of rifampicin.

https://doi.org/10.1515/9783110642070-006

RIF at 24 h of incubation with *Staphylococcus aureus*. Cytotoxicity assays showed a reduction of the toxicity when RIF was loaded into PHBV microparticles while maintaining its antibacterial activity [6].

In turn, Cunha et al. used fucoidan as a matrix of microparticles containing two first-line antitubercular antibiotics – isoniazid (INH) and rifabutin (RFB). These drug carriers were produced by spray-drying for an application as inhalable tuberculosis therapy. Drug association efficiencies were 81% for RFB and 95% for INH, resulting in loading capacities of 3.9% and 8.6%, respectively. The produced microparticles showed adequate aerodynamic properties for pulmonary delivery of antibiotics with possibility to reach the respiratory zone [7].

In a study carried out by Franco and coworkers, zein microparticles loaded with amoxicillin and ampicillin were prepared by supercritical antisolvent coprecipitation. These two antibiotics belong to the group of aminopenicillins, which have a broad spectrum of antibacterial activity in the treatment of Gram-positive and some of Gram-negative bacteria. In this investigation, researchers successfully encapsulated antibiotics into microparticles, obtaining 99.5–99.8% encapsulation efficiency for all the samples. An *in vitro* release study revealed that the release of the drugs was significantly prolonged, confirming that zein/antibiotic microparticles can be used for controlled-release systems [8]. Montes' group performed encapsulation of ampicillin and amoxicillin into ethyl cellulose microparticles by a supercritical antisolvent process using CO_2 as the antisolvent. The drug release profiles were examined in simulated gastric fluid (SGF), pH 1.2, and simulated intestinal fluid (SIF), pH 6.8. The release of antibiotics from the precipitates was slower than from a solution of the pure drug, and sustained release was more marked in SIF than in SGF. In addition, the presence of ampicillin and amoxycillin on the microparticle surface caused faster release of the drugs [9, 10].

Kasten et al. developed azithromycin (AZI)-loaded low-density polycaprolactone microparticles by the double emulsion/solvent evaporation method. AZI is an antimicrobial macrolide with wide antibacterial spectrum, including most major respiratory pathogens. The obtained microparticles, which showed drug loading up to 23.1% and allowed the control of the antibiotic release over 24 h, can be used for the local treatment of lung infections. They estimated the pulmonary deposition profiles using an *in silico* model, and the results indicated that a significant fraction of the microparticles may be deposited in the deeper lung regions [11]. Tung et al. also successfully fabricated polymeric microparticles containing AZI to bitter taste masking using two different methods: spray-drying and solvent evaporation [12].

Another research group prepared vancomycin-loaded levan microparticles using the precipitation method. Vancomycin is a first-generation bactericidal glycopeptide antibiotic, and its pharmacological action is due to its ability to inhibit bacterial cell wall biosynthesis. According to the authors, the drug encapsulation capacity of microparticles changed between 47.7% and 74.7%. The antibiotic release from the particles presented a biphasic behavior. During 5 h, 10–20% of the drug was released rapidly from the particles. After this time, approximately more than 70% of the drug released from the particles at constant speed. The obtained microparticles showed an antibacterial effect against *Bacillus subtilis* ATCC 6633 [13]. There is also another study about vancomycin encapsulated chitosan-alginate polyelectrolyte microparticles as a controlled drug delivery system. In this case, lyophilized microparticles showed the controlled release of vancomycin with an average release of 22 μg per day for 14 days [14].

Very interesting research was presented by Wu et al. The authors designed minocycline-calcium-dextran sulfate complex microparticles as local delivery systems for periodontitis treatment. Minocycline is a semisynthetic, second-generation tetracycline antibiotic with anti-inflammatory properties, commonly used in the treatment of bacterial infections (Figure 6.2). The loading efficiency of minocycline into these complex precipitates was about 97%, whereas the loading content was about 45%. The published results reported that the obtained microparticles achieved sustained release of minocycline for at least 9 days at pH 7.4 and 18 days at pH 6.4. Moreover, the investigation showed that complex microparticles demonstrated potent antimicrobial effects against *Streptococcus mutans* and *Aggregatibacter actinomycetemcomitans* in agar disk diffusion and biofilm assays [15].

Flores's group prepared gentamicin-loaded poly(lactic-co-glycolic acid) (PLGA) microparticles using a double emulsion solvent extraction/evaporation technique. They proposed the application of these microparticles to prevent and provide protection against bone infections. Gentamicin is an aminoglycoside antibiotic with a broad-spectrum activity, water solubility, and stability (Figure 6.3). The microparticles containing gentamicin demonstrated a sustained release for 25–30 days with initial burst effect, which related to a rapid antibacterial activity. According to the biological evaluation, the obtained microparticles exhibited cytocompatibility

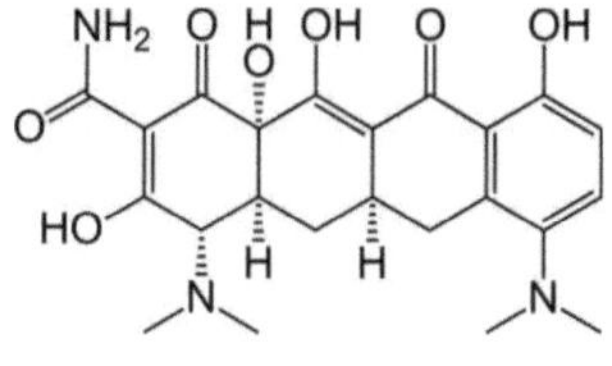

Fig. 6.2: Chemical structure of minocycline.

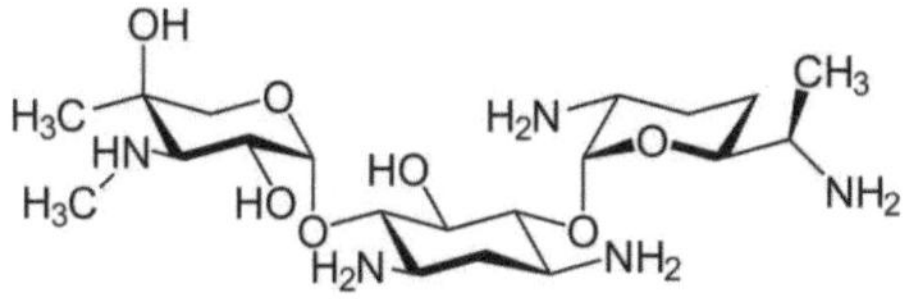

Fig. 6.3: Chemical structure of gentamicin.

and nonhemolytic properties and antibacterial activity against *Staphylococcus aureus* [16]. Aquino et al. designed alginate/pectin microparticles containing gentamicin by supercritical assisted atomization for the treatment of wound bacterial infections. According to the authors, gentamicin loading was varied between 20% and 33% (w/w), while encapsulation efficiency almost amounted to 100%. They conducted stability tests in storage accelerated conditions. The results showed that microparticles had a good stability and constant water content after 90 days of storage. The release behavior of antibiotic-loaded microparticles was examined using vertical Franz-type diffusion cells. The release of pure gentamicin solution proceeded by diffusion process, and the total content of the drug was released in less than 3 h. In turn, gentamicin/alginate/pectin microparticles showed sustained release of antibiotic until 6 days, achieved by the swelling and erosion of the polymers. All formulations presented preservation of gentamicin activity after 6 days and the drug activity was higher at 12 and 24 days. What is important, the rate and extent of bactericidal activity over 24 days were confirmed by a time-killing assay [17].

Mustafa et al. performed encapsulation of kanamycin sulphate (KS) into microparticles and nanoparticles made of PLGA with d-α-tocopheryl polyethylene glycol 1000 succinate (vitamin-E-TPGS) for intramuscular administration to reduce dosing frequency and side effects. The encapsulation efficiency of KS microparticles was about 69%, while for KS nanoparticles, it was about 73%. The biphasic release profiles were observed. The therapeutic agent was released from all formulations with initial burst effect with later sustained release. In case of microparticles, the drug release was about 20% within the first 24 h; up to 95% was released after 10 days. The antibiotic release from nanoparticles was improved, and it was about 16% in the initial 24 h, with relatively 97% of the drug released after 13 days. During the early phases, the drug released mainly through diffusion in the polymer matrix, while the later phases were related to both diffusion and degradation of the polymer matrix. Based on published data, an *in vivo* test presented that nanoparticles produced detectable blood levels for 5 days in comparison to 4 days of microparticles and 8 h of pure drug [18].

Raval et al. encapsulated cefixime into poly(lactide-co-glycolide) (PLGA) microparticles by spray-drying technique. Cefixime is a semisynthetic, third-generation cephalosporin used mainly in infections of the upper respiratory tract, urinary tract, and gonorrhea (Figure 6.4). Based on published data, the concentration of polymer and drug has an impact on the encapsulation efficiency and drug-release rate. All the prepared formulations exhibited high encapsulation efficiency, above 80%. The controlled and sustained release of the cefixime from microparticles continued for 74 h. Antibacterial studies were carried out using a standard agar diffusion method and it was observed that these microparticles effectively inhibited the growth of microorganisms [19].

Wright's group produced nano and microparticles from PLGA with an entrapment novel antibiotic – Ramizol for subcutaneous and intramuscular administration. For this purpose, they used two methods: emulsification solvent evaporation

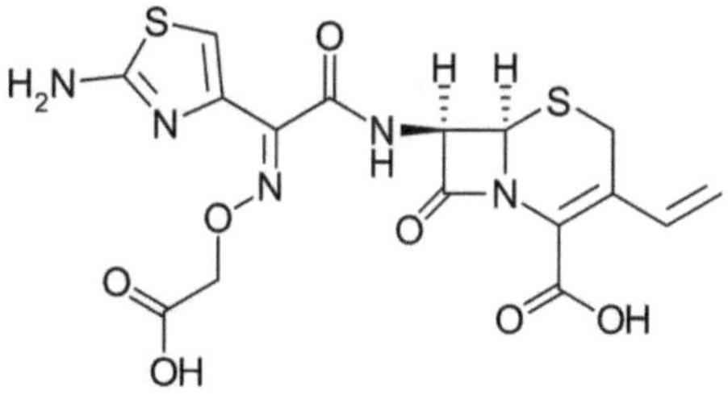

Fig. 6.4: Chemical structure of cefixime.

techniques and nanoprecipitation, but the first one turned out to be more appropriate for the preparation of drug carriers to the target site. The selection of formulation method had an impact on drug loading and encapsulation efficiency. In the case of emulsification solvent evaporation, loading levels of the prepared particles were about 9% and encapsulation efficiency was about 92%. The antibiotic incorporated into all formulations fabricated by this technique was released in two steps: initial with burst effect followed by a sustained-release phase. Pharmacokinetic studies in rats showed prolonged absorption and improved bioavailability of Ramizol after subcutaneous and intramuscular dosing compared to the oral administration [20].

Another research group encapsulated daptomycin into poly(methyl methacrylate) (PMMA) and PMMA-Eudragit RL 100 (EUD) microparticles by a double emulsion-solvent evaporation method. The addition of EUD to the formulations caused an improvement in antibiotic encapsulation efficiency and release behavior. The encapsulation efficiency of these microparticles amounted to 90%. The antibiotic release was observed only after adding EUD to the microparticles, and it was proportional to the amount of EUD. The *in vitro* drug release increased up to 52.2% in the case of PMMA-EUD 30% after 24 h. The microparticles with and without daptomycin showed similar biocompatibility profiles. The antibacterial activity of the formulations was assessed by isothermal microcalorimetry against two clinically relevant methicillin-susceptible and methicillin-resistant *Staphylococcus aureus* strains (MSSA and MRSA, respectively). The results indicated that prepared PMMA-EUD microparticles containing drug showed the highest antibacterial activity against both strains. It was reported that 150 mg/mL of daptomycin was released from PMMA-EUD 30% microparticles after 24 h based on bioassay [21].

Thomas' group incorporated ciprofloxacin into PLGA micro- and nano-particles using the double emulsion solvent evaporation method. Ciprofloxacin is synthetic antibiotic included in fluoroquinolones. Due to its broad spectrum of antimicrobial activity, it remains effective in a wide variety of infections (Figure 6.5). The aim of this study was to increase the efficiency antibiotic against biofilms of *Staphylococcus aureus* and *Pseudomonas aeruginosa*. The drug loading levels were 4.3% and 7.5% for nano- and micro-particles, respectively. Hence, microparticles exhibited 96% of encapsulation efficiency, while for nanoparticles, this value was

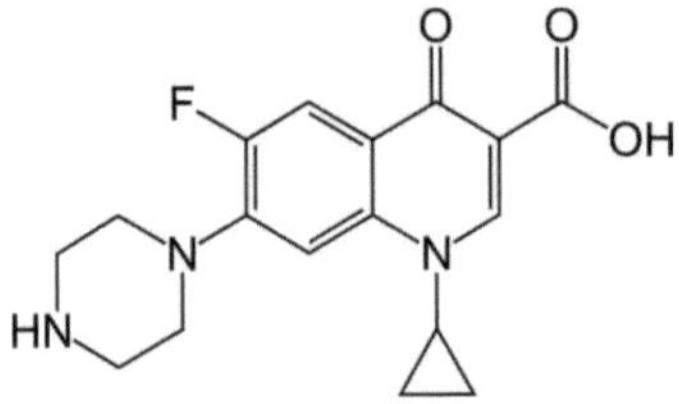

Fig. 6.5: Chemical structure of ciprofloxacin.

55%. All formulations presented an initial burst release of ciprofloxacin followed by a sustained drug release. Based on published data, approximately 50–60% of antibiotic was released during the initial 24 h and total drug release was observed within 5 days. An antibiofilm study of ciprofloxacin treatments was examined. In case of free ciprofloxacin during 2 days, only a slight reduction in the number of culturable cells of *P. aeruginosa* and *S. aureus* was observed compared to controls. After 6 days of treatment with free antibiotic, complete eradication of culturable cells was achieved. In contrast, treatment with prolonged-release drug with micro- and nanoparticles over 6 days showed a significant reduction in culturable *P. aeruginosa*, and the eradication of culturable *S. aureus* was equally effective as the continuous 6 days of treatment with free ciprofloxacin [22].

Martín and coworkers produced three types of microspheres made from sodium alginate, chitosan-coated, and hydrogel-containing nystatin (Nys) using emulsification/internal gelation method. The main goal of this study was to fabricate effective antifungal mucoadhesive systems for the treatment of oral candidiasis. Nys is a polyene antibiotic obtained from *Streptomyces noursei* and possesses a broad spectrum with both antifungal and fungistatic activity. It is used in the treatment of cutaneous, vaginal, and oral fungal infections. In this investigation, all prepared systems presented an explicit fungicidal activity against *Candida albicans* strains up to 48 h. *In vivo* studies showed that Nys was not found in systemic circulation, providing the safety of the treatment. Sodium alginate microparticles and coated chitosan microparticles indicated a two-step release profile. The first step was related to rapid burst release of the drug, followed by a slower sustained-release phase. Approximately 60% of the drug was released from these microparticles after 2 h. In the case of hydrogel with microparticles, Nys was released in a more constant manner and only 25% of the drug was released from this system after 2 h [23].

6.2 Microencapsulation of anticancer agents

Cancer is a group of diseases caused by an abnormal growth of cells, which tend to proliferate in an uncontrolled way and, in some cases, to metastasize. The cancerous cells are not caused by a single mutation or protein abnormality, they are a result of a combination of numerous abnormalities such as genetic or functional change, mutation in somatic cells, variations in size and shape of cells, larger and darker nucleus, and abnormal number of chromosomes arranged in a disorganized way [24, 25].

Cancer incidence increases rapidly each year around the world. The reasons for this phenomenon are complex but reflect the aging and growth of population, as well as social and economic development. According to the "World Cancer Report 2018" prepared by the International Agency for Research on Cancer, the global burden of cancer increased to 18.1 million new cases and 9.6 million new deaths worldwide in 2018 [26]. Worldwide, 1 in 5 men and 1 in 6 women suffer from cancer during their lifetime, as well as 1 in 8 men and 1 in 11 women die of the disease. It is estimated that the number of cancer cases will increase from 18.1 million to 21.5 million in 2025 and to 29.5 million in 2040. More and more cases are also reported in children and young people.

The treatment of cancer includes surgery, chemotherapy, radiation, and hormonal therapy. A main disadvantage of anticancer drugs is their insufficient selectivity for tumor tissue, which leads to severe side effects and low cure rates. Therefore, there is a great need to develop effective therapy targeting solely abnormal cells. A microparticulate drug delivery system is one of the solutions used to provide the site-specific delivery along with controlled and sustained release of compounds, without causing side effects on normal cells. Polymers used in encapsulation of anticancer drugs procedure must fulfill several requirements such as nontoxicity, biocompatibility, biodegradability, and ease of processing.

Doxorubicin (DOX) is an anthracycline originally isolated from the pigment-producing bacteria *Streptomyces peucetius* early in the 1960s (Figure 6.6) [27]. DOX is an antibiotic that is one of the most commonly used and highly effective anticancer drugs in chemotherapy treatment of neoplastic diseases like leukemia and various solid tumors such as breast, ovary, pancreas, lung, and skin cancers [28–30]. It is also the protocol therapy for AIDS-related Kaposi's sarcoma. However, the clinical use of this broad-spectrum drug is limited due to its serious nonspecific toxicity to normal tissues, leading to severe side effects such as cardiac and renal toxicity and

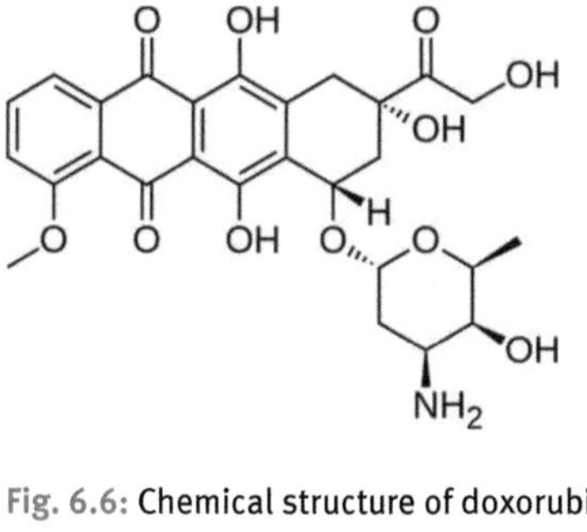

Fig. 6.6: Chemical structure of doxorubicin.

the development of multiple drug resistance [31, 32]. During the last years, several scientific publications have been dedicated to the preparation of tumor-targeted microparticulated formulations and to investigating their therapeutic effects in combination with low cytotoxicity.

Janes et al. [33] prepared chitosan nanoparticles loaded with DOX by dissolving chitosan or chitosan-DOX complex in acetic acid solution and adding tripolyphosphate (TPP) aqueous solution under magnetic stirring, which led to the immediate formation of the nanoparticles. The addition of the negatively charged TPP caused electrostatic interactions with positively charged chitosan and DOX, resulting in the distribution of DOX to the chitosan network. The entrapped drug maintained its cytostatic effect and was delivered into the cancerous cells via an endocytic mechanism in its active form. DOX-loaded chitosan microspheres were also developed by Park and coworkers [29] using emulsification and cross-linking methods. Aqueous chitosan solution, prepared by dissolving in an acetic acid solution, was mixed with DOX and, afterwards, with TPP. This aqueous solution was added to a mixture of liquid paraffin and petroleum ether containing sorbitan sesquioleate as an emulsifier, which resulted in the formation of a water-in-oil (W/O) emulsion. Subsequently, glutaraldehyde saturated toluene was dropped into the flask for the cross-linking reaction. The use of glutaraldehyde as a saturated solution in toluene predominantly causes more cross-linking to the surface of the microsphere than the interior. Prepared microspheres exhibited strong antitumor activity to VX2 cells in the rabbit auricle model.

Lin and coworkers encapsulated DOX into microparticles based on poly(D,L-lactide-*co*-glycolide) (PLGA) [34]. These microparticles were fabricated using the spray-drying technique. Moreover, Pluronic P105 (PLU) and poly(L-lactide) (PLLA) were individually added to PLGA solution in order to improve the drug release characteristics. The authors showed that the addition of PLU delayed the release of DOX in comparison to pure PLGA microparticles. A further improvement in drug release characteristics was observed in the case of DOX-loaded PLLA/PLGA microparticles. Prepared microparticles were used in a short-term cancer cell cytotoxicity test, which was determined using a MTT [3-(4,5-dimethylthiazol-2-yl)-2,5-diphenyltetrazolium bromide] assay. It was found that delivering DOX in polymeric microparticulate form enhanced the cytotoxicity of drug to glioma C6 cancer cells.

Another recent study involving the encapsulation of DOX into a PLGA polymer matrix was carried out by Babos et al. [35]. The authors entrapped DOX and sorafenib together into poly(D,L-lactide-co-glycolide) and polyethylene glycol-poly(D,L-lactide-*co*-glycolide) (PEG-PLGA) matrix in order to enable the anticancer drugs to exert a synergistic activity. Sorafenib is a novel, tyrosine kinase inhibitor drug indicated as a treatment for advanced renal cell carcinoma, hepatocellular carcinoma (HCC), and thyroid cancer (Figure 6.7). Although it exhibits therapeutic activity, it can also lead to several adverse effects such as hand-foot syndrome, facial erythema, subungual splinter hemorrhages, diarrhea, and hypertension due to oral administration and its nonspecific uptake by normal cells [36, 37].

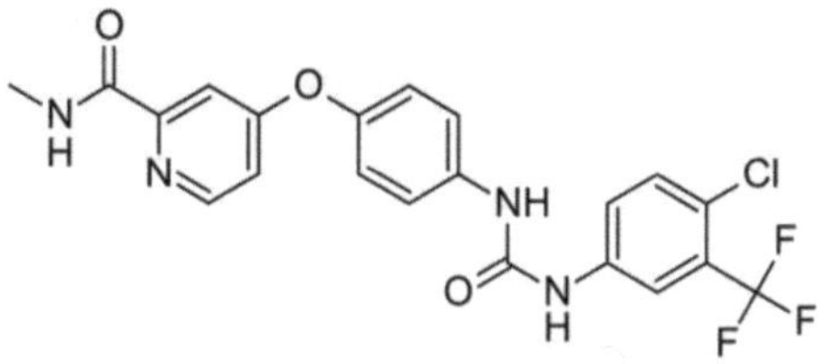

Fig. 6.7: Chemical structure of sorafenib.

Babos and coworkers used the water-in-oil-in-water (W/O/W) double emulsion solvent evaporation process to obtain their dual-agent nanocomposites. DOX hydrochloride solution constituted the inner water phase, which was added to the organic phase that was composed of PLGA with sorafenib. The first emulsification was carried out by sonication. In order to perform the W/O/W emulsion, the formed W/O emulsion was pipetted into the outer water phase that consisted of PVA in phosphate buffer and subsequently sonicated. Nanoparticles were obtained by evaporation of organic solvents. The authors investigated their drug release profile. In human blood plasma, DOX was released continuously within 6 days, while sorafenib was released quickly during 24 h. However, the tumor microenvironment is generally acidic; thus, these drugs were characterized by accelerated DOX and sustained sorafenib release in acidic conditions. *In vitro* cellular studies on the human cancer cell line HT-29 determined the higher cellular uptake of PEG-PLGA nanocomposites, which correlated with the higher cytotoxicity of these nanoparticles. Chen et al. [38] fabricated PLGA microspheres with sorafenib and ferrofluid of iron oxide for magnetic resonance imaging (MRI) transcatheter delivery to liver tumors. These microparticles were synthesized via a double emulsion solvent evaporation method. The published *in vitro* studies using McA-RH7777 rat hepatoma cell culture and *in vivo* with orthotropic rodent HCC models showed potent therapeutic responses of sorafenib-PLGA microspheres. Iron-oxide ferrofluid coencapsulation permits *in vivo* MRI visualizing of selective delivery of these microspheres to HCC in rodent models.

Gao's research group encapsulated sorafenib in PLGA nanoparticles via single-step nanoprecipitation [39]. The addition of the anionic lipid dioleoylphosphatidic acid (DOPA) and D-α-tocopherol polyethylene glycol 1000 succinate (TPGS) stabilized the structure of the developed nanoparticles. PLGA, sorafenib, TPGS, and DOPA in dimethyl sulfoxide (DMSO) were mixed as the oil phase and added to water dropwise under gentle stirring. Self-assembled lipid-coated PLGA-sorafenib nanoparticles were modified by the addition of AMD3100 to the emulsion. The authors indicated that AMD3100 exhibits anticancer activity when it is combined with chemotherapy and serves as a ligand to target HCC. The authors achieved targeted codelivery of sorafenib together with AMD3100 into neovessels and HCC, which showed a synergistic anticancer effect in HCC both *in vitro* and *in vivo*. The developed nanoparticles demonstrated high *in vitro* cellular uptake and cytotoxicity, which was conducted using both murine HCC (HCA-1) and human HCC (JHH-7) cell lines. *In vivo* pharmacokinetic studies and tumor uptake were carried out with the use of C3H mice with orthotopic HCA-1 tumors. PLGA nanoparticles with sorafenib and AMD3100 efficiently targeted the drugs into HCC and triggered tumor apoptosis, which resulted in the inhibition of tumor growth and metastasis.

Paclitaxel is a diterpenoid compound first isolated from the bark of *Taxus brevifolia* (Figure 6.8). It is one of the most widely used anticancer agents and is first-or second-line treatment for several types of cancers, including ovarian, breast, lung, pancreatic, and cervical cancer, as well as Kaposi's sarcoma [40–42]. Due to its excellent therapeutic effects against a wide spectrum of cancers, paclitaxel achieved great commercial success as the bestseller among various anticancer agents [42]. There exist serious formulation problems of paclitaxel due to its poor solubility in water. Paclitaxel is a complex diterpenoid product with an extended fused ring system and lack of functional ionizable groups. Therefore, manipulation of pH does not enhance its solubility [43]. Paclitaxel is currently formulated in a vehicle composed of 1:1 blend of Cremophor EL (polyethoxylated castor oil) and ethanol or dextrose solution. Paclitaxel in Cremophor EL is called Taxol. Cremophor EL has been reported to be responsible for severe side effects such as hypersensitivity reactions, neutropenia, neuropathy, hypotension, and cardiotoxicity [40, 44]. Therefore, a solution enhancing the therapeutic efficacy and simultaneously reducing the side effects of this anticancer agent is constantly being explored. One way to overcome these problems is to encapsulate the drug.

Ruan et al. fabricated paclitaxel microspheres in a copolymer of poly(lactic acid)-poly(ethylene glycol)-poly(lactic acid) (PLA-PEG-PLA) [45]. Paclitaxel-loaded PLA-PEG-PLA microspheres were obtained using the oil-in-water (O/W) single-emulsion solvent extraction/evaporation technique. Paclitaxel was dissolved in an organic solvent (dichloromethane (DCM), in some cases mixed with acetone) with the PLA-PEG-PLA or PLGA polymer solution, which was then poured rapidly into a PVA solution while being stirred. The resulting O/W emulsion was further stirred overnight to completely extract/evaporate the organic solvent, leaving behind solid microspheres.

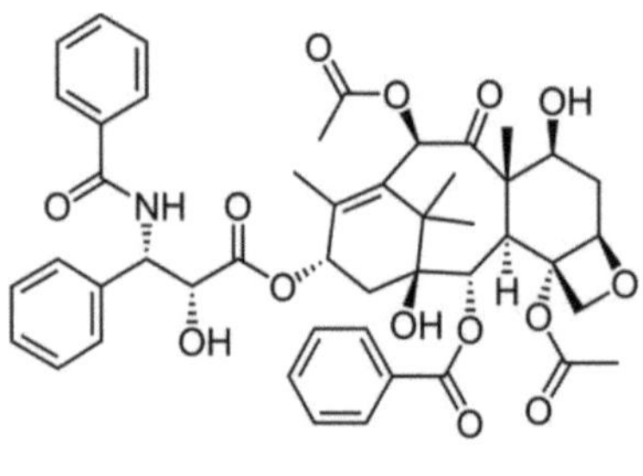

Fig. 6.8: Chemical structure of paclitaxel.

The final product was obtained by centrifugation, washing, and freeze drying. The presented results showed that the incorporation of hydrophilic PEG segment within the hydrophobic PLA in microspheres facilitated paclitaxel release due to the enhanced porosity. Moreover, the addition of water-soluble solvent acetone in the organic solvent phase further increased the porosity of the paclitaxel-loaded PLA-PEG-PLA microspheres and fastened the drug release. The authors indicated that PLA-PEG-PLA microspheres might be promising for the clinical administration of highly hydrophobic drugs such as paclitaxel.

Zhang and coworkers developed paclitaxel-loaded poly(lactide)-tocopheryl polyethylene glycol succinate (PLA-TPGS) nanoparticles [46]. First, the authors synthesized the PLA-TPGS copolymers via the ring-opening polymerization method. Solvent extraction/evaporation method was used for the fabrication of nanoparticles with high encapsulation efficiency ranging between 80% and 90%. Afterwards, the PLA-TPGS nanoparticles with paclitaxel were subjected to *in vitro* cellular uptake with the use of cancer cell lines HT-29 and Caco-2 and cell viability MTT assays. The results present that paclitaxel-loaded PLA-TPGS nanoparticles of 89:11 PLA:TPGS ratio achieved the best effects on drug encapsulation efficiency, cellular uptake, and cancer cell mortality.

Pulmonary delivery of paclitaxel-loaded alginate microparticles for lung chemotherapy was investigated by Alipour et al. [47, 48]. The authors demonstrated that aerosolized delivery provides chemotherapeutic effectiveness of paclitaxel, as well as continuous and direct exposure of drug to the lungs with lower side effects for other organs. Alginate microparticles with paclitaxel were prepared using an emulsification/gelation method. A sodium alginate-hydroxypropyl methylcellulose (HPMC) solution (9:1 ratio) and Tween 85 were mixed. Paclitaxel was added to this mixture and sonicated. Subsequently, iso-octane was added to the dispersion and sonicated to get an emulsion. In order to achieve solidification of emulsified alginate, calcium chloride solution was added. Microparticles were collected by filtration and washed with isopropyl alcohol for further hardening. The obtained microparticles were characterized by a volume diameter of 3 μm, mass median aerodynamic diameter of 5.9 μm, fine particle fraction of 13.9%, and encapsulation efficiency of 61%. The results of the *in vitro* cytotoxicity effect of paclitaxel-loaded microparticles on lung cancer cell lines (A549 and Calu-6) showed effective cell growth inhibition. The authors also performed a study on jugular vein cannulated rats. The use of a noninvasive pulmonary route – endotracheal administration – resulted in the minimization of drug loss during application, efficient concentration, and longer residence time of drug in the lung tissue.

6.3 Microencapsulation of vaccines

One of the leading global health problems is infectious diseases caused by pathogenic microorganisms. Vaccine is a preparation providing effective, safe protection from infectious diseases by developing active acquired immunity to particular disease. A vaccine typically contains an agent that resembles a disease-causing microorganism, which stimulates the body's immune system to recognize it as a threat, destroy it, and further recognize and destroy this agent in the future. Despite the World Health Organization's Expanded Program on Immunization, which resulted in a dramatic rise in worldwide vaccination rates over the past 40 years, infectious disease remains the leading cause of death in developing countries [49]. Several vaccines require repeated administrations, which poses a serious problem for patients in developing nations. Therefore, microparticulate delivery system, providing prolonged and controlled drug release, may eliminate the need for repeat immunizations. Encapsulated vaccine enables targeting to antigen-presenting cells such as macrophages and dendritic cells, as well as inducing stronger immune response [50].

Tetanus is an infection caused by, commonly occurring in soil and dust, the bacterium *Clostridium tetani*. The bacteria generally enter the body through contaminated wounded skin. It affects the nervous system, leading to painful muscle spasms, which begins in the jaw muscles and subsequently progresses to the rest of the body. A vaccine preventing tetanus is tetanus toxoid (TT), a high-molecular-weight protein. This vaccine is on the World Health Organization's List of Essential Medicines, which contains the most effective and safe medicines needed in a health system [51]. During childhood, five doses are recommended, with a sixth given during adolescence. Despite the significant reduction in mortality on tetanus, there were about 57,000 deaths globally in 2015, which mostly occurred in developing countries [52].

Gupta et al. evaluated the immunogenicity of TT encapsulated in PLA and PLGA microspheres [53]. TT-containing microspheres were fabricated with the use of W/O/W double emulsion procedure. Fabricated microspheres were suspended in an aqueous vehicle composed of sorbitol, carboxymethyl cellulose, and Tween 80 immediately before injection in order to perform the immunization experiments. The author's aim was to develop a formulation with a controlled TT release, which could replace the two to three doses required for primary immunization but would also need a booster dose after 12–18 months. Immunogenicity was investigated in mice and guinea pigs for 1 year. TT-loaded microsphere formulations elicited significantly higher immunoglobulin G (IgG) antibody levels in mice than soluble TT. Furthermore, a single injection of TT-encapsulated microspheres provides strong immunological memory due to significantly high anamnestic response to a low-dose booster 1 year after priming. However, the analyzed microsphere formulations released a small fraction of antigenic TT throughout *in vitro* release studies due to the denaturation of TT during encapsulation and hydration of microspheres. To overcome this shortcoming, Jiang and coworkers revealed two principal TT instability mechanisms during the release from PLGA microspheres, namely protein aggregation mediated by formaldehyde and acid-induced protein unfolding and epitope damage [54]. The authors systematically identified additives in the PLGA matrix, which can efficiently inhibit TT aggregation and retain TT antigenicity. TT was encapsulated by an oil-in-oil emulsion and solvent extraction method. TT with or without stabilizers was micronized by adding aqueous antigen solution into PLGA in acetonitrile. The suspension was homogenized and dropwise added to cottonseed oil containing Span 85. After stirring, petroleum ether was poured into the cottonseed oil bath to extract the acetonitrile from the polymer. Subsequently, the microspheres were sieved, collected, and lyophilized. It was found that coencapsulation of lysine, sorbitol, and trehalose inhibited formaldehyde-mediated aggregation pathway and $MgCO_3$ bypassed acid-induced damage, which resulted in an unparalleled stability of TT encapsulated in PLGA microspheres and slow, continuous release of high doses of the antigen.

Tafaghodi et al. investigated the systemic and mucosal immune responses in rabbits following intranasal immunization with encapsulated TT and CpG-ODN in alginate microspheres [55]. The addition of oligodeoxynucleotides (ODN) containing immunostimulatory CpG motifs (CpG-ODNs) boosts the immune response to produce more antibodies and longer-lasting immunity. The presented results showed that the encapsulation efficiency of TT and CpG-ODN was determined as 47.7% and 34.2%, respectively. The release of TT and CpG-ODN in a simulated model with nasal cavity was 14.2% and 36.7%, respectively, after 4 h. The authors exhibited that the highest serum IgG and antitoxin, as well as nasal lavage IgA titers, were observed in groups immunized with TT-loaded alginate microspheres. Furthermore, there was an indicated lack of alginate matrix toxicity, studied by a standard hemolysis test, and local irritation in human volunteers.

Hepatitis B is an infectious disease affecting the liver caused by hepatitis B virus (HBV), which is transmitted through contact with infected blood or body fluids. It can cause both acute and chronic infections; however, many people have no symptoms during the initial infection. Acute hepatitis B causes hepatitis, vomiting, jaundice, and rarely death, while chronic infection induces liver cirrhosis and HCC [56]. HBV infection can be prevented by a protective three-dose vaccination series recommended by the World Health Organization [57]. This disease continues to cause epidemics in Asia and Africa.

Shi and coworkers performed encapsulation of the HBV surface antigen (HBsAg) into the PLGA microsphere by double the emulsion method, which requires five major steps: primary emulsion, second emulsion, hardening, washing/filtration, and drying [58]. The *in vitro* release kinetics of HBsAg from the PLGA microspheres indicated that the antigen is relatively stable under conditions of temperature and pH that imitate *in vivo* conditions. The *in vivo* immunogenicity of the HBsAg-loaded PLGA microspheres was compared with conventional hepatitis B vaccine in C3H mice. The results indicated that a single injection of a mixture of aluminum and obtained HBsAg PLGA microparticles results in equal or better immune responses than two injections of conventional aluminum-formulated HBsAg vaccine. HBsAg-loaded PLGA microspheres were also prepared via double emulsion microencapsulation technique by Feng et al. [59]. The authors obtained two kinds of microspheres differing in their lactide/glycolide ratio exhibiting rapid and prolonged release of HBsAg antigen, which has the application as priming dose and "autobooster dose," respectively. The results of *in vivo* immunization of BALB/c mice showed that a single subcutaneous injection of HBsAg-PLGA microspheres had the capacity to induce a long-lasting immune response in a manner comparable to that of three injections of the conventional HBsAg-aluminum vaccine. HBsAg-loaded microspheres induced a strong anti-HBsAg antibody response.

6.4 Microencapsulation of protein

Bioactive proteins and peptides are finding increasing application in pharmaceutical, supplement, and food industries. They exhibit a range of biological activities, including nutritional, antioxidant, antimicrobial, flavor, antihypertension, antidiabetes, and anticancer [60, 61].

Protein and peptide drugs are becoming very important agents in the diagnostic as well as the therapeutic sector, and their mode of activity is classified into four groups (Figure 6.9) [62]. The growth in science, especially in molecular biology, led to the development of new protein therapeutics, such as antigens, hormones, anticoagulants, blood factors, bone morphogenetic proteins, engineered protein scaffolds, enzymes, Fc fusion proteins, growth factors, interferons, interleukins, and thrombolytics [63]. Therapeutic protein drugs are now used to treat a wide variety of clinical indications, including cancers, diabetes, infectious diseases, and many other disorders, because they possess several advantages over the conventional ones [64].

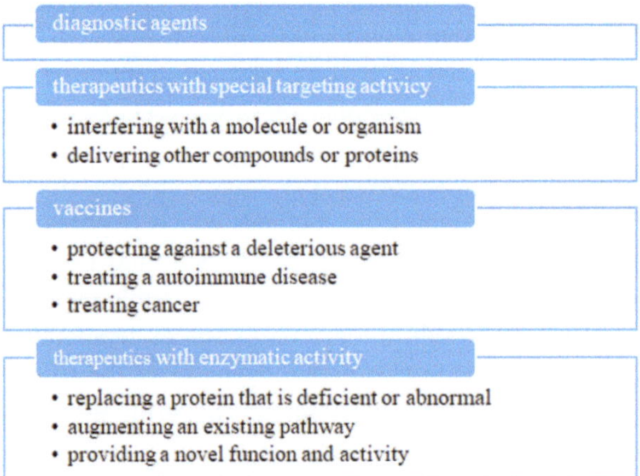

Fig. 6.9: Overview of the function/mode of activity-based classification of therapeutic proteins and peptides [62].

These drugs are highly versatile, providing a wide variety of pharmaceutical targets and high specificity. Moreover, therapeutic protein and peptide drugs show low toxicity and low side effect. The first commercial human protein therapeutic derived from recombinant DNA technology was human insulin (Humulin, Eli Lilly), created in 1982 [65]. At present, there are about 380 protein pharmaceutical products available on the market, and many are at the stage of clinical trials [62].

Unfortunately, despite the several advantages, proteins have several significant limitations as therapeutics. The oral delivery of many bioactive proteins in commercial products is often challenging, not only in the case of drugs but also in functional foods and supplements. The problems can appear during manufacturing, transport, storage, as well as inside the gastrointestinal (GIT) tract after ingestion [60, 66]. Proteins and peptides may be exposed to temperature, light, oxygen, and humidity, which in turn leads to changes in their molecular structure and significant reduction in their activity. Other environmental factors that are found during the manufacturing process, such as pH, metal ions, and adsorption, may decrease the biological activity of proteins. Many bioactive proteins may lose their biological activity after oral delivery, and they are generally characterized by a relatively short circulating half-life time, due to the high susceptibility to aggregation, degradation, and hydrolysis by enzymes. This is especially so within the GIT tract as the highly acidic and protease-rich environment encourages the degradation of protein [60, 64, 67]. In addition, hydrolysis of proteins is promoted by proteases in the mouth, stomach, and small intestine [68]. Other drawbacks of proteins as therapeutics are low solubility, immunogenicity, high molecular weights, low lipophilicity, and presence of some charged functional groups. For these reasons, we observed low bioavailability of protein therapeutics due to difficulty in their absorption in various sites of the GIT tract. Moreover, protein therapeutics are equally difficult to administer through the body due to weak noncovalent interactions such as van der Waals forces [69].

Various strategies have been developed to improve the delivery and effectiveness of protein, including microencapsulation techniques. Microencapsulation can protect the bioactive protein during storage and after ingestion but then releases them at the appropriate site of action within the human body and can release them slowly to realize a higher constant serum concentration for a prolonged time [70].

Various encapsulation techniques have been developed to accommodate varying properties of proteins, including microparticles (microcapsules and microspheres), microemulsions, emulsions, nanoemulsions, multiple emulsions, and liposomes [60, 67]. Numerous types of delivery systems with different structural designs to protein encapsulation have several advantages and disadvantages (Table 6.1). Some of the patents related to the oral delivery of encapsulated therapeutic proteins and peptides are shown in Table 6.2.

Microparticles have been widely used for protein encapsulation, but it remains a challenge to design uniform particles with well-controlled spherical morphology and

Table 6.1: Numerous types of delivery systems with different structural designs to protein encapsulation [59, 71–73].

Delivery system	Structural design	Advantages/ Disadvantages	Examples	Ref.
Microparticles	Small particles with encapsulated active agents with a diameter range of 1–1,000 µm	– Protection of the liable protein/peptide *in vitro* and *in vivo* degradation – Sustained and controlled release of drug for long period of time – Beneficial way of delivering active substances are otherwise difficult to deliver due to limited – solubility in water – Improvement in bioavailability – Reducing adverse effects	Chitosan-based insulin-loaded microparticles	[74]
Microcapsules	Active agent (core) surrounded by a polymeric shell		Poly(lactic-co-glycolic acid) (PLGA)-based insulin loaded microparticles	[75]
Microspheres	Active agent dispersed in a polymeric matrix		Alginate-based bovine serum albumin-loaded microparticles	[76]
Emulsion systems	Mixture of two or more liquids in which one is dispersed in the other as microscopic or ultramicroscopic droplets – Water-in-oil emulsion – Oil-in-water emulsion – Double emulsion method	– Possibility to produce reverse micelles by maintaining optimum conditions – The protein drugs are prone to denaturation and structural perturbations at the water/organic solvent interface – Mechanical forces employed in the creation of the emulsion, such as homogenization and sonication, which may cause protein degradation and loss of bioactivity	Phospholipid-based anhydrous reverse micelles for oral peptide delivery	[77]
Liposomes	Sphere-shaped vesicles consisting of one or more phospholipid bilayers	– Hydrophobic and hydrophilic character – Tend to have a relatively low encapsulation – efficiency – The stability of liposomes may be improved by coating them with biopolymer layers	Wheat germ agglutinin-carbopol-modified liposomes	[78]

Table 6.2: Some of the patents related to the oral delivery of encapsulated therapeutic proteins and peptides [72].

Protein/peptide	Drug dosage form	Patent no.	Ref.
Insulin	Liposomes	US 4582820	[79]
	Polymeric hydrogel	EP 0918543	[80]
	Microparticles	WO 2008132727	[81]
	Microemulsion	US 7605123	[82]
Insulin/human growth factor/myelin basic protein/collagen S antigen/ transforming growth factor beta	Microspheres	US 6613332	[83]
Proteins/peptides/oligonucleotides	Microemulsion	US 6280770	[84]
Erythropoietin	Liposomes	US 5597562	[85]
Prolamine	Microspheres	US 5679377	[86]
Interferon/interleukin/antisense oligonucleotide/vaccine/glucagon	Microparticles	US 6355270	[87]
Granulocyte colony stimulating factor/interferon/indinavir	Microcapsules	US 7097851	[88]
Hemoglobin/pepsin/ immunoglobulin/lipase/peroxidase/ myoglobin	Microparticles	US 7217410	[89]
Pituitary growth hormone/calcitonin/ tumor necrosis factor/interferon	Nanoparticles	WO 2009087633	[90]

size distribution and high protein encapsulation efficiency. It results in poor reproducibility in large-scale production and the delivery system may fail to be approved in the market. In the case of encapsulation of protein drug, it is also difficult to maintain the bioactivity of protein/peptide drugs during the preparation and storage because there are several reasons responsible for protein denaturation. During the encapsulation process, proteins can lose their therapeutic effect because many parameters can affect microencapsulated products and their final characteristics, for example, as a result of mechanical shear force when proteins lose their three-dimensional structure, as well as in the wake of coagulation during the contact of protein with oil/water interface, or cross-linking of protein. Moreover, a hydrophobic interaction between protein and hydrophobic wall material is possible, which leads to the denaturation of proteins [70, 91].

6.5 Cell encapsulation

The concept of cell encapsulation has received much attention in the past three decades; however, it was first introduced by Bisceglie et al. in 1933 [92–94]. He studied the effect of vascularization absence on tumor cells by encapsulating the cells and transplanting them in pigs. The results showed that the cells survived a long time and they were not destroyed by the immune system [92]. It was also then concluded that encapsulation involves the protection of living cells from the effects of the host immune system. Since that time, such devices have been produced in different conformations, with application to a wide range of therapeutic treatments, including hemophilia B, anemia, dwarfism, renal and liver failure, pituitary disorders, diabetes, and central nervous system insufficiency [93, 95, 96].

Cell encapsulation technology is based on the immobilization of cells within a generally semipermeable and polymeric membrane [92, 94, 97]. This membrane protects the inner cells from the host's immune system (antibodies and other immunologic moieties) and permits the entry of nutrients, oxygen, and therapeutic molecules (Figure 6.10).

It is widely understood that biocompatible materials, which do not interact with cells, have to be employed. Many different materials, natural and synthetic, are applied to encapsulate cells [98, 99]. However, finding a suitable material to form a capsule is still challenging. It is also essential that such scaffold should have mechanical stability to allow for exchange of nutrients, oxygen, and metabolic waste. Among different materials used, alginates are the most studied [97]. In the past years, it was considered that capsule geometry can affect the foreign body response and can also influence the health of transplanted cells or tissues. Since that time, encapsulation of cells has been applied in two families of geometrics: macro- and micro-capsules [94]. There are a lot of persistent challenges with both: Where it could be implanted? Is the exchange of metabolites, nutrients, and

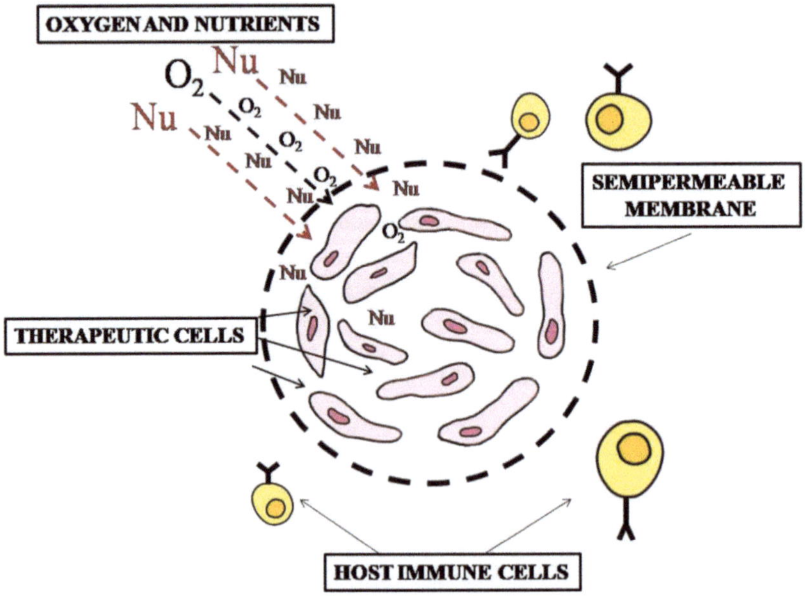

Fig. 6.10: Scheme of the cellular encapsulation technology.

oxygen fast enough? Are they biocompatible? What about the injected cell density? We know already some of the answers; however, it is still the key of clinical application. All encapsulated cell-based therapeutics consist of two elements: the material/capsule and the transplanted cells/tissues [94, 100]. Taking this into account, it is also worth considering the suitability of the transplanted cells or tissue type for encapsulation technology. A wide range of cell lines have been investigated, but mostly, porcine donors are used for xenogenic tissues, including islets or hepatocytes [101–104]. This is possible due to their similarity to humans. Nevertheless, regardless of the source, the most important is the ability of cells or tissues to survive encapsulation technology. Some of the cell sources used in encapsulation technology are presented in Table 6.3.

Why does cell encapsulation still continue to hold significant promise for medicine and also biotechnology? The encapsulation of cells in place of therapeutic products permits the delivery of the aforementioned products for a longer period of time, as cells release the products continuously. In addition, cell encapsulation allows the transplantation of nonhuman cells, which can be an alternative source to the limited donor tissues. Moreover, cells with genetic modification can be immobilized to express any desired protein *in vivo* without modification of the host [92, 97, 105]. These strategies can overcome the present difficulties relating to whole organ graft rejection, usage of immunomodulatory protocols, or chronic administration of immunosuppressants, cell migration, and necrosis resulting from the harsh environment [106]. In addition, the small size of the capsules (from 100 μm to 500 μm) allows their implantation in close contact to the blood stream, which could be beneficial in specific applications [92, 97]. Determining the optimal site of implantation is still a matter of intense research. Figure 6.11 shows the main implantation sites in cell encapsulation technology.

In general, intraperitoneal implantation results in poorer functionality of the devices due to an increased inflammatory response that takes place in the peritoneum [97, 105, 107]. Capsules implanted intraperitoneally showed decreased viability and insulin secretion rates than capsules implanted subcutaneously or under the kidney capsule. Furthermore, the immune response against capsules implanted

Table 6.3: Cell sources for cell immobilization technology.

Cell type	Application
Fibroblasts	Metabolic deficiencies
Myoblasts	Cancer
Kidney cells	Hemophilia
Pancreatic islets	Diabetes
Hepatocytes	Liver transplantation
Chondrocytes	Bone and cartilage regeneration
Tumor cells	Cancer vaccine

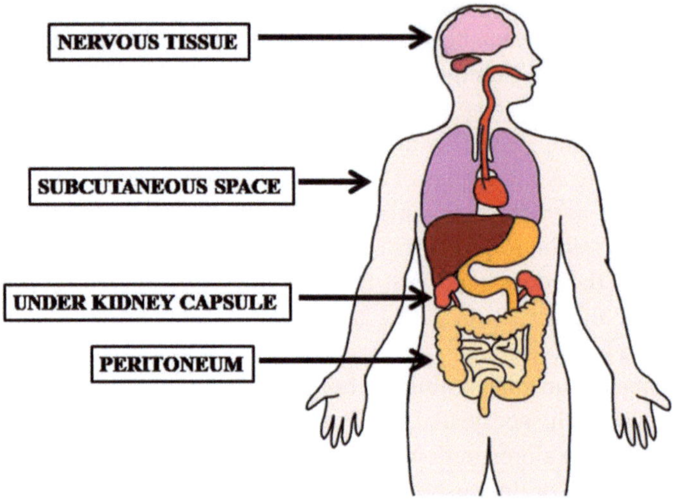

Fig. 6.11: Implantation sites in cell encapsulation technology.

intraperitoneally was more severe that against capsules implanted subcutaneously or under the kidney capsule. On the other hand, the ability to monitor the implanted devices in cell encapsulation technology is a huge challenge [97, 108]. Once macro- or micro-capsules are transplanted, invasive recovery surgery is necessary to assess their functionality. There are some research groups who used radiopaque microcapsules containing either barium sulfate or bismuth sulfate, but they are toxic both for the encapsulated cells and for the recipient [108].

Cells can be encapsulated for a variety of applications, mostly cell-based therapies. Cell-based therapies have numerous advantages in regenerative medicine. Cells can be grafted to differentiate and replace damaged tissues, protect other donor cells from apoptosis, promote cell activation and growth, participate in genetic and non-genetic material transfer, and provide immunomodulation [98]. Encapsulation technology used in such cell-based therapies protects donor cells by isolation from host's immune system. Depending on the desired therapeutic use of the encapsulated cells, the residence time for cells within the devices may vary [92]. In order for this to be possible, the material used to encapsulate the cells must be able to undergo controlled degradation *in vivo*. Nondegradable polymer can be used if the therapeutic use of microencapsulated stem cells is insulin production in the case of diabetic patients. A degradable polymer can be used if it is required for the stem cells to be released to support tissue regeneration.

The most widely researched microencapsulated cells are pancreatic beta cells within alginate/poly-L-lysine-based hydrogel microcapsules. Cell encapsulation technology is also currently under study for the treatment of neurodegenerative disorders or treatment of chronic neuropathic pain. Other therapies, including for Parkinson's disease, metachromatic leukodystrophy, and mucopolisaccharidosis type VII, should also be highlighted [109, 110, 111].

There are some clinical trials available based on cell encapsulation. The analysis of completed and closed clinical trials is presented in Table 6.4. We excluded studies that were withdrawn or terminated or with unknown status. We also narrowed our search in the ClinicalTrials.gov database using the following terms: "cell encapsulation," "cell microencapsulation," "cell macroencapsulation," "encapsulation of cells," "cellular encapsulation," and "bioencapsulation of cells."

Unfortunately, it is difficult to confirm any effect of the above treatment since no final results can be found.

One of the most important fields in cell encapsulation nowadays is cancer therapy [112, 113]. Some studies on animal models have shown that microencapsulation of engineered cells allows for deliver sustained levels of drugs.

Encapsulating therapeutic cells provides numerous advantages and helps overcome a number of caveats in the current cell-based therapies, especially by enabling

Table 6.4: Summary of selected clinical trials, based on ClinicalTrials.gov.

Experiment	Application	Procedure/ Intervention	Capsule type	Final results/ outcomes
Encapsulated Mesenchymal Stem Cells for Dental Pulp Regeneration – completed	Periapcal periodontitis	Regenerative endodontic procedure/ conventional root canal treatment	Biological scaffold (plasma-derived biomaterial)	Unknown
Bet Cell Therapy in Diabetes Type 1 – unknown	Type 1 diabetes	Transplantation of encapsulated beta cells	Alginate + human beta cells	Unknown
Efficacy and Safety of L-Asparaginase Encapsulated in RBC Combined with Gemcitabine or FOLFOX in 2nd Line for Progressive Metastatic Pancreatic Carcinoma – completed	Cancer	Drug: ERY001 Drug: Gemcitabine Drug: 5-fluorouracil/ oxaliplatin/ leucovorin (folfox)	L-asparaginase encapsulated in erythrocytes	Unknown
MVX-ONCO-1 in Patients With Solid Tumor – Recruiting	Cancer	Administrated subcutaneously with advance metastatic solid tumor/ implantation loaded capsules	Capsules + irradiated autologous tumor cells	Ongoing

transplantation without immunosuppression. However, cell encapsulation technology still faces several significant challenges, including prevention of uncontrolled cell growth and proliferation and standardization of polymers in terms of physicochemical properties. For rigorous quality control, environment and operational discipline is essential for allowing this technology to enter human clinical trials and become a real clinical therapeutic strategy.

6.6 Acknowledgements

Financial support from the National Science Centre (NCN, Poland) (grant no. UMO 2016/21/D/ST8/01705) is gratefully acknowledged.

6.7 References

[1] Kohane, D. S., Microparticles and nanoparticles for drug delivery, Biotechnol Bioeng **96** (2007) 203–209.

[2] Oh, J. K., Drumright, R., Siegwart, D. J., Matyjaszewski, K., The development of microgels/nanogels for drug delivery applications, Prog Polym Sci **33** (2008) 448–477.

[3] Siepmann, J., Siepmann, F., Microparticles used as drug delivery systems, Progr Colloid Polym Sci **133** (2006) 15–21.

[4] Kumar, B. P., Chandiran, I. S., Bhavya, B., Sindhuri, M., Microparticulate drug delivery system: a review, Indian J Pharm Sci Res **1** (2011) 19–37.

[5] Xiong, M. H., Bao, Y., Yang, X. Z., Zhu, Y. H., Wang, J., Delivery of antibiotics with polymeric particles, Adv Drug Deliv Rev **78** (2014) 63–76.

[6] Durán, N., Alvarengal, M. A., Da Silva, E. C., Melo, P. S., Marcato, P. D., Microencapsulation of antibiotic rifampicin in poly(3-hydroxybutyrate-co-3-hydroxyvalerate), Arch Pharm Res **31** (2008) 1509–1516.

[7] Cunha, L., Rosa da Costa, A. M., Lourenço, J. P., Buttini, F., Grenha, A., Spray-dried fucoidan microparticles for pulmonary delivery of antitubercular drugs, J Microencapsul **35** (2018) 392–405.

[8] Franco, P., Reverchon, E., De Marco, I., Production of zein/antibiotic microparticles by supercritical antisolvent coprecipitation, J Supercrit Fluids **145** (2019) 31–38.

[9] Montes, A., Gordillo, M. D., Pereyra, C., Martínez de la Ossa, E. J., Co-precipitation of amoxicillin and ethyl cellulose microparticles by supercritical antisolvent process, J Supercrit Fluids **60** (2011) 75–80.

[10] Montes, A., Gordillo, M. D., Pereyra, C., Martínez de la Ossa, E. J., Polymer and ampicillin co-precipitation by supercritical antisolvent process, J Supercrit Fluids **63** (2012) 92–98.

[11] Kasten, G., Silva, L. F., Lemos-Senna, E., Development of low-density azithromycin-loaded polycaprolactone microparticles for pulmonary delivery, Drug Dev Ind Pharm **42** (2016) 776–787.

[12] Tung, N. T., Tran, C. S., Nguyen, T. L., Hoang, T., Trinh, T. D., Nguyen, T. N., Formulation and biopharmaceutical evaluation of bitter taste masking microparticles containing azithromycin loaded in dispersible tablets, Eur J Pharm Biopharm **126** (2018) 187–200.

[13] Sezer, A. D., Kazak Sarılmışer, H., Rayaman, E., Çevikbaş, A., Öner, E. T., Akbuğa, J., Development and characterization of vancomycin-loaded levan-based microparticular system for drug delivery, Pharm Dev Technol **22** (2017) 627–634.

[14] Unagolla, J. M., Jayasuriya, A. C., Drug transport mechanisms and in vitro release kinetics of vancomycin encapsulated chitosan-alginate polyelectrolyte microparticles as a controlled drug delivery system, Eur J Pharm Sci **114** (2018) 199–209.
[15] Wu, L., Chen, W., Li, F., Morrow, B. R., Garcia-Godoy, F., Hong, L., Sustained release of minocycline from minocycline-calcium-dextran sulfate complex microparticles for periodontitis treatment, J Pharm Sci **107** (2018) 3134–3142.
[16] Flores, C., Degoutin, S., Chai, F., Raoul, G., Hornez, J. C., Martel, B., Siepmann, J., Ferri, J., Blanchemain, N., Gentamicin-loaded poly(lactic-co-glycolic acid) microparticles for the prevention of maxillofacial and orthopedic implant infections, Mater Sci Eng C Mater Biol Appl **64** (2016) 108–116.
[17] Aquino, R. P., Auriemma, G., Mencherini, T., Russo, P., Porta, A., Adami, R., Liparoti, S., Della Porta, G., Reverchon, E., Del Gaudio, P., Design and production of gentamicin/dextrans microparticles by supercritical assisted atomisation for the treatment of wound bacterial infections, Int J Pharm **440** (2013) 188–194.
[18] Mustafa, S., Devi, V. K., Pai, R. S., Comparative study of kanamycin sulphate microparticles and nanoparticles for intramuscular administration: preparation in vitro release and preliminary in vivo evaluation, J Microencapsul **33** (2016) 679–688.
[19] Raval, J. P., Naik, D. R., Patel, P. S., Spray-dried cefixime encapsulated poly(lactide-coglycolide) microparticles: characterization and evaluation of in vitro release kinetics with antibacterial activity, Dry Technol **30** (2012) 865–872.
[20] Wright, L., Rao, S., Thomas, N., Boulos, R. A., Prestidge, C. A., Ramizol® encapsulation into extended release PLGA micro- and nanoparticle systems for subcutaneous and intramuscular administration: in vitro and in vivo evaluation, Drug Dev Ind Pharm **44** (2018) 1451–1457.
[21] Ferreira, I. S., Bettencourt, A., Bétrisey, B., Gonçalves, L. M., Trampuz, A., Almeida, A. J., Improvement of the antibacterial activity of daptomycin-loaded polymeric microparticles by Eudragit RL 100: an assessment by isothermal microcalorimetry, Int J Pharm **485** (2015) 171–182.
[22] Thomas, N., Thorn, C., Richter, K., Thierry, B., Prestidge, C., Efficacy of poly-lactic-co-glycolic acid micro- and nanoparticles of ciprofloxacin against bacterial biofilms, J Pharm Sci **105** (2016) 3115–3122.
[23] Martín, M. J., Calpena, A. C, Fernández, F., Mallandrich, M., Gálvez, P., Clares, B., Development of alginate microspheres as nystatin carriers for oral mucosa drug delivery, Carbohydr Polym **117** (2015) 140–149.
[24] Martincorena, I., Campbell, P. J., Somatic mutation in cancer and normal cells, Science **349** (2015) 1483–1489.
[25] Lekka, M., Discrimination between normal and cancerous cells using AFM, Bionanoscience **6** (2016) 65–80.
[26] Bray, F., Ferlay, J., Soerjomataram, I., Siegel, R. L., Torre, L. A., Jemal, A., Global cancer statistics 2018: GLOBOCAN estimates of incidence and mortality worldwide for 36 cancers in 185 countries, CA Cancer J Clin **68** (2018) 394–424.
[27] Arcamone, F., Di Marco, A., Gaetani, M., Scotti, T., Isolation and antitumor activity of an antibiotic from Streptomyces, G. Microbiol **9** (1961) 83–90.
[28] Rocconi, R. P., Straughn, J. M., Leath, C. A., Kilgore, L. C., Huh, W. K., Barnes, M. N., Alvarez, R. D., Pegylated liposomal doxorubicin consolidation therapy after platinum/paclitaxel-based chemotherapy for suboptimally debulked, advanced-stage epithelial ovarian cancer patients, Oncologist **11** (2006) 336–341.
[29] Park, J. M., Lee, S. Y., Lee, G. H., Chung, E. Y., Chang, K. M., Kwak, B. K., Lee, J., Design and characterisation of doxorubicin-releasing chitosan microspheres for anti-cancer chemoembolisation, J Microencapsul **29** (2012) 695–705.

[30] Deepa, K., Singha, S., Panda, T., Doxorubicin nanoconjugates, J Nanosci Nanotechnol **14** (2014) 892–904.
[31] Polegato, B. F., Minicucci, M. F., Azevedo, P. S., Carvalho, R. F., Chiuso-Minicucci, F., Pereira, E. J., Matsubara, L. S., Acute doxorubicin-induced cardiotoxicity is associated with matrix metalloproteinase-2 alterations in rats, Cell Physiol Biochem **35** (2015) 1924–1933.
[32] Tacar O, Sriamornsak P, Dass C. R., Doxorubicin: an update on anticancer molecular action, toxicity and novel drug delivery systems. J Pharm Pharmacol **65** (2013) 157–70.
[33] Janes, K. A., Fresneau, M. P., Marazuela, A., Fabra, A., Alonso, M. J., Chitosan nanoparticles as delivery systems for doxorubicin, J Control Release **73** (2001) 255–267.
[34] Lin, R., Shi Ng, L., Wang, C. H., In vitro study of anticancer drug doxorubicin in PLGA-based microparticles, Biomaterials **26** (2005) 4476–4485.
[35] Babos, G., Biró, E., Meiczinger, M., Feczkó, T., Dual drug delivery of sorafenib and doxorubicin from PLGA and PEG-PLGA polymeric nanoparticles, Polymers **10** (2018) 895.
[36] Autier, J., Escudier, B., Wechsler, J., Spatz, A., Robert, C., Prospective study of the cutaneous adverse effects of sorafenib, a novel multikinase inhibitor, Arch Dermatol **144** (2008) 886–892.
[37] Robert, C., Soria, J. C., Spatz, A., Le Cesne, A., Malka, D., Pautier, P., Armand, J. P., Cutaneous side-effects of kinase inhibitors and blocking antibodies, Lancet Oncol **6** (2005) 491–500.
[38] Chen, J., Sheu, A. Y., Li, W., Zhang, Z., Kim, D.-H., Lewandowski, R. J., Larson, A. C., Poly (lactide-co-glycolide) microspheres for MRI-monitored transcatheter delivery of sorafenib to liver tumors, J Control Release **184** (2014) 10–17.
[39] Gao, D. Y., Lin, T. T., Sung, Y. C., Liu, Y. C., Chiang, W. H., Chang, C. C., Chen, Y., CXCR4-targeted lipid-coated PLGA nanoparticles deliver sorafenib and overcome acquired drug resistance in liver cancer, Biomaterials **67** (2015) 194–203.
[40] Singla, A. K., Garg, A., Aggarwal, D., Paclitaxel and its formulations, Int J Pharm **235** (2002) 179–192.
[41] Spencer, C. M., Faulds, D., Paclitaxel, Drugs **48** (1994) 794–847.
[42] Sofias, A. M., Dunne, M., Storm, G., Allen, C., The battle of "nano" paclitaxel, Adv Drug Deliv Rev **122** (2017) 20–30.
[43] Meerum Terwogt, J. M., Nuijen, B., Ten Bokkel Huinink, W. W., Beijnen, J. H., Alternative formulations of paclitaxel, Cancer Treat Rev **23** (1997) 87–95.
[44] Huang, S. T., Wang, Y. P., Chen, Y. H., Lin, C. T., Li, W. S., Wu, H. C., Liposomal paclitaxel induces fewer hematopoietic and cardiovascular complications than bioequivalent doses of Taxol, Int J Oncol **53** (2018) 1105–1117.
[45] Ruan, G., Feng, S. S., Preparation and characterization of poly (lactic acid)–poly (ethylene glycol)–poly (lactic acid)(PLA-PEG-PLA) microspheres for controlled release of paclitaxel, Biomaterials **24** (2003) 5037–5044.
[46] Zhang, Z., Feng, S. S., The drug encapsulation efficiency, in vitro drug release, cellular uptake and cytotoxicity of paclitaxel-loaded poly(lactide)-tocopheryl polyethylene glycol succinate nanoparticles, Biomaterials **27** (2006) 4025–4033.
[47] Alipour, S., Montaseri, H., Khalili, A., Tafaghodi, M., Non-invasive endotracheal delivery of paclitaxel-loaded alginate microparticles, J Chemother **28** (2016) 411–416.
[48] Alipour, S., Montaseri, H., Tafaghodi, M., Preparation and characterization of biodegradable paclitaxel loaded alginate microparticles for pulmonary delivery, Colloids Surf B Biointerfaces **81** (2010) 521–529.
[49] Greenwood, B., The contribution of vaccination to global health: past, present and future, Philos Trans R Soc Lond B Biol Sci **369** (2014) 20130433.
[50] Lin, C. Y., Lin, S. J., Yang, Y. C., Wang, D. Y., Cheng, H. F., Yeh, M. K., Biodegradable polymeric microsphere-based vaccines and their applications in infectious diseases, Hum Vaccin Immunother **11** (2015) 650–656.

[51] World Health Organization, Essential Medicines: WHO Model List, 20th edition. 2017.
[52] The selection and use of essential medicines: report of the WHO Expert Committee, 2017 (including the 20th WHO Model List of Essential Medicines and the 6th WHO Model List of Essential Medicines for Children). Geneva: World Health Organization; 2017 (WHO technical report series; no. 1006). Licence: CC BY-NC-SA 3.0 IGO.
[53] Gupta, R. K., Alroy, J., Alonso, M. J., Langer, R., Siber, G. R., Chronic local tissue reactions, long term immunogenicity and immunologic priming of mice and guinea pigs to tetanus toxoid encapsulated in biodegradable polymer microspheres composed of poly lactide-co-glycolide polymers, Vaccine **15** (1997) 1716–1723.
[54] Jiang, W., Schwendeman, S. P., Stabilization of tetanus toxoid encapsulated in PLGA microspheres, Mol Pharm **5** (2008) 808–817.
[55] Tafaghodi, M., Tabassi, S. A. S., Jaafari, M. R., Induction of systemic and mucosal immune responses by intranasal administration of alginate microspheres encapsulated with tetanus toxoid and CpG-ODN, Int J Pharm **319** (2006) 37–43.
[56] Shepard, C. W., Simard, E. P., Finelli, L., Fiore, A. E., Bell, B. P., Hepatitis B virus infection: epidemiology and vaccination, Epidemiol Rev **28** (2006) 112–125.
[57] World Health Organization, Geneva, Switzerland. Global health sector strategy on viral hepatitis 2016–2021. Towards ending viral hepatitis (No. WHO/HIV/2016.06). 2016.
[58] Shi, L., Caulfield, M. J., Chern, R. T., Wilson, R. A., Sanyal, G., Volkin, D. B., Pharmaceutical and immunological evaluation of a single-shot hepatitis B vaccine formulated with PLGA microspheres, J Pharm Sci **91** (2002) 1019–1035.
[59] Feng, L., Qi, X. R., Zhou, X. J., Maitani, Y., Cong Wang, S., Jiang, Y., Nagai, T., Pharmaceutical and immunological evaluation of a single-dose hepatitis B vaccine using PLGA microspheres, J Control Release **112** (2006) 35–42.
[60] McClements, D. J., Encapsulation, protection, and delivery of bioactive proteins and peptides using nanoparticle and microparticle systems: a review, Adv Colloid Interface Sci 253 (2018) 1–22.
[61] Zhang, Z., Zhang, R., Zou, L., McClements, D. J., Protein encapsulation in alginate hydrogel beads: Effect of pH on microgel stability, protein retention and protein release, Food Hydrocoll **58** (2016) 308–315.
[62] Usmani, S. S., Bedi, G., Samuel, J. S., Singh, S., Kalra, S., Kumar P., Ahuja, A. A., Sharma, M., Gautam, A., Raghava, G. P. S., THPdb: Database of FDA-approved peptide and protein therapeutics, PLoS One 12 (2017) e0181748.
[63] Carter, P. J., Introduction to current and future protein therapeutics: a protein engineering perspective, Exp Cell Res 317 (2011) 1261–1269.
[64] Black, K. A., Priftis, D., Perry, S. L., Yip, J., Byun, W. Y., Tirrell, M., Protein encapsulation via polypeptide complex coacervation, ACS Macro Lett **3** (2014) 1088–1091.
[65] Goeddel, D. V., Kleid, D. G., Bolivar, F., Heyneker, H. L., Yansura, D. G., Crea, R., Hirose, T., Kraszewski, A., Itakura, K., Riggs, A. D., Expression in Escherichia coli of chemically synthesized genes for human insulin, Proc. Natl Acad Sci USA 76 (1979) 106–110.
[66] Muheem, A., Shakeel, F., Jahangir, M. A., Anwar, M., Mallick, N., Jain, G. K., Warsi, M. H., Ahmad, F. J., A review on the strategies for oral delivery of protein and peptides and their clinical perspectives, Saudi Pharm J **24** (2016) 413–428.
[67] Ye, C., Chi, H. A review of recent progress in drug and protein encapsulation: Approaches, applications and challenges, Mater Sci Eng C Mater Biol Appl **83** (2018) 233–224.
[68] Moreno, F. J. Gastrointestinal digestion of food allergens: effect on their allergenicity, Biomed Pharmacother **61** (2007) 50–60.
[69] Almeida, A. J., Souto, E., Solid lipid nanoparticles as a drug delivery system for peptides and proteins, Adv Drug Deliv Rev **59** (2007) 478–490.

[70] Ma, G., Microencapsulation of protein drugs for drug delivery: strategy, preparation, and applications, J Control Release **193** (2014) 324–340.

[71] Wong, C. Y., Al-Salami, H., Dass, C. R., Microparticles, microcapsules and microspheres: a review of recent developments and prospects for oral delivery of insulin, Int J Pharm **537** (2017) 223–244.

[72] Gupta, S., Jain, A., Chakraborty, M., Sahni, J. K., Ali, J., Dang, S., Oral delivery of therapeutic proteins and peptides: a review on recent developments, Drug Deliv **20** (2013) 237–246.

[73] Tan, M. X. L., Danquah, M. K., Drug and protein encapsulation by emulsification: technology enhancement using foam formulations, Chem Eng Technol **35** (2012) 618–626.

[74] Rekha, M. R., Sharma, C. P., Synthesis and evaluation of lauryl succinyl chitosan particles towards oral insulin delivery and absorption, J Control Release **135** (2009) 144–151.

[75] Cozar-Bernal, M. J., Holgado, M. A., Arias, J. L., Munoz-Rubio, I., Martin-Banderas, L., Alvarez-Fuentes, J., Fernandez-Arevalo, M., Insulin-loaded PLGA microparticles: flow focusing versus double emulsion/solvent evaporation, J Microencapsul **28** (2011) 430–441.

[76] Suksamran, T., Opanasopit, P., Rojanarata, T., Ngawhirunpat, T., Ruktanonchai, U., Supaphol, P., Biodegradable alginate microparticles developed by electrohydrodynamic spraying techniques for oral delivery of proteins, J Microencapsul **26** (2009) 563–570.

[77] Wang, T., Wang, N., Hao, A., He, X., Li, T., Deng, Y., Lyophilization of water-inoil emulsions to prepare phospholipid-based anhydrous reverse micelles for oral peptide delivery, Eur J Pharm Biopharm **39** (2010) 373–379.

[78] Makhlof, A., Fujimoto, S., Tozuka, Y., Takeuchi, H., In vitro and in vivo evaluation of WGA-carbopol modified liposomes as carriers for oral peptide delivery, Eur J Pharm Biopharm **77** (2011) 216–224.

[79] Teng, L. L., inventor; Research Corporation, assignee. Orally administered biologically active peptides and proteins. United States patent US 4582820. April 15, 1986.

[80] Plate, N. A., Valuev, L. I., Valueva, T. A., Staroseltseva, L. K., Ametov, A. S., Knyazhev, V. A., Henis, J. M. S., inventor; Orex Pharmaceutical Development Corp., assignee. Polymer composition for oral administration of peptides and proteins. European Patent EP 0918543. April 21, 2004.

[81] Shimoni, E., Ramon, O., Kopelman I. J., Mizrahi, S., Salzman, M., Nahmias, Y., Oren, A., inventor; Technion Research and Development Foundation Ltd., assignee. Oral delivery of proteins and peptides. World Intellectual Property Organization WO/2008/132727. November 6, 2008.

[82] Radhakrishnan, B., Aggarwal, D., Ferro, M., James, K. D., Malkar, N. B., Miller, M. A., Pavliv, L., Polowy, K., Puskas, M., Ekwuribe, N. N.; Biocon Ltd., assignee. Fatty acid formulations for oral delivery of proteins and peptides, and uses thereof. United States patent 7605123. October 20, 2009.

[83] Michael, J. C., Litwin, A., inventor; The University of Cincinnati, assignee. Oral administration of therapeutic proteins. United States patent US 6613332. September 2, 2003.

[84] Pather, S. I., Gupte, S. V., Khankari, R. K., Hontz, J., Robinson, J. R., Eichman, J. D, Kumbale, R., inventor; Cima Labs Inc., assignee. Microemulsions as solid dosage forms for oral administration. United States patent 6280770. August 28, 1996.

[85] Nomura, H., Maruyama, K., inventor; Kirin-Amgen, Inc., assignee. Oral dosage form of biologically active proteins. United States patent US 5597562. January 8, 1997.

[86] Bernstein, H., Morrel, E., Mathiowitz, E., Schwaller, K., Beck, T. R. Alkermes Controlled Therapeutics, Inc., assignee. Protein microspheres and methods of using them. United States patent US 5679377. October 21, 1997.

[87] Ferrari, M., Dehlinger, P., Martin, F. J., Grove, C. F., Friend, D. R., inventor; The Regents of the University of California, assignee. Particles for oral delivery of peptides and protein. United States patent US 6355270. March 12, 2002.

[88] Takada, K., inventor. Oral formulation for gastrointestinal drug delivery. United States patent 7097851. August 29, 2006.
[88] Suslick, K. S., Toublan, F. J., Boppart, S. A., Marks, D. L., inventors; The Board of Trustees of the University of Illinois, assignee. Surface modified protein microparticles. United States patent US 7217410. May 15, 2007.
[90] Vol, A., Gribova, O., inventor; Oshadi Drug Administration Ltd., assignee. Methods and composition for oral administration of protein and peptide therapeutic agents. World Intellectual Property Organization WO2009087633. July 16, 2009.
[91] Paulo, F., Santos, L. Design of experiments for microencapsulation applications: a review, Mater Sci Eng C Mater Biol Appl **77** (2017) 1327–1340.
[92] Orive, G., Hernandez, R. M., Rodriguez Gascon, A., Calafiore, R., Chang, T. M., de Vos, P., Hortelano, G., Hunkeler, D., Lacik, I., Pedraz, J. L., History, challenges and perspectives of cell microencapsulation, Trends Biotech **22** (2004) 87–92.
[93] de Vos, P., Historical perspectives and current challenges in cell microencapsulation, Cell Microencapsulation Methods Protocols **1479** (2017) 3–21.
[94] Smith, K., Johnson, R., Papas, K., Update on cellular encapsulation, Xenotranplant **25** (2018) e12399.
[95] Liu, H. W., Ofosu, F. A., Chang, P.l., Expression of human factor IX by microencapsulated recombinant fibroblasts, Hum Gene Ther **4** (1993) 291–301.
[96] Koo, J., Chang, T. S. M., Secretion of erythropoietin from microencapsulated rat kidney cells, Int J Artif Organs **16** (1993) 557–560.
[97] Murua, A., Portero, A., Orive, G., Hernandez, R. M., Cell microencapsulation technology: towards clinical application, J Controll Rel **132** (2008) 76–83.
[98] Aijaz, A., Perera, D., Olabisi, R., Polymeric materials for cell microencapsulation. Cell Microencapsulation Methods Protocols (2016) 79–93.
[99] Tibbitt, M. W., Anseth, K. S., Hydrogels as extracellular matrix mimics for 3D cell culture, Biotechnol Bioeng **103** (2009) 655–663.
[100] Anderson, J. M., Rodriguez, A., Chang, D. T., Foreign body reaction to biomaterials, Semin Immunol **20** (2008) 86–100.
[101] Dorling, A., Riesbeck, K., Warrens, A., Clinical xenotransplantation, Tissue Antigens **86** (2015) 239–253.
[102] Kolstoe, M. M., Iversen, T., Ostensvik, O., Abdelghani, A., Secic, I., Nesbakken, T., Specific pathogen-free pig herds also free from campylobacter?, Zoonoses Public Health **62** (2015) 125–130.
[103] Gao, G., Yonezawa, T., Hubbell, K., Dai, G., Cui, X., Inkjet-bioprinted acrylated peptides and PEG hydrogen with human mesenchymal stem cells promote robust bone and cartilage formation with minimal printhead clogging, Biotechnol J **10** (2015) 1568–1577.
[104] Stiegler, P. B., Stadlbauer, V., Schaffellner, S., Halwachs, G., Lackner, C., Hauser, O., Iberer, F., Tscheliessenigg, K., Cryopreservation of insulin-producing cells microencapsulated in sodium cellulose sulfate, Transplant Proc **38** (2006) 3026–3030.
[105] de Vos, P., Bucko, M., Gemeiner, P., Navratil, M., Svitel, J., Faas, M., Strand, B. L., Skjak-Braek, G., Morch, Y. A., Vikartovska, A., Lacik, I., Kollarikova, G., Orive, G., Poncelet, D., Pedraz, J. L., Ansorge-Schumacher, M. B., Multiscale requirements for bioencapsulation in medicine and biotechnology, Biomaterials **30** (2009) 2559–2570.
[106] Hill, E., Boontheekul, T., Mooney, D. J., Regulating activation of transplanted cells controls tissue regeneration, Proc Natl Acad Sci USA **103** (2006) 2494–2499.
[107] Thanos, C. G., Bintz, B. E., Emerich, D. F., Stability of alginate-polyornithine microcapsules in profoundly dependent on the site of transplantation. J Biomed Mater Rea A **81** (2007) 1–11.

[108] Barnett, B. P., Kraitchman, D. L., Lauzon, C., Magee, C. A., Walczak, P., Gilson, W. D., Arepally, A., Bulte, J. W. M., Radiopaque alginate microcapsules for X-ray visualization and immunoprotection of cellular therapeutics, Mol Pharm **3** (2006) 531–538.
[109] Yasuhara, T., Date, I., Intracerebral transplantation of genetically engineered cells for Parkinson's disease: toward clinical application, Cell Transplant **16** (2007) 125–132.
[110] Sajadi, A., Bensadoun, J. C., Schneider, B. I., Lo Bianco, C., Aebischer, P., Transient striatal delivery of GDNF via encapsulated cells leads to sustained behavioral improvement in a bilateral model of Parkinson disease, Neurobiol Dis **22** (2006) 119–129.
[111] Sieving, P. A., Caruso, R. C., Tao, W., Coleman, H. R., Thompson, D. J. S., Fullmer, K. R., Bush, R. A., Ciliary neurotrophic factor (CNTF) for human retinal degeneration: phase I trial of CNTF delivered by encapsulated cell intraocular implants, Proc Natl Acad Sci USA **103** (2003) 3896–3901.
[112] Shah, K., Encapsulated stem cells for cancer therapy, Biomatter **3** (2013) 1–7.
[113] Aftab, S., Shah, A., Nadhman, A., Kurbanoglu, S., Avsil, S., Dionysios, O., Shyam, D., Tejrej, S. S., Aminabhavi, M., Nanomedicine: an effective tool in cancer therapy, Int J Pharm **540** (2018) 132–149.

Gaetano Palumbo

7 Smart coatings for corrosion protection by adopting microcapsules

7.1 Introduction

For decades, iron and steel have been two of the most important materials in our daily lives due to their good mechanical properties, availability and relatively reasonable cost. However, iron does not exist in nature as iron, but as a compound such as iron oxide. In order to get a usable material, these compounds must be refined and processed from their natural state. The conversion process (reduction, eq. 7.1) of iron from iron ore is undertaken by heating, therefore providing energy, with carbon under careful control, in order to prevent the reverse reaction.

$$2Fe_2O_3 + 3C \rightarrow 4Fe + 3CO_2 \tag{7.1}$$

According to the second law of thermodynamics [1, 2], there is a strong tendency for the system to move from order to disorder: its energy tends to be transformed from high energy states into lower levels energy states available, in order to reach the state of complete randomness and be unavailable for further work. Corrosion occurs due to this tendency of metals to recombine with components of the environment to reach its low energy state. In the case of iron, the metal tends to be oxidized into a reddish oxide, commonly known as rust, leading to a gradual destruction or deterioration of the metal [3–7]. All metals, with a few exceptions such as gold and silver, show this natural tendency to return to their lower energy state to form oxide and hydrate and therefore are prone to corrosion. Corrosion degradation is one of the main reasons for large industrial losses. One of the most suitable methods to protect metal for the external environment is the application of coatings on the metal's surface. Coatings are designed to protect the metal by physically isolating and preventing the diffusion of aggressive species towards the metal surface. However, coatings are prone to degrade, either because of the long exposure time – processes commonly known as weathering and ageing (e.g. moisture, UV radiation, thermo-oxidation, etc.) – or mechanical factors (e.g. damage due to impacts, scratches, defects, etc.). These chemical and mechanical degradation processes will eventually lead to the formation of microcracks and premature failure of the coating system. When the coating fails, the corrosion of the substrate is greatly accelerated. Repairs and maintenance of failed coatings are well-known to be both expensive and time consuming. Therefore, an *active protection* in addition to the time-limited passive protection is required. In order to prolong the service life of the metallic structures, scientists have developed a new class of intelligent polymeric coatings able to fully or partially regenerate their structural integrity resulting from external damage, without the help of any external

https://doi.org/10.1515/9783110642070-007

factors. These new class of coatings are known as “smart coatings”. This chapter aims to cover the latest state-of-the-art developments in the field of smart coating, focusing the attention exclusively on the use of microcapsules loaded with functional active species autonomically activated by different external stimuli, namely mechanical damage and microcrack propagation as a result of thermal cycling or UV exposure, pH change, etc.

7.2 Basic principles: electrochemical nature of corrosion

Corrosion can be defined as a deterioration of the metal by reaction with its environment [3–7]. It is possible to distinguish the corrosion processes in two major areas [3–7]: dry and wet corrosion. In dry corrosion, the corrosion reactions take place in high-temperature dry gaseous environments, and so are referred to as gaseous or oxidation corrosion. While in wet corrosion, the process involves the exposure of the metal to an electrolyte (normally a liquid). As most of the corrosion processes occur in aqueous environments and the electrolyte involved is, usually, an aqueous solution, wet corrosion is also commonly referred to as aqueous corrosion. Which is, in turn, classified on the base of the apparent morphology in six others types of corrosion [3–7]: general corrosion (uniform, quasi-uniform, nonuniform, galvanic); localized corrosion (pitting, crevice); metallurgical corrosion (intergranular, dealloying); mechanically assisted corrosion (erosion, fatigue, wear); microbiological corrosion and environmentally induced cracking (stress-corrosion cracking, embrittlement, etc.).

A metal surface is composed of a network of local galvanic cells short-circuited by the metal itself (Figure 7.1). As long as the metal is dry, corrosion processes are not observed. The corrosion process is triggered when an electrical circuit is established between these local cells (Figure 7.2), involving the generation and transport of electrons from the metal (negative or anodic sites) to active species present in the aqueous solution in contact with the metal (positive or cathodic sites). This established circuit leads to a continuous degradation of the metal [8–10]. As the process occurs through two distinguished reactions taking place at the surface of the metal, i.e. anodic and cathodic reactions, corrosion in aqueous media is often defined as an electrochemical process. The anodic process or oxidation-dissolution reaction (eq. 7.2 and eq. 7.3) involves the oxidation of the metal, and thus the generation of the electrons as a result of the changing from the metal, or some of the metals present in the alloy state, into metallic ions.

$$\mathrm{Me} \rightarrow \mathrm{Me}^{n+} + ne^- \tag{7.2}$$

$$\mathrm{Me} + n\mathrm{OH}^- \rightarrow \mathrm{Me(OH)}n + n\mathrm{H}^+ \tag{7.3}$$

where n is the valence of the ion and the number of electrons produced, Me^{n+} and $Me(OH)_n$ are the metallic ion and hydroxide, respectively. Equations 7.2 and 7.3 refer to anodic reactions that involve soluble and insoluble corrosion products, respectively.

As the system does not produce any net charge [8–10], the electrons generated by the electrochemical anodic reaction must be consumed by another electrochemical reaction, called the cathodic process. Therefore, the anodic and cathodic processes are inseparable. Corrosion of the metal is a process that requires the presence of three important elements: anodic and cathodic reactions and a conductive liquid, the electrolyte, which is the medium surrounding the anode and the cathode, closing the circuit. In aerated or an in contact with air aqueous solution, where a significant amount of dissolved oxygen is available for reduction, the following cathodic reactions are very common [3–10]:

- Reduction of oxygen dissolved reactions in acid and in neutral or basic environments:

$$O_2 + 4H^+_{(aq)} + 4e^- \rightarrow 2H_2O_{(l)} \tag{7.4}$$

$$O_2 + 2H_2O_{(l)} + 4e^- \rightarrow 4OH^-_{(aq)} \tag{7.5}$$

- In acid and neutral and/or basic deaerated environments, the main cathodic process involves the evolution of molecular hydrogen, as result of the proton and water reactions:

$$2H^+_{(aq)} + 2e^- \rightarrow H_2 \tag{7.6}$$

$$2H_2O_{(l)} + 2e^- \rightarrow 2OH^- + 2H_2 \tag{7.7}$$

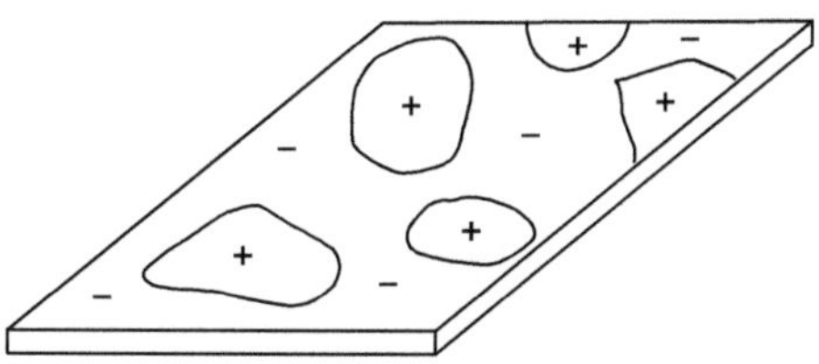

Fig. 7.1: Surface network of local galvanic cells short-circuited by the body of the metal itself.

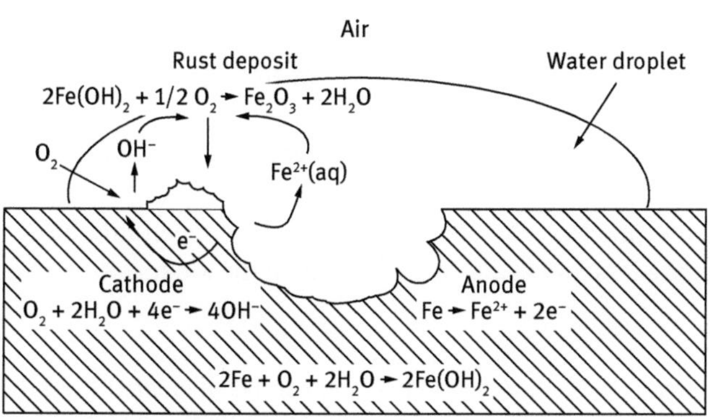

Fig. 7.2: The electrochemical corrosion of iron under a drop of water.

If, for example, the oxidized metal is iron and the electrolyte is water, the presence of oxygen lead to the formation of a corrosion product layer of ferrous hydroxide that, in the presence of excess of oxygen reacts to form the final reddish corrosion product layer of hematite (Fe_2O_3) (eq. 7.8–7.10), commonly known as rust [3–10]:

$$Fe \rightarrow Fe^{2+} + 2e^- \tag{7.8}$$

$$2Fe + O_{2(g)} + 2H_2O_{(l)} \rightarrow 2Fe^{2+} + 4OH^- \rightarrow 2Fe(OH)_2 \tag{7.9}$$

Thus, the overall reaction, which proceeds through a series of intermediate steps, is given by:

$$2Fe(OH)_2 + 1/2\, O_{2(g)} \rightarrow Fe_2O_3 + 2H_2O_{(l)} \tag{7.10}$$

7.2.1 Thermodynamics of corrosion

As stated above, metals try to lower their energy by spontaneously reacting with the external environment to form a compound or solution with a greater thermodynamic stability [1, 2]. The tendency of a metal to corrode into its own corrosion product is thermodynamically indicated by the free energy change of the reaction or Gibbs free energy, ΔG. When ΔG is negative (i.e. reaction products having a lower energy than the reactants) the reaction will proceed spontaneously. The more negative the free energy value is, the greater the tendency of the reaction to proceed. The free energy change associated with an electrochemical reaction may be related to the electrode potential and is given by the following equation:

$$\Delta G = -nF\,(\text{emf}) = -nF\,(E_{cat} - E_{an}) \tag{7.11}$$

where n is the number of electrons or equivalents exchanged in the reaction, F is Faraday's constant (96,500 C/mole) and E_{cat}/E_{an} = electrode potentials (V), respectively form the cathodic and anodic reactions. Therefore, there are three possibilities for a process in which the initial state of the system is given by G_I and the possible final state is given by G_F:

i. $G_F - G_I = 0 \rightarrow \Delta G = 0$, the process is at equilibrium and no net change in components will occur.
ii. $G_F - G_I > 0 \rightarrow \Delta G > 0$, the process will not occur spontaneously as the final energy state is higher.
iii. $G_F - G_I < 0 \rightarrow \Delta G < 0$, the process may occur spontaneously depending on the activation energy of the process and at a rate dependent on the kinetics system of the process.

In addition, the electrode potentials, E, for a single reaction can be calculated from the Nernst equation, that for e generic reaction (eq. 7.12) is described by the following equation:

$$a\text{Rd} + b\text{H}^+ + ne^- \leftrightarrow c\text{Ox} + d\text{H}_2\text{O} \tag{7.12}$$

$$E = E^{\circ} - \frac{RT}{nF}\ln\frac{\{Ox\}^{c}\{H_2O\}^{d}}{\{Rd\}^{a}\{H^{+}\}^{b}} = E^{\circ} - \frac{0.05916}{n}\log\frac{[Ox]^{c}[H_2O]^{d}}{[Rd]^{a}[H^{+}]^{b}} \tag{7.13}$$

where: R is the universal gas constant (8.314 J/K mole), T is Kelvin temperature (298 K), E° is the standard electrode potential (V) and {i} is the activity of the species. At low concentrations, the activities are replaced with the concentrations. At 25 °C, RT/F can be treated like a constant and in changing the natural logarithm in the one of base 10, we have the coefficient 0.05916. The standard electrodes potential are calculated at standard condition for each metal (Table 7.1). The electrochemical series consists of the arrangement of metals in order of electrode potential, the more negative the single potential is, the more active the metal is, i.e. there is a greater tendency of the metal to corrode [3–7]. If we set up a galvanic cell, e.g. zinc and iron with a generic electrolyte, based on the potential listed in Table 7.1, zinc will act as the anode and is therefore prone to corrode, whereas iron will be the cathode.

Table 7.1: Standard electrode potentials in aqueous solution at 25 °C [11].

Electrode reactions	E° (V) at 25 °C vs. SHE
$Au^{3+}_{(aq)} + 3e^{-} \rightarrow Au$	+1.498
$O_{2(g)} + 4H^{+}_{(aq)} + 4e^{-} \rightarrow 2H_2O_{(l)}$	+1.230
$Cu^{2+}_{(aq)} + 2e^{-} \rightarrow Cu$	+0.337
$2H^{+}_{(aq)} + 2e^{-} \rightarrow H_2$	0.000
$Ni^{2+}_{(aq)} + 2e^{-} \rightarrow Ni$	−0.250
$Fe^{2+}_{(aq)} + 2e^{-} \rightarrow Fe$	−0.440
$Zn^{2+}_{(aq)} + 2e^{-} \rightarrow Zn$	−0.763
$2H_2O_{(l)} + 2e^{-} \rightarrow H_{2(g)} + 2OH^{-}_{(aq)}$	−0.830
$Al^{3+}_{(aq)} + 3e^{-} \rightarrow Al$	−1.662
$Mg^{2+}_{(aq)} + 2e^{-} \rightarrow Mg$	−2.363

The application of thermodynamics to corrosion can be generalized by means of potential-pH plots or a Pourbaix diagram. Figure 7.3 and Figure 7.4 show the Pourbaix diagram for the solution systems Ni-H_2O and Cu-H_2O at 25 °C, respectively [12]. The Pourbaix diagram grants useful thermodynamic information on the feasibility of a generic corrosion reaction. The diagram provides a graphical representation of the relations between the pH of the aqueous system and the equilibrium potential of the electrochemical species involved. Such diagrams are temperature-dependent and are constructed using a Nernst equation (eq. 7.13) on the basis of the solubility for various metal compounds. The continuous lines delimit the stability domains of the metal species involved in the electrochemical process whereas the two dashed lines (a and b) indicate the proton and water reduction reactions. The area between the dashed lines delimits the stability domain of water. So, below line A water is unstable due to hydrogen evolution and above line b water is unstable due to the evolution of oxygen. Horizontal lines describe reactions that are dependent only on potential (e.g. eq. 7.8), vertical lines describe reactions that are dependent only on pH (eq. 7.9), whereas the angled lines correspond to reactions that depend on both potential and pH (eq. 7.9).

In the Pourbaix diagram it is possible to distinguish three regions:

i. Corrosion: where the metal will dissolve into its soluble species.
ii. Immunity: where the metal is thermodynamically stable and therefore, will not corrode.
iii. Passivity: where the metal is covered by a solid compound thermodynamically stable and it prevents its dissolution activity.

If, for example, the oxidized metal is nickel in equilibrium with its own specie (Ni^{2+}) at concentration of 10^{-4} M in acid solution, using the Pourbaix diagram it is possible to predict the species thermodynamically stable at different pH. The standard potential of nickel calculated in standard condition is $E° = -0.250$ V (Table 7.1), with the Nernst equation it is possible to calculate the potential of the following anodic and cathodic reaction in acid deaerated solution.

The anodic reaction corresponding to the dissolution reaction of the nickel is:

$$Ni \rightarrow Ni^{2+} + 2e^- \tag{7.14}$$

$$E_{Ni^{2+}/Ni} = E°_{Ni^{2+}/Ni} + \frac{0.059}{n}\log[Ni^{2+}] = -0.250 + \frac{0.059}{2}\log 10^{-4} = -0.368\ V \tag{7.15}$$

The cathodic reaction in acid solution is represented by eq. 7.16. From Table 7.1 $E°_{H^+/H_2} = 0$ V with partial pressure $P_{H_2} = 1$ atm, the Nernst equation will be:

$$E_{H^+/H_2} = E°_{H^+/H_2} + \frac{0.059}{n}\log\frac{[H^+]}{P_{H_2}} = 0 - 0.059\ pH = -0.059\ pH \tag{7.16}$$

From pH = 1 to pH = 6, the equilibrium potential of the species H^+/H vary from −0.059 to −0.368 V (eq. 7.16), and −0.368 V for the species Ni^{2+}/Ni (eq. 7.15), respectively. As the corrosion reaction involves the generation and consuming of the electrons, the electrons flow from the specie with equilibrium potential more negative (nickel; −0.368 V) to the one with equilibrium potential less negative (hydrogen). The mixed potential of the two reactions will be a value between the nickel and hydrogen equilibrium potential, and for pH ranging from 1 to approximately 6 it is located in the region of Ni^{2+} stability and below the dashed line A, in the region of H_2 stability (Figure 7.3). Therefore, in an acid solution ions of nickel are thermodynamically stable and nickel is prone to be corroded (oxidized) into Ni^{2+}, (eq. 7.1), while H^+ is reduced into hydrogen gas (eq. 7.6). However, when increasing the pH, e.g. from 6 to 8, the equilibrium potential of the species Ni^{2+}/Ni will be less negative than the hydrogen one, therefore

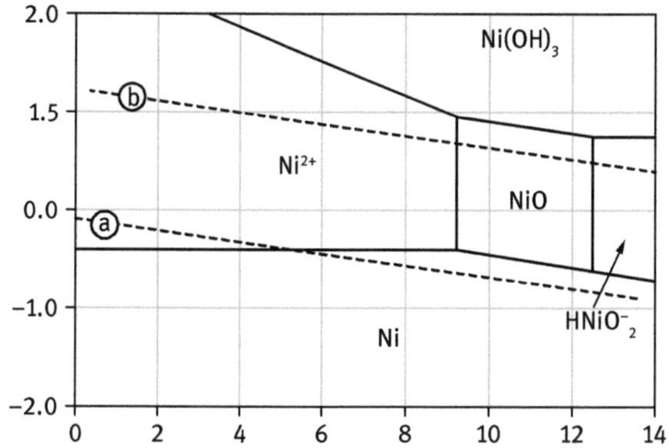

Fig. 7.3: Pourbaix diagram for Ni-H_2O at 25 °C.

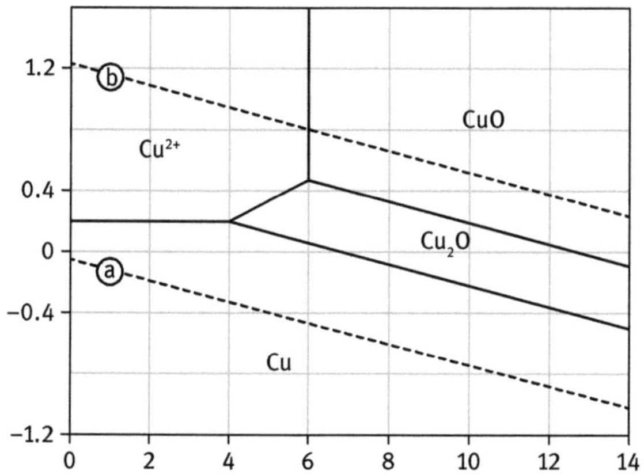

Fig. 7.4: Pourbaix diagram for Cu-H_2O at 25 °C.

in this case, the electrons will flow from hydrogen (oxidized) to Ni^{2+} (reduced). Hence, under these conditions nickel behaves as a noble metal and it is no longer susceptible to corrosion. For a further increase of pH, e.g. higher than 8, the thermodynamic stable specie is the oxide of nickel. In this case the metal surface will be covered by a protective oxide layer and the metal is said to be in the passive state. This oxide can protect the metal by physically blocking the aggressive species and preventing or reducing further corrosion phenomenon. This is the reason why nickel and nickel-based alloys are widely used in modern industry: both are ductile and tough and easily processed using conventional methods and can withstand a variety of extreme corrosive environments [5, 13]. Another interesting metal worth mention and extensively used in the industry due to its conductive and mechanical properties is copper and its alloys. Figure 7.4 shows the Pourbaix diagram for the system Cu-H_2O. At first sight of the diagram, it is clearly noticeable that the equilibrium potential of the species

Cu^{2+}/Cu (+0.337 V; Table 7.1) is above the line of the equilibrium potential of the species H^+/H_2 (dashed line A) for a wild range of pH. The H^+ ions are thermodynamically stable in contact with copper; therefore, metallic copper is thermodynamically stable in deaerated solution with respect to the other dissolved species. However, in the presence of oxygen the equilibrium potential of the species O_2/H_2O is higher than that of the Cu^{2+}/Cu (+1.230 and +0.337 V, respectively; Table 7.1), thus copper is prone to corrode in acid media. At neutral and alkaline pH values, the thermodynamically stable species is the copper oxide (Cu_2O). As in the case of nickel, the metal's surface will be passivated with a protective layer. Due to these excellent corrosion properties, copper is broadly used as metallic coating on metals such as iron and iron-based alloys, Figure 7.7(b).

7.3 Corrosion protection

Every year, corrosion processes cause $US billions of damage [6, 14]. The National Association of Corrosion Engineers [15], jointly with the Federal Highway Agencies (FHWA) [16], conducted a study on the cost of corrosion and found that the direct cost was estimated to be around 276 billion $US – approximately 3.1% of the national GDP. Therefore, understanding the mechanisms of the corrosion processes can facilitate analysis of material loss and help to develop a predictive model of material lifetime. Even with a well-designed system, together with a carefully choice of the materials used, there is no absolute way to eliminate this process. It is possible nevertheless to mitigate and control the effects of these corrosion processes. As the system does not produce any net charge [8–10], in order to ensure the electroneutrality of the system the rate at which the electrons are generated must be equal to the rate at which the electron are consumed. It is therefore possible to control the rate of the corrosion process by controlling the rate of one these reactions. The control can be achieved by a number of different methods and sometimes, when the structure is placed in severe environments, with multiple methods applied together such as, cathodic protection, corrosion inhibitors, coating, etc.

7.3.1 Cathodic protection

Cathodic protection (CP) together with protective coatings is one of the most effective methods widely used in the oil and gas industry and provides an effective way of preventing corrosion on substrates that are immersed or buried in an electrolyte. Cathodic is broadly used to protect large infrastructure, pipelines and reinforced concrete structures, with its first application dating back to 1824 [6].

CP can be achieved by supplying an external current to the buried or submerged structure. According to Le Châtelier's Principle [17], if a change occurs in one of the factors under which a system is in equilibrium, the system will tend to adjust itself so as to annul, as far as possible, the effect of that change. Taking as an example the dissolution reaction of iron (eq. 7.8), by supplying additional electrons (external current) the equilibrium will move to the left in order to decrease the "concentration" of electrons, and so reducing the dissolution rate of the metal.

Figure 7.5(a), shows the sketch of a pipe cathodically protected, in which a direct current (dc) source is connected with its positive terminal to the auxiliary electrode buried some distance from the structure to be protected, and its negative terminal to the structure itself; in this way the pipe becomes the cathode and the auxiliary electrode the anode with the current flowing from the auxiliary electrode through the soil to the structure. As long as the current is applied the metal cannot corrode. The most common auxiliary electrode used is composed either of scrap iron or graphite. Scrap iron is relatively cheap but is consumed at the rate of 15 lb/A per year, and therefore needs to be restored periodically. Conversely, graphite is more efficient and is consumed at the rate 2 lb/A per year, but is more expensive and fragile compared to scrap iron and must be installed with greater care [5].

There are, however, cases in which is not economical to install power lines and is preferable to protect the structure by setting up a galvanic cell using a "sacrificial anode". Basically, the sacrificial anode, with a potential more negative (so, more active in the galvanic series [11]) than the metal being protected, is consumed in this galvanic process (Figure 7.5). It is important to point out that the sacrificial anode, coupled with the metal of the structure, must provide a sufficient driving voltage to generate sufficient current in order to adequately protect the structure. One of the most-used sacrificial anodes is magnesium, a metal that with a standard potential of $E°= -2.38$ V, coupled for example with iron pipe $E°= -0.41$ V, provides the crucial driving voltage required to protect the pipe. Besides magnesium, others metals are used as anodes and are capable of ensuring an adequate driving voltage, such as zinc ($E°= -0.76$ V) or aluminum ($E°= -1.66$ V).

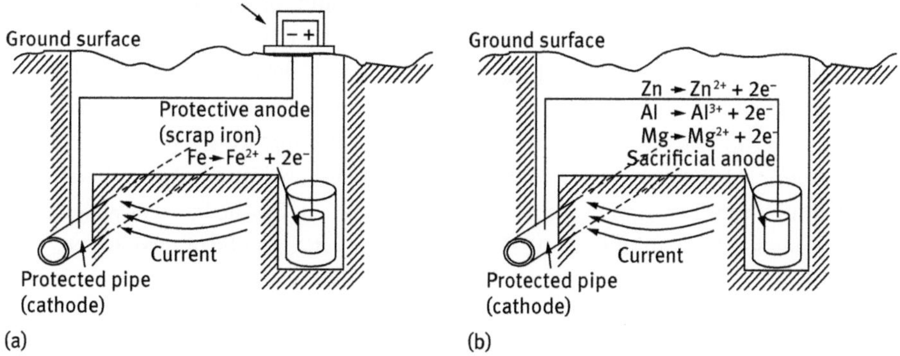

Fig. 7.5: Sketch of the pipe cathodically protected: (a) with a direct current source; (b) with a sacrificial anode.

When the structure to protect is in an extremely corrosive environment, e.g. seawater, CP is sometimes not really economic. In fact, large infrastructures, i.e. offshore, long oil pipelines, would require a prohibitively high current. Therefore, in order to increase the level of protection of the system, CP is often used together with other protection methods. The most common method, and rather economical, is the use of organic coatings. The combined use of CP system and barrier coatings reduce the amount of current needed for CP, however, in the case of failure of the coating the CP system will protect the bare metal from corrosive attack.

7.3.2 Anodic protection

Anodic protection is a method of corrosion control relatively newer than CP, but nevertheless less frequently used. Anodic protection, similar to the CP, is an electrochemical means of corrosion control but is based on a different electrochemical principle. As its name suggests, this form of protection is achievable by the formation of a passive film on the surface of the anodic electrode (the metal to be protected) with the application of an electrical current (Figure 7.7), therefore, bringing the structure into the regime of thermodynamic immunity in the Pourbaix diagram. One of the reasons why this method is less frequently used than CP is that only certain metals in specific environments can be anodically protected, whereas all metals can be protected with the CP. Anodic and CP, are sometimes confused even though they are essentially two different techniques. In practice, the difference involves the electrode that is protected: the cathode is protected in CP, and the anode is protected in anodic protection.

7.3.3 Corrosion inhibitors

Another practical method to protect the metal dissolution from aggressive environments is the use of corrosion inhibitors. Corrosion inhibitors are chemical substances that are added to the corrosive environment in small amounts, reducing the corrosion rate to an sufficient level [18]. It is generally accepted that the inhibitor interacts with the metal surface, inhibiting the corrosion reactions that take place at the cathodic or anodic sites of the corrosion cell, e.g. either by blocking the reduction of the oxygen to form hydroxide ions at the metal's surface (eq. 7.5) or by keeping the metal from dissociating into ions (eq. 7.2). By its adsorption on the metal surface, the inhibitors form a protective film or barrier layer, which isolates the metal from the corrosive environment and therefore is able to inhibit the corrosion processes [18–20] (Figure 7.6). On the basis of which reaction is suppressed, the inhibitors are classified as cathodic, anodic or mixed-type.

7.3.4 Corrosion protection by coatings

The most common and relatively economical method used to prevent the corrosion of metallic structures is via coating. This provides long-term protection under a broad

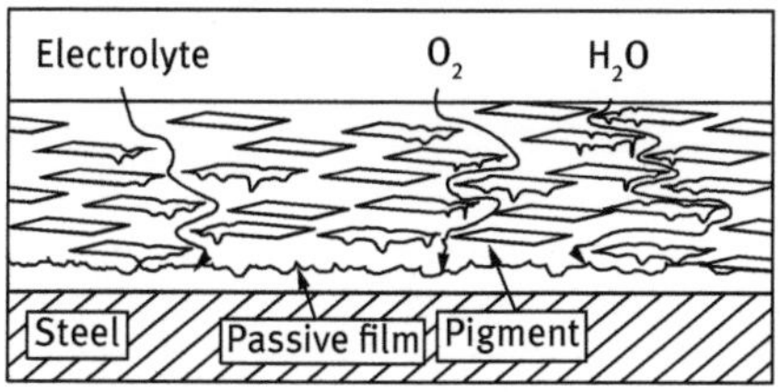

Fig. 7.6: Corrosion protection provided by inhibitors pigments.

range of corrosive conditions by isolating the structure from the aggressive external environment. Based on the types of coatings produced, coatings can be classified in three different ways [21–23]:

i. nonmetallic inorganic coatings such as conversion layers, anodized layers, glass and ceramic chemical vapor deposition (CVD) and physical vapor deposition layers;
ii. metallic coatings;
iii. organic coating such as paints and polymer sheets.

While the organic coatings essentially act as a barrier protecting the metal from the surrounding environment, metallic coatings provide their protective action by functioning either as a sacrificial anode or as cathodic protector, therefore changing the surface properties of the substrate. This chapter focuses only on the latter two types of coatings.

7.3.4.1 Metallic coatings

Metallic coatings basically utilize the same mechanism as the classic cathodic and anodic protections described above. A less noble metal, so more active in the galvanic series [11] than the substrate to protect, is used as coating and therefore acts as an anode under corroding conditions, whereas the substrate supports the cathodic reaction (Figure 7.7a). The most common coatings utilising this mechanism are zinc, aluminum, manganese, cadmium, and their alloys that, due to their more negative standard potential, are used to protect iron and steel substrates. The deposition is usually obtained either by wet chemical processes or by new alternative techniques such as thermal spraying, hot dipping, CVD, etc.

It is also possible to employ metal coatings more noble than the metal of the structure e.g. copper (Figure 7.7b). Several metals can be used as metallic coatings in addition to copper, such as nickel, tin, chromium, lead, etc., under the condition that the applied layer covers the entire surface of the substrate with a total absence of defects and pores. In fact, the presence of pores and defects in the coating may lead to the corrosion of the metal, as the result of the diffusion of water throughout the coating. The diffused water will establish the electric connection required to start the corrosion process between the substrate, the less noble metal (the anode electrode) and the coating, in which, as more noble metal, will act as the cathode. Conversely, with a less

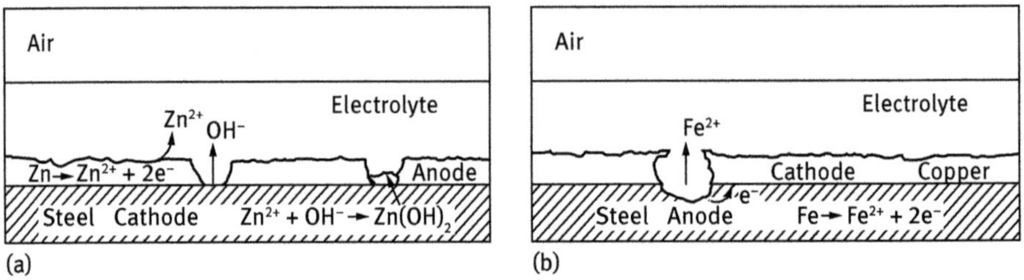

Fig. 7.7: Schematic illustration of the cathodically protection coating (a) and anodic protection coating (b).

noble metal, in the presence of defects, the exposed portion of the substrate will trigger a galvanic process between the substrate (in this case, the cathode) and the surrounding coating (anode), with the formation of corrosion products. The corrosion products formed by this galvanic action may also lead to the formation of a second barrier that will fill up the defect itself [5] (Figure 7.7a). Other noble coatings, such as gold, silver, platinum etc., due to the their high cost are usually used in the field of jewelry and watches, etc., when the aesthetic aspect of the surface is the dominant factors, or are also used in electrical equipment due to their good conductive properties.

7.3.4.2 Organic coating

Three elements are necessary for the corrosion process to occur: cathode, anode and a conductive media that allows the electrons to flow from the anode to the cathode establishing an electric circuit [8–10]. It is possible to interrupt this electric circuit by providing a protective barrier of coating between the metal and the external environment. Therefore, the main function of an organic coating is to insulate the metal from the external environment, by decreasing the diffusion of water, oxygen and aggressive ionic species towards the surface of the metal.

As the cathodic and anodic reaction are inseparable, preventing the oxygen from reaching the metal/coating interface deprives the cathodic reaction of the necessary fuel, thus reducing the corrosion rate of the metal. However, for years it was believed that the corrosion prevention provided by the coatings was only owing to the effect of the barrier that hindered the permeation of water and oxygen.

It is important to point out that – although water and oxygen are necessary to the corrosion processes to take place – their permeation is not the rate-controlling step [24]. In fact, most of the coatings (with the exception of a few) are permeable in some degree to water and oxygen, and therefore it is not the barrier that prevents them from diffusing through the coating. For example, the corrosion of steel in pure water is very slow, due to the formation of ferrous hydroxide, $Fe(OH)_2$. Ferrous hydroxide has a low solubility in water, and therefore precipitates, forming a corrosion product layer on the surface that retards the diffusion of the oxygen and water necessary to continue the corrosion process [23, 24]. Conversely, the presence of water and oxygen, combined with aggressive species, e.g. chloride ions (Cl^-), leads to an autocatalytic process dissolution of the metal with formation of localized corrosion spots. Thus, it is important to prevent the permeation of water and oxygen, but also the diffusion of these aggressive species [25–27]. This method is accomplished by using resistive coatings.

The theory of ionic resistance or resistance inhibition, suggests that by creating a path of high electrical resistance between the anode and the cathode areas, the flow of current available for cathode and anode is reduced. In other words, the resistance to ionic current flow isolates anodic and cathodic areas and thereby corrosion is slowed down.

In summary, the coating efficiency and its service life, strongly depend upon the composition (high electrical resistance), structure and chemical bonds of the metal/polymer interface as well as the precautions taken during its application on the substrate [24, 28, 29]. Defects or pores may occur during the application of the film (Figure 7.8) as well as the result of abrasion or damages. Damage could be mechanical, e.g. the impact of solid objects or chemical by exposure to acids or solvents and high-temperature exposure or exposure to light [10]. The presence of defects lower the diffusion barrier effect provided by the coating, thereby promoting the penetration of the electrolyte through the coating itself. In particular, the reduction of oxygen generates hydroxyl ions that will promptly attack the chemical bonds between the film and the substrate [23, 30–34]. Therefore, above all, the coating must provide a good adhesion with the substrate. Poor adhesion could lead to the formation of blistering and coating failure (Figure 7.10).

On the basis of their composition, coatings may be divided into solvent-borne, water-borne and solvent-free (solid). For the water and solvent based coatings, their deposition is usually done by spray-gun. However, in the last few decades, new technology are developed, e.g. electrostatic-sprays, utilizing solvent-free powder coating. In particular, powder coatings provide more environmental benefits than traditional coatings, meeting the stringent environmental regulations limiting the emissions of VOCs [36, 37]. Regardless of the physical form – either liquid or solid – the major constituents of coating are binders, pigments, fillers and additives [21, 24, 38, 39]. The binder, usually a polymer, is the matrix of the coating in which all the others components are integrated to form a dense, tight film, Figure 7.9. The binder can be a resin (epoxy, polyurethane etc.) or inorganic compound.

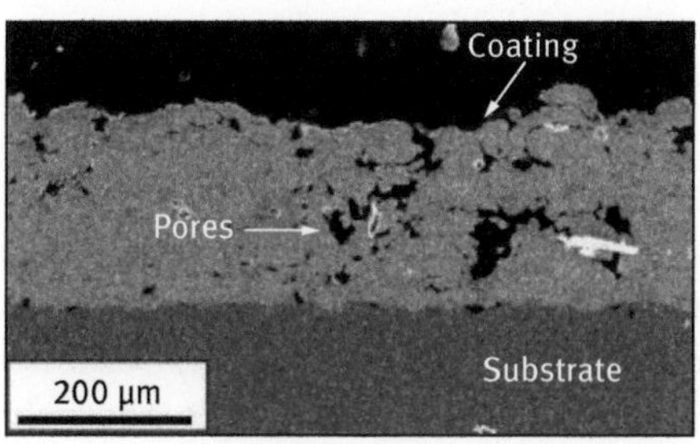

Fig. 7.8: Pores or defects generated during the application of the coating (adapted from [35] and reprinted with permission from Elsevier).

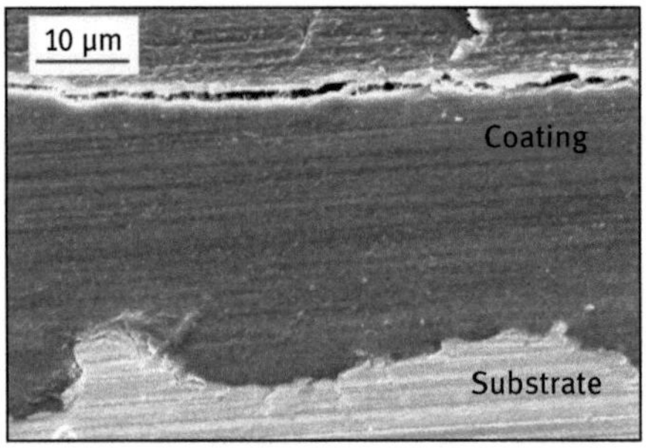

Fig. 7.9: Compact coating free of defects and pores.

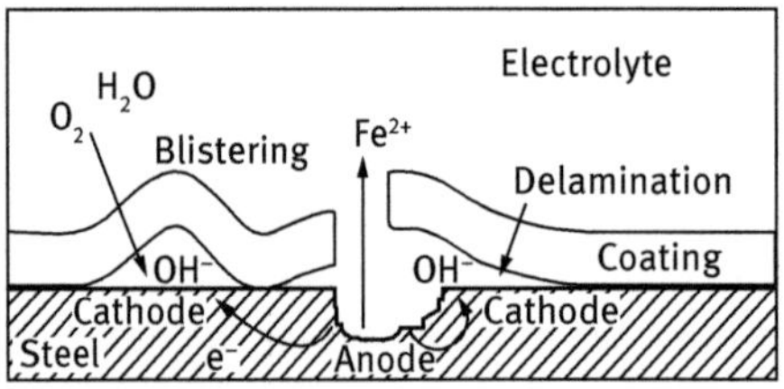

Fig. 7.10: Schematic illustration of the blistering and delamination mechanism.

Pigments are basically dry powder added to the coating either to improve the corrosion protection properties, e.g. chemical and moisture vapor transfer resistance, or purely for aesthetic reasons, e.g. by providing color to the coating. There are different ways by which the pigments can enhance the corrosion action of the coating and they can be classified by the way they work: chemical (active pigments) or physically (barrier pigments). One of the most effective methods is achieved by using flake-shaped pigments parallel to the surface of the metal (Figure 7.11). They can increase the barrier effect of the system by retarding the diffusion through the coating (i.e. barrier pigments). Figure 7.12 shows the effects of the pigments on the oxygen and water permeability through the coating and how the permeability of both decreases by increasing the percentage of pigments on the coating. Occasionally, water permeation is desirable to a certain degree in order to activate inhibitive pigments (i.e. active pigments) (Figure 7.6). Sometimes the application of a single layer is not enough to protect the substrate, especially in aggressive environments, and in some cases the application of a second or third layer is required. Figure 7.11 shows a schematic example of a multilayered coating, in which each layer aimed for specific functionalities, namely: adhesion to the metal or adhesion between layers, water resistance, corrosion inhibition, and et cetera.

The first layer of this multilayered system, the primer, is usually applied on the clean metal surface, providing long-lasting adhesion properties between the metal and the finishing coating, better than if it were used alone. In fact, in a multilayered system the use of a poor quality primer may compromise the efficiency of the entire system, even in the presence of a high performance finishing coating on top, and it will eventually lead to the formation of blistering and thus disbonding of the coating. After the primer layer, an intermediate coat or undercoat might be present. The main

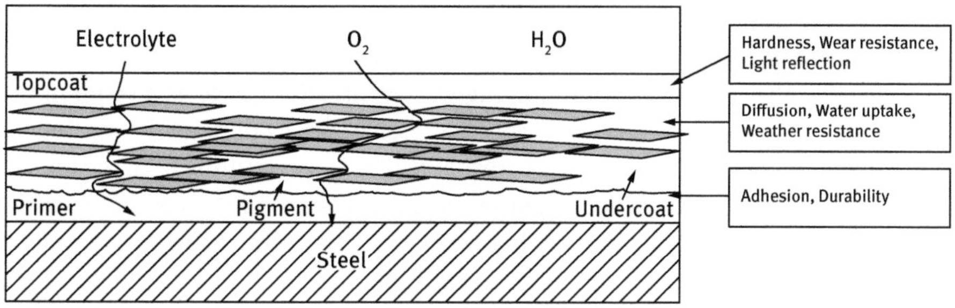

Fig. 7.11: Barrier protection provided by the coating with flake-shaped pigments in a multi-layered system.

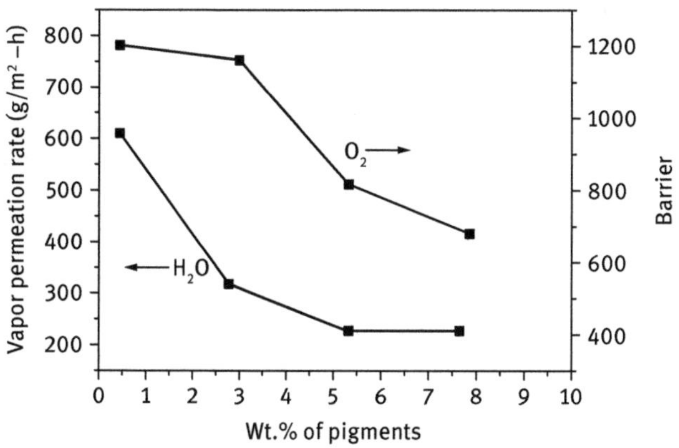

Fig. 7.12: Permeability of the H_2O and O_2 as function of the pigments content in the coating (adapted from [40]).

function of the undercoats is essentially to increase the thickness, thus the protection, of the overall system with the pigments embedded on the undercoat itself. The thickness can range from a few µm to up 1 mm, depending on the application of the structure. The finishing coat, sometimes called topcoat, is the last layer to be applied and provides the first line of defense against the external environment. The main function of the topcoat can be vary; they can provide toughness and wear resistance, enhancing the resistance towards light, chemical agents, etc. of the surface or be merely decorative.

7.4 Smart coatings

Coatings are designed to provide long-lasting protection against most chemical aggressive environments. There are different reasons for the failure of a coating: misapplication, defects, or simply a long period of exposure to aggressive elements, e.g. high temperatures, light, acids, etc. Furthermore, with the exception of a few coatings, most of them are permeable to the water to some degree. In fact, during its service life, the properties of the coating will eventually change as a result of the penetration of the water, oxygen and other aggressive species through the coating. In addition, defects or the combination of mechanical damage resulting from the impact of solid objects may lead to the formation of microcracks on the coating surface, providing pathways by which the aggressive species may diffuse towards the metal surface. Whatever the reason, the failure will eventually lead to disbonding of the coating and flake formation from the metal coating interface (Figure 7.10).

Maintenance and repair of the damaged coatings are known to be very expensive, especially on large structures, e.g. offshore oil rigs, ships, etc. Moreover, the conventional repair procedures are generally time consuming and it is mainly appropriate to repair the external damage rather than the internal microcracks. Dry and Sottos performed the first successful attempt at a self-healing material in 1993, using fibres as healing agent carriers [41]. In 2001, White et al. published in *Nature* [42] a landmark paper entitled "Autonomic healing of polymer composites" with the first successful attempt at an encapsulated healing agent combined with a dispersed encapsulated catalyst agent within a polymeric matrix. In the last few decades, as witnessed by the increasingly number of scientific papers [43, 44], textbooks published [22, 45–47] and four international conferences on self-healing materials held in Netherlands (2007) [48], USA (2009) [49], UK (2011) [50] and Belgium (2013) [51], self-healing materials have become an intense field of research. Since then, different approaches have been researched and tested; to date the most common approach and the most successful is the one proposed by White and Sottos [42], which involves the encapsulation of a self-healing reagent within microcapsules incorporated in the coatings (Figure 7.13).

The encapsulation of reactive agent has been widely used in the agriculture, food [52] and pharmaceutical industries [53, 54] as a delivery device. In the context of corrosion protection it has proved to be an effective means of increasing the life-span of the coating, enhancing the compatibility of the active agent with the coating matrix and sustaining, prolonging and controlling the release rate of the agent itself. In particular, as will be discussed later, the process of encapsulation has proved to be very useful in the case of corrosion inhibitors because it allowed coating scientists to solve some problems related to their chemical activity and solubility in water [18, 55–58].

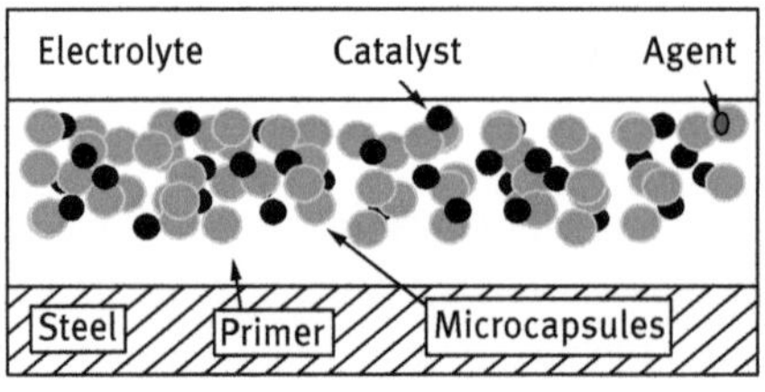

Fig. 7.13: Schematic representation of the microcapsules incorporated within the coating matrix.

This new class of coatings are commonly referred as "smart coatings" due to their ability to respond to physical, chemical or mechanical external stimuli, namely temperature, UV light, and pH changes, and they can be considered as an alternative mean to further enhance the corrosion protection and thereby the durability of metallic structures. Based on the type of microcapsules content, three major release mechanisms were proposed: mechanical rupture of the microcapsule wall, diffusion through the wall, and erosion. In mechanical rupture, upon rupture resulting from microcrack or sudden damage of the coating, the encapsulated self-healing agent is released into the damaged region, and by acting as a healant prevents further propagation of the crack, allowing the recovery of the barrier property of the coating (Figure 7.14). In particular, corrosion inhibitor microcapsules present all of three mechanisms above mentioned, depending on the type and final use of the coating. The release can be achieved either upon rupture of the capsules themselves [44, 59] or by localized pH changes as a result of the corrosion process. Changes towards acidic and alkaline pH may lead the inhibitors either to diffuse through the shell of the capsule into the coating or by a spontaneous leakage due to the erosion of the embedded capsule [60–62]. In both cases the released agent can react with the environment, forming a thin protective layer on top the metal surface [44, 59, 63]. There are basically two different methods to incorporate the microcapsules into the coating, mostly depending on the final use of the coating: blending them within coating matrix [43, 64, 65], or electrolytically co-depositing the microcapsules on the substrate with metal ions in order to produce metal-matrix composites (MMCs) [66–77] (Figure 7.15).

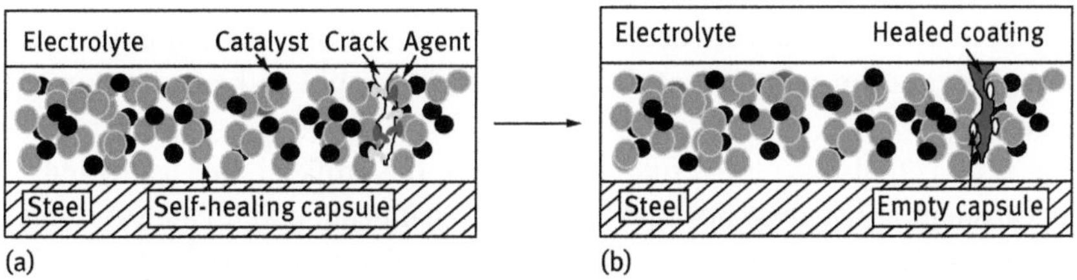

Fig. 7.14: Schematic representation of self-healing coating; (a) before and (b) after healing.

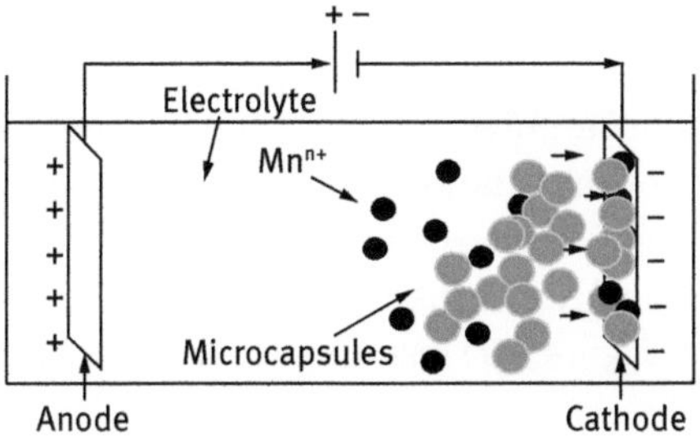

Fig. 7.15: Schematic representation of the microcapsules incorporated within the coating matrix applied by electrolytic codeposition with metal ions.

7.4.1 Self-healing coatings

As White et al. showed [42, 78–82], it is possible to design polymer composite to self-heal autonomically from external damage by embedding liquid-containing microcapsules in a polymeric matrix. One of the first successful attempts to incorporate microencapsulated agent into commercial epoxy coating for corrosion prevention was made by Kumar and co-workers [83] in 2003. The corrosion study as well as the self-healing behavior was carried out by using poly(urea-formaldehyde) (PUF) microcapsules loaded with several types of self-healing agents (tung oil and spar varnish) and corrosion inhibitors (camphor and alkylammonium salt of 2-benzothiazolylthio succinic acid in xylene) incorporated into the coating. The coating was then applied on a steel hand scribed and exposed for 12 weeks to a cycle treatment of salt-fog/UV. The results demonstrated that, after scribing the coating surface, the undercutting corrosion and the growth of the artificial scribe were reduced, but not fully recovered, resulting in the release of the healing agent and corrosion inhibitor. Although the experimental results displayed only partial recovery of the artificial damage, nevertheless this approach showed the effectiveness of the microcapsule as a suitable route to increase the service life of the coating. Suryanarayana et al. [84] were able to achieve full recovery of the artificial damage using PUF microcapsules as shell material loaded with linseed oil as self-healing agent. The finding reveals that after exposed both self-healing and the control scribed coatings specimens to the spray chamber, no sign of undercutting corrosion was found for the self-healing coating, unlike the control coating. Figure 7.16 displays the healing process of the initial scratch shrinking gradually with the time until it is completely healed, showing no visual corrosion evidence. Linseed oil, upon rapture of the microcapsule, was released into the crack and through oxidation with atmospheric oxygen was able to heal the damage efficiently.

White et al. [79] also achieved full recovery from the artificial damage by individually encapsulating reactive siloxanes and tin, as self-healing and catalyst agents, respectively, in polyurethane (PU) microcapsules. Both microcapsules were then incorporated into epoxy-based coatings and characterized by means of morphological and electrochemical techniques. The results showed that upon the artificial damage a polymerization reaction took placed within the damage, resulting in the release of healant and catalyst agents, preventing further propagation and therefore restoring the barrier properties of the scribed coating.

Samadzadeh et al. [85] synthesized by *in situ* polymerization, tung oil liquid core PUF microcapsules, without the presence of any catalyst agent. The self-healing action was evaluated through corrosion testing of damaged and healed coated steel systems compared to control samples in 3.5% of NaCl. The morphological results confirmed full recovery of the artificial scribe (Figure 7.17) resulting from the release

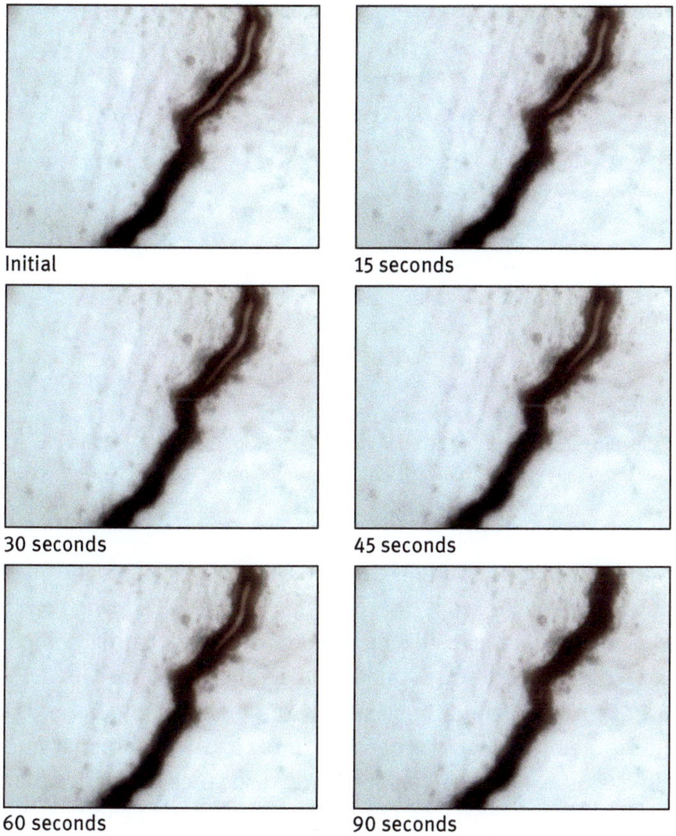

Fig. 7.16: Optical microscope photos of self-healing coating films process as a function of the healing time (reprinted from [84] with permission from Elsevier).

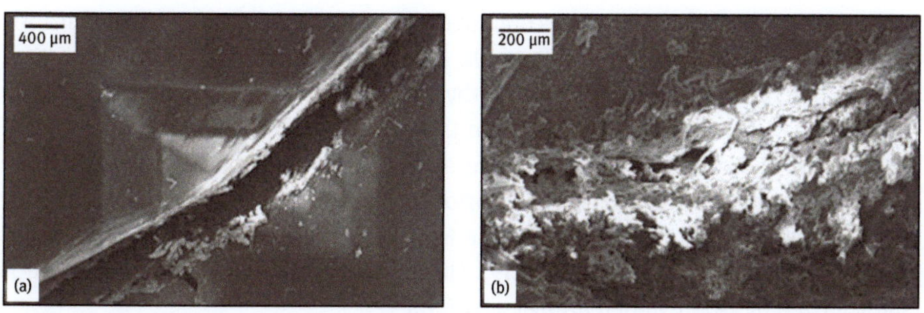

Fig. 7.17: Scanning electron microscope (SEM) image of artificial scratch after exposure of the specimens for 14 days in 3.5% of NaCl solution. (a) control epoxy coating, (b) healed scratch on self-healing coating (reprinted from [85] with permission of Elsevier).

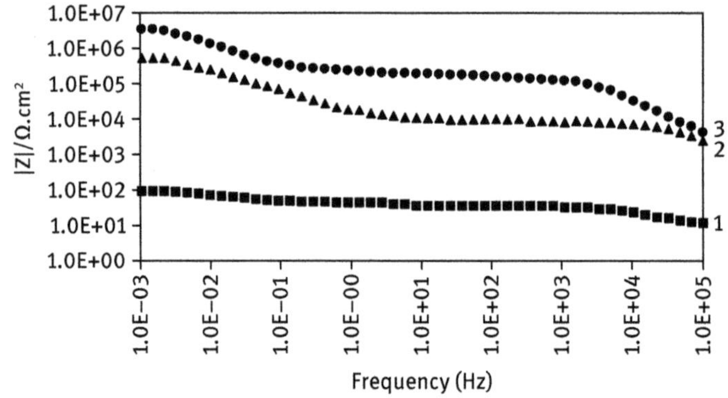

Fig. 7.18: EIS plot of samples after 14 days of immersion in 3.5 wt. % NaCl aqueous solution. (a) scratched neat epoxy coating, (b) scratched Tung oil microcapsule incorporated epoxy coating; (c) neat epoxy coating without scratch (reprinted from [85] with permission from Elsevier).

of the healing agent tung oil. The recovered barrier property was also confirmed by electrochemical impedance spectroscopy (EIS) results. Figure 7.18 shows the EIS data for the scratched tung oil microcapsule incorporated epoxy coating compared with the neat epoxy coating with and without the scratch performed in 3.5% of NaCl after 14 days of exposure. Generally speaking, the higher the value of the impedance |Z|, the better the anticorrosive properties. It is clearly visible that microencapsulated scratched coating shows better corrosion resistance properties than the scratched neat epoxy coating, confirming the effectiveness of the healing agent by the full recovery of the artificial damage. Moreover, the corrosion resistance of the microencapsulated scratched coating displays high value of impedance, almost comparable to the unscratched neat epoxy coating, indicating that even after a long exposure in an aggressive media the recovered damage area still provides good barrier properties.

Following the promising results obtained with the one-step intelligent self-healing approach proposed by Samadzadeh et al., Huang and co-workers [86, 87] investigated the anticorrosive property of examethylene diisocyanate (HDI) PU microcapsules dispersed in epoxy resin coating. Even though the work was more focused on the optimization of the process parameters, the anticorrosion performance undertaken in 1 M of NaCl solution showed the excellent corrosion protection achieved. No undercutting corrosion signs for the self-healing coating were noticed after exposing the specimens for 48 h in the test solution (Figure 7.19). On the other hand the control specimen is characterized by severe undercutting corrosion.

All the approaches mentioned so far are based on the idea of closing the gap generated during the damage and therefore restoring the barrier effect loss. More recently, new interesting and different approaches were proposed. These approaches tend to isolate rather than fill the local damaged area, by incorporating microcapsules loaded

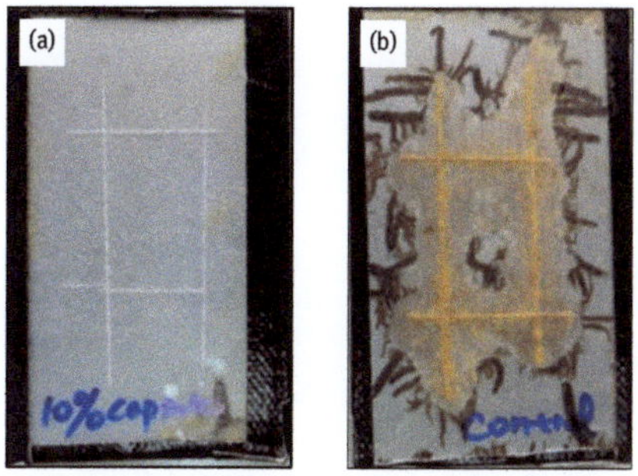

Fig. 7.19: Corrosion test results after exposure in 1 M NaCl solution for 48 h. (a) Self-healing coating, (b) control epoxy coating (modified and reprinted from [87] with permission from Elsevier).

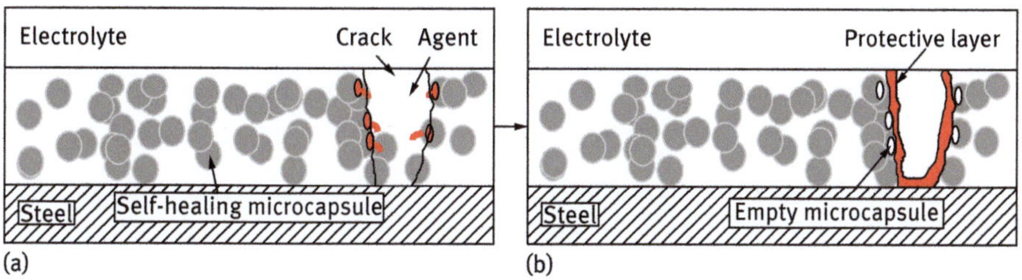

Fig. 7.20: Schematic representation of the anticorrosive hydrophobic layer of the self-healing coating; (a) before and (b) healed.

with environment-reactive healing agent (Figure 7.20). García and co-workers [44] used silyl ester as a healing agent in a single healing component. Silyl ester was encapsulated in PUF microcapsules incorporated into an epoxy resin coating and then applied to aluminum alloy. Fourier transform infrared spectroscopy (FTIR) and contact angle measurements showed the formation of a tightly adhered hydrophobic barrier layer on top of the metallic surface caused by the reaction of the silyl ester with the water/humidity presented in the coating. The corrosion protection performance of the formed layer were also characterized by means of EIS, comparing the performance of a scratched coatings with and without the presence of the encapsulated agent. The result displayed in Figure 7.21 clearly shows a well-defined distinction between the two scratched coatings. An increase in the module of impedance upon a long exposure time for the coating containing the encapsulated silyl ester was observed, confirming the formation of the hydrophobic protective layer. This work demonstrated that the silyl esters were able not only to wet the damaged surface but also the cut edge of the coating, avoiding further leakage of the inhibitor from the coating, Figure 7.21.

Latnikova et al. [88] used microcapsules loaded with alkoxysilanes in PU. The applied coating was then tested in 0.1 M NaCl solution for 6 h. Accordingly, alkoxysilanes possessing a long hydrophobic tail were found to display better corrosion resistance than those without a tail, as a result of the water-repelling properties of the tails [88]. Huang et al. [89] successful incorporated 1H,1H,2H,2H-perfluorooctyl triethoxysilane (POTS) containing microcapsules in epoxy resin matrix.

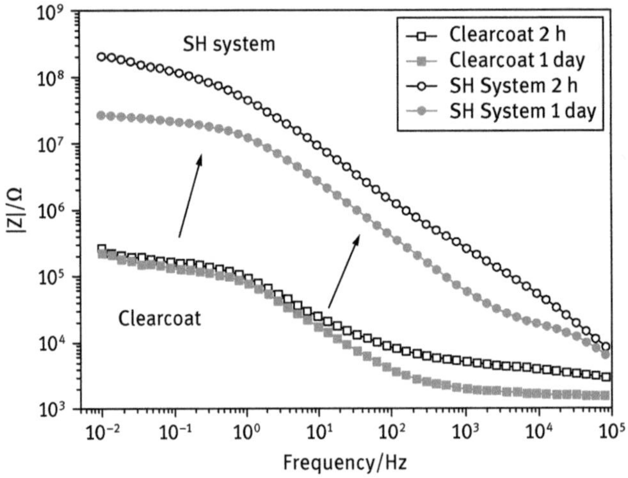

Fig. 7.21: EIS plot of scratched coating at different immersion time in the presence and absence of the encapsulated agent (reprinted from [44] with permission from Elsevier).

Figure 7.23 shows the optical and SEM-EDX analysis for the neat scribed epoxy and the self-healing scribed coatings after exposing the specimens in 1 M NaCl for 48 h. The presence of the undercutting corrosion on the scribed neat epoxy coating denoting the poor corrosion resistance provided by this coating as result by the diffusion of the electrolyte through the scribe. Conversely, no sign of undercutting corrosion was found on the self-healing coating (Figure 7.23a,b). Moreover, SEM analysis on the scribed area clearly shows that the gap generated by the artificial scribe has been partially sealed, resulting in leakage of the silane agent (Figure 7.23c,d). EDX analysis undertaken on the bulk of the coating and in the scribed area for the self-healing coating showed that both areas share the same composition, i.e. silicon (S) and fluorine (F). As only the healing agent is comprised of these two elements, the analysis confirmed that leakage occurred from the microcapsule. On the contrary, the EDX analysis performed on the scribed neat epoxy coating only displayed the presence of iron (Fe), sodium (Na) and chlorine (Cl) from the electrolyte solution, without any signs of silicon (S) and (F) fluorine. The presence of sodium and chlorine indicated the diffusion of the electrolyte through the artificial scribe, while the presence of iron could be ascribed to the formation of iron oxide Fe_2O_3 on the metal surface, indicating that corrosion process occurred (eq. 7.10). Both the presence of these elements and the absence of any traces of the latter two elements lend the conclusion that no leakage process occurred. The corrosion resistance of the scribed self-healing coating was also investigated by the electrochemical technique shown Figure 7.22; the figure shows the current density vs. the potential of the specimen – the higher the value of current density the higher the corrosion rate of the metal. It is clearly visible that after 48 h exposure in an aggressive media the scribed control self-healing coating displays lower current density compared with the control coating. The lower current, thus the lower corrosion rate, is the result of the barrier effect provided by the coating, which isolates the metal from the external electrolyte, impeding the current flow. This electrochemical experiment once more confirmed that the self-healing coating has recovered its barrier properties during the immersion process.

All those approaches, in addition to the aforementioned corrosion protection effect, present another interesting advantage that is not present with other methods so far discussed. These methods use single self-healing microcapsules, rather than the conventional two microcapsules method, i.e. a self-healing agent containing microcapsules and the catalyst microcapsules necessary to accelerate the polymerization of the healing agent.

As will be detailed below, this two-step self-healing approach may pose a serious drawback. In fact, the presence of a second phase decreases the likelihood of having both phases present at the same time and in the required ratio in the region of the scratch, hindering the restoration of the barrier properties of the coating itself [44].

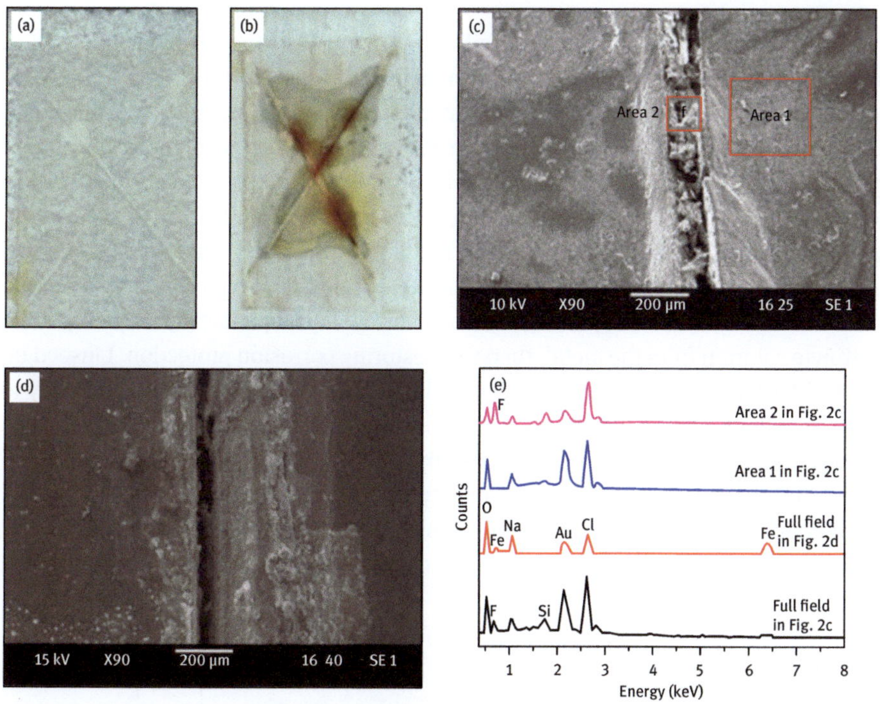

Fig. 7.22: Optical images of the scribed coatings: (a) self-healing coating, (b) neat epoxy control coating. Scanning electron microscope (SEM) of the (c) self-healing scribed coating and (d) neat epoxy control scribed coating. (e) EDX analysis of the areas of the control coating and self-healing coating. The figures were taken after the the specimens were exposed for 48 h in 1 M NaCl solution (reprinted from [88] with permission from Elsevier).

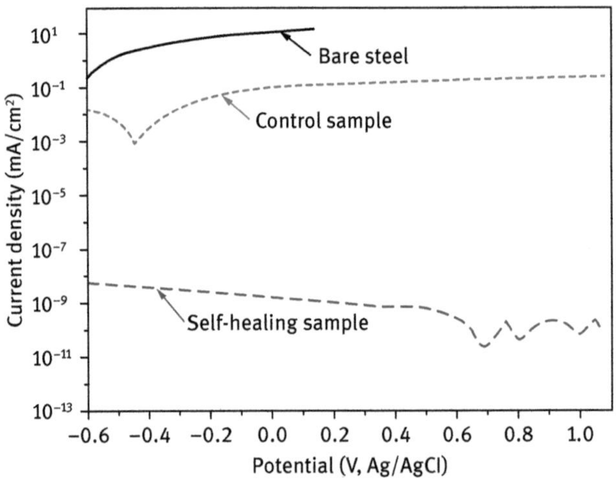

Fig. 7.23: Electrochemical plot comparing the behavior of the bare steel panel, scribed control and self-healed specimens (adapted from [88] and reprinted with permission from Elsevier).

However, this probability can be enhanced by increasing the concentration of microcapsules within the coating, which in turn, can negatively affecting the adhesion between the metal and the coating [85, 90]. What instead makes these materials a catalyst-free self-healing additive is their ability, once released, to react with the environment and form a hydrophobic film on the surface of the metal or to reduce the growth of the damage. For example, the self-healing approach proposed by García and Huang utilising silane is able to react with the hydroxyl groups present on the metal surface, producing a hydrophobic layer able to passivate the surface, according to the reaction showed in Figure 7.24 [44, 88, 89, 91]. The new formed hydrophobic layer will repel the electrolyte solution from the metal, therefore restoring corrosion protection. Linseed oil [84, 90, 92, 93] and tung oil [85] have comprehensibly demonstrated their capability to form a protective layer on the metal surface by their oxidation in contact with air, and thus no catalyst is needed. Isocyanates also have proved their effectiveness as catalyst-free healing agents due to the higher reactivity with water [87, 89] (Figure 7.25).

One major disadvantage overlooked by all the studies so far mentioned is that once the system has recovered itself from the damage by the released healing

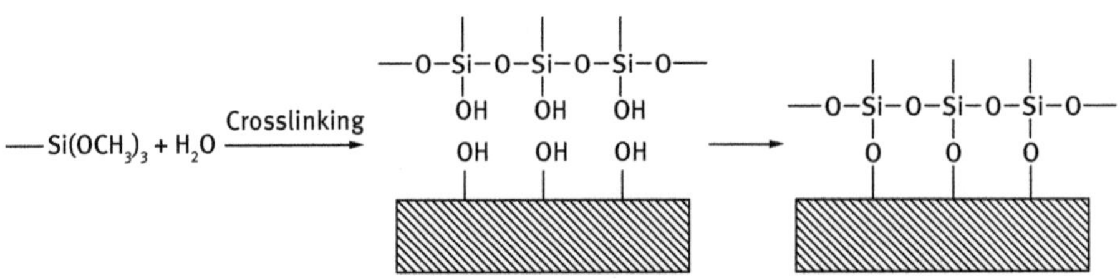

Fig. 7.24: Reaction of silanes agent with water from ambient medium. The surface becomes passive and hydrophobic resulted by binding of silanes with the metal.

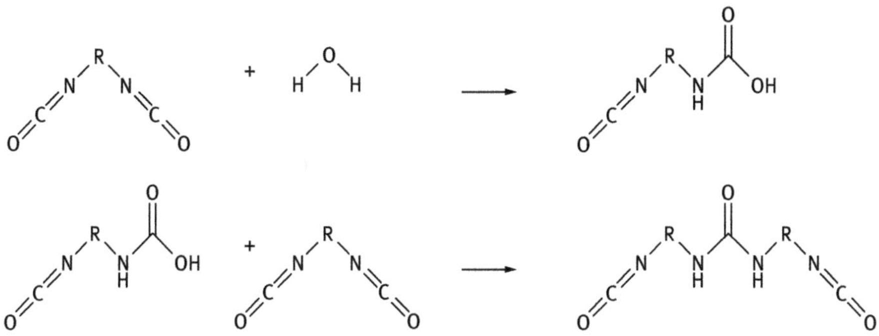

Fig. 7.25: Reaction of diisocyanate monomer with water.

agent,corrosive species may still be trapped between the healed coating layer and the substrate. Therefore, further electrochemical reactions can take place at the coating/ substrate interfaces and that eventually may lead to disbonding of the coating itself. This shortcoming can be addressed by the combination of self-healing and self-releasing microcapsules containing inhibitors. While the active component in ruptured embedded self-healing microcapsules is released into damaged coating (Figure 7.15), allowing the recovery of the barrier property, the released inhibitor forms a protective layer due to the reaction of the inhibitor and the entrapped aggressive species (Figure 7.27).

7.4.2 Self-releasing inhibitor coatings

As discussed so far, there are essentially three suitable routes to protect the metal for the external environment and therefore hinder its degradation:

i. by using the electrochemical properties of the metal, i.e. cathodic and anodic protection;
ii. by isolating the structures by means of an appropriate coating;
iii. by the incorporation of corrosion inhibitors into coating formulations in order to confer active protection when the coating barrier property fails.

As has been stated above, regardless of the best coating system designed it is just a matter of time before oxygen, water and others species diffuse through the coating towards the metal substrate. The inhibitor protection action is achieved via the formation of a passive layer on the metal surface. For isolating the metal from the external environment, chromate conversion coatings (CCCs) were widely used as corrosion inhibitors, providing additional long-term corrosion protection for both the bare and coated metal by not only providing a thick barrier of oxide/hydroxide layer, but also improving the adhesion between the metal and the primer [94, 95].

Furthermore, CCCs can provide self-healing properties to the system in the case of chemical and mechanical damage. In fact, hexavalent chromium [Cr(VI)] contained in certain types of CCCs can leach from the coating into the surrounding solution and migrate to the coating damage, where it may prevent further corrosion activity and therefore act as a self-healing material [96–98].

However, chromate compounds are highly toxic and carcinogenic due to their high potential to oxidase organic molecules [99, 100]. As a result of these drawbacks they were banned in Europe in 2007 and worldwide shortly afterwards by the Environmental Protection Agency (EPA) [101, 102].

Because of this, scientists have focused their research on finding more environmentally friendly, but nevertheless effective organic corrosion inhibitors in replacement of the chromate-based inhibitors in industry. Organic inhibitors, e.g. salicylaldoxime, tolyltriazole (TTA), benzotriazole (BTA), etc. have shown great potential as corrosion inhibitors of commonly used aerospace aluminum alloys [103]. However, the addition of corrosion inhibitors to coating formulations is not straightforward. Due to the functional groups present on their structure they can react with the coating material, causing coating degradation and/or inhibitor deactivation.

Another issue that must be taken into consideration during the formulation of a corrosion-resistant coating is the solubility of the inhibitors themselves. In fact, corrosion inhibitors can only provide the effectiveness if their solubility is in the right range, the so-called "window of solubility" [18, 55–58]. Corrosion inhibitors with too low solubility cause a lack of sufficient inhibiting ions available to diffuse from the matrix to the metal surface to protect. Conversely, upon higher solubility inhibitors may leach out from the coating and are thereby lost, providing the ability to protect the substrate for only a limited period of time. The leaching is the result of the diffusion of the inhibitors through the coating, and being replaced by water molecules that in turn diffuse into the coating from the coating surface, lead to osmotic blistering (Figure 7.10) [32–34]. All these factors can modify the leaching rate, i.e. the amount of inhibitor that is available for corrosion control, thus they cannot be used directly as pigments in the organic coating.

One approach to extend the life-span of the coating is to incorporate high amounts of inhibitors, but this may not be economically, environmentally and practically viable. A strategy that has attracted considerable interest among corrosion scientists in the last decade is the encapsulation of corrosion inhibitors in hosting systems into the coating matrix. These containers will store the inhibitors, preventing the inhibitors themselves from interacting with the matrix, where they can act as slow release pigments similar to chromate pigments, extending the corrosion protection effectiveness of the coating. Sometimes – either due to their toxicity or otherwise potential harm when used alone – it is preferable to encapsulate these types of inhibitors.

Therefore, the encapsulation process provides a potential environmentally acceptable and controlled means of prolonging inhibitor activity in coatings. In addition, controlling their release leads to a reduction in the threshold levels of inhibitors commonly used to prevent the corrosion process. Once encapsulated, the inhibitor-loaded microcapsules can be incorporated in the coating in the same manner by which the classic inhibitors are used. In order to solve these shortcomings, Khramov et al. proposed to encapsulate organic inhibitors such as mercaptobenzothiazole (MBT) and mercaptobenzimidazole

(MBI) in cyclic oligosaccaharides and then incorporates them in hybrid organo-ceramic coatings [104, 105]. The study compared the anticorrosive performance of the coating applied on aluminum alloy AA 2024-T3, with the corrosion inhibitors directly dispersed in the matrix with the encapsulated one. The coating were characterized by means of EIS, an electrochemical technique widely used to test the anticorrosion properties of coatings over a long exposure time (Figure 7.26) [10, 106–108]. The results showed that for short immersion times the coatings displayed similar behaviors. However, upon higher immersion times the nonencapsulated inhibitors showed a remarkable decrease whereas the encapsulated inhibitors exhibited stable anticorrosion properties, as the result of the slow and long-term releasing rate of the inhibitors in coating.

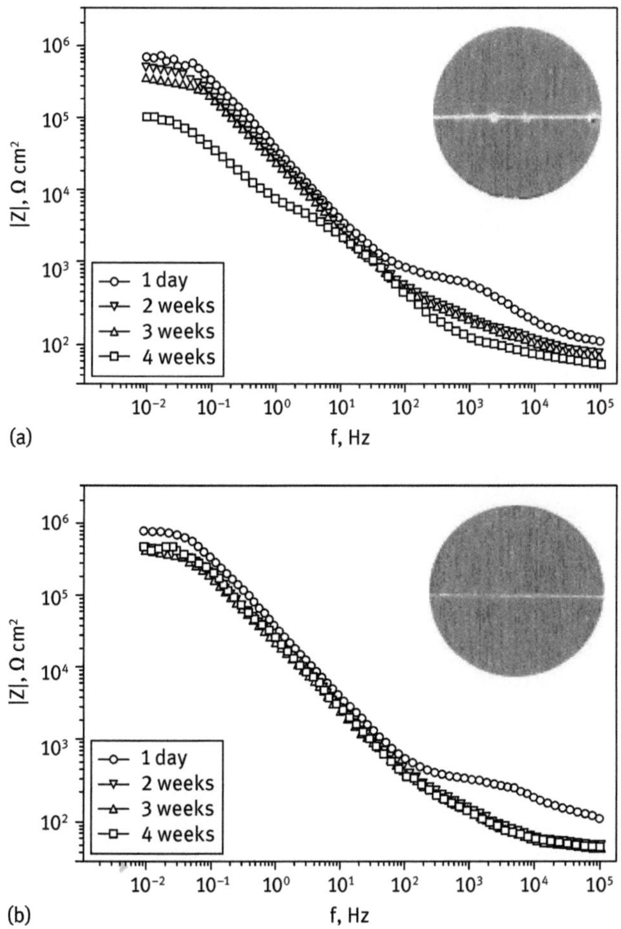

Fig. 7.26: EIS plots for the scribed coatings at different immersion times in dilute Harrison's solution (a) without inhibitor and (b) with MBI/β-cyclodextrin complex (adapted from [105] and reprinted with permission from Elsevier).

The cyclodextrin is a truncated cup-shape structure with a hydrophilic exterior surface and a hydrophobic interior cavity in which the inhibitor is allocated, forming an inclusion complex with the cyclodextrin itself. Although the method proposed by Khramov and co-workers back in 2004, strictly speaking, does not involve the encapsulation of the inhibitors within closed structures (capsule), nevertheless the novelty of this study clearly shows the effectiveness of this approach as a suitable route to ensure long-lasting release rate of the inhibitors within coating thereby, opening new opportunities for increase corrosion protection of the substrate.

Kuang and co-workers encapsulated the inhibitor thiurea within microcapsules by using thermal phase separation method together with vigour Finestran drying bath method. The microcapsules were comprised a shell polyvinyl alcohol (PVA) sealed by alginate natrium [58]. The electrochemical results showed that the releasing times of the encapsulated inhibitor almost tripled and the corrosion rate, initially decreased with the release of the microcapsules, remained stable after the inhibitor was released completely upon higher immersion times.

Another interesting approach was proposed by Yang and van Ooji [103]. In an attempt to decrease the negative effects associated with highly soluble corrosion inhibitors, they used a plasma polymerization technique to encapsulate triazole inhibitor particles in a one ultrathin polymer layer comprised either of a single shell of perfluorohexane or of two ultrathin polymer layers comprised of an interior shell of perfluorohexane and an exterior shell of pyrrole. The authors found out that the inhibitor release and degradation performances of capsule were closely related to the shell material.

High hydrophobic capsules may render it difficult to disperse the capsule into paints, however high hydrophilicity leads to an increase in adsorption of water and formation of blisters. Plasma polymerization is a technique generally used to change the hydrophilicity, hydrophobicity as well as other anticorrosion properties of materials [109, 110]. The inhibitors were then applied on aluminum copper alloy as a water-based epoxy coating and immersed in 3.5% NaCl solution. The result shown that the introduction of the outer layer of pyrrole enhanced the compatibility of the capsules with the coating and that the inhibitors were able to reduce the corrosion rate of the metal by acting as cathodic inhibitors, i.e. suppressing the cathodic reaction. The presence of the second layer increased the thickness of the microcapsule, allowing the inhibitor to be slowly released by diffusion through the microcapsule to the coating, therefore lowering the corrosion rate of the substrate.

Polyurethanes (PUs) are widely used in coating industries due to their excellent chemical, mechanical and physical properties. However, a direct addition of an inhibitor like MBI or MBT will yield the deactivation of the inhibitors themselves as a result of the reaction between the isocyanate groups of the PU coating and the thiol groups of the inhibitors, through a thiol-isocyanate coupling reaction. Marathe et al. [111] encapsulated 2-MBI and 2-MBT inhibitors in UF microcapsules dispersed in acrylic-based PU coatings in order to enhancing anticorrosive properties of PU coatings. The anticorrosive properties of the self-releasing inhibitor PU coating were carried out by exposing the specimens in 0.5 M HCl and 3.5% NaCl solutions for up to 240 h. The electrochemical and immersion results showed that self-releasing PU coatings performed better than the nonencapsulated coating microcapsules. Moreover, the anticorrosion property of the coating increased increasing the loading of UF microcapsules in the coating.

7.4.2.1 pH-sensitive self-release microcapsules

There is no doubt that self-healing materials would offer enormous possibilities, in particular for the applications with long-term reliability in severe environments. However, it would be preferable to detect corrosion processes at an early stage, so that action can be taken to prevent and/or to correct the problem to avoid failures. Over time, many corrosion sensors or probes based on electrochemical or nonelectrochemical theory have been developed and applied to corrosive environments in industry [112–115]. One suitable route to prevent the failure of the structure is the using of a corrosion control coating filled with pH-sensitive carriers [60–62, 116–121]. The main cathodic reaction in the presence of oxygen involves the reduction of oxygen with formation of hydroxide ions that lead an increase in the local pH (eq. 7.5). In this sense, these pH-sensitive carriers loaded with inhibitors act at early corrosion stages by releasing the inhibitor *on demand* into the system, resulting in pH changes, stabilising the local pH gradients and preventing the propagation of the corrosion process. Once the pH returns to a more suitable value, i.e. a pH value in which the capsules are stable, the release of the inhibitor is stopped (Figure 7.27). However, at present the literatures contains only studies related to the use of nanocontainers as pH-trigger release inhibitors, e.g. inhibitor-loaded silica nanoparticles and inhibitor-loaded halloysite nanotubes, and only few of them are related to the use of microcapsules [116–118].

The literature reports some interesting attempts to formulate corrosion control coatings based on pH indicators incorporated in the coating that are able to detect the increased pH (alkaline) associated with the local cathodic activity under the coating, through colour changes by visual inspection [60–63, 119, 122, 123] or fluorescence changes [124–126]. This concept is easy to understand using a simple laboratory demonstration shown in Figure 7.28. Placing an iron nail in agar in which ferricyanide ions ($[Fe(CN)_6]^{3-}$) and phenolphthalein (PhPh) indicator have been previously placed, the ends of the nail will turn blue and the middle of the nail will turn pink. The pH indicators show which part of the exposed nail tends to be acid (anodic) or basic (cathodic). The blue colour (Prussian blue), is the result of the reaction between the ferricyanide ions indicator with the iron ions released during the oxidation reaction (anodic regions, eq. 7.8), while the pink colour is due to the reaction hydroxide ions released during the oxygen reduction reaction (cathodic regions, eq. 7.5) with phenolphthalein. Zhang and co-workers proved the effectiveness of this approach back in 1999 by mixing several color changing pH indicators, namely phenolphthalein (PhPh) or bromothymol blue, to a transparent acrylic-based coating, thereby conferring sensing functionality to

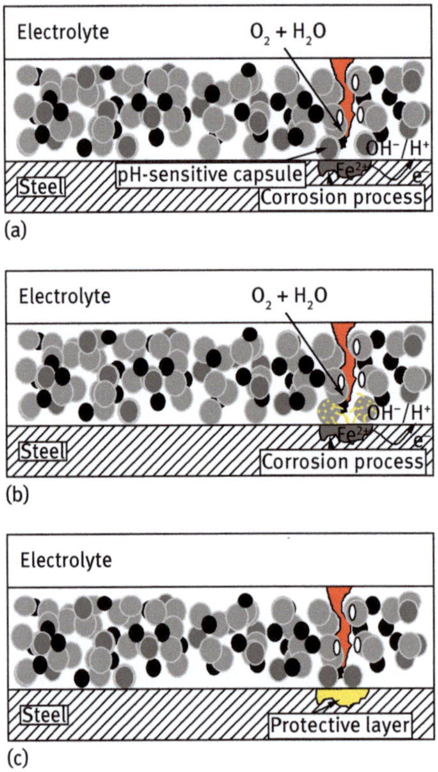

Fig. 7.27: Schematic representation of microcapsule-based on pH-sensitive release releasing agent. (a) Onset corrosion process, (b) respond, (c) formation of a protective layer.

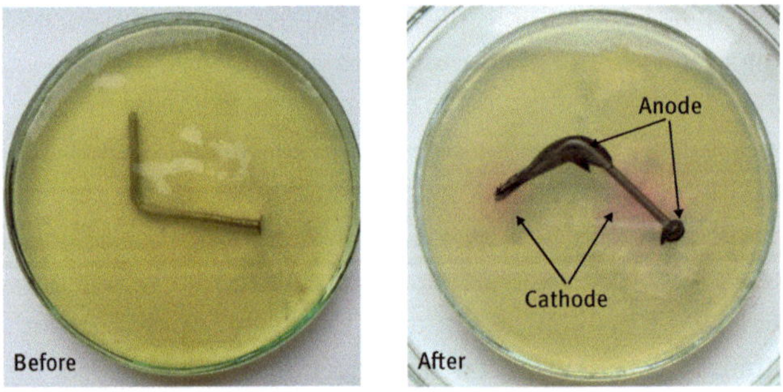

Fig. 7.28: Iron nail in agar gel enriched with ferricyanide ions $[Fe(CN)_6]^{3-}$ and phenolphthalein indicator. The blue color (Prussian blue) indicates the anodic regions while the pink color indicates the cathodic regions.

the coating [122]. At that time this approach showed great promise from an industrial standpoint and soon others scientists followed this interesting route [121]. PhPh is a good pH indicator because of its intense colour that displays at a pH higher than 8.2 (pink) and also that it is colourless at pH values below 8.2, which is useful for the indication of corrosion process and avoiding false-positive corrosion indications [60–62, 119–122, 127]. However, unlike with the inhibitors, some technical challenges associated with the solubility and possible interaction between the indicators and coating system had to be considered. Once again these negative effects were overcome by encapsulating the indicator into containers [60–62, 119, 120].

To date, to the knowledge of the author, the literature reports few successful attempts to discuss the incorporation of encapsulated pH indicators microcapsules in functional coatings used as coating sensing [60–62, 120, 128]. The microcapsules filled with inhibitors described above are actually only inert containers that do not respond to any trigger mechanism but they are simply used to increase the compatibility between the coatings and to reduce the release rate. The main difference between the self-releasing and pH-sensitive microcapsules, essentially, is the release mechanism; in the first the healing contents is released by its diffusion through the microcapsule shell whereas, the latter will release their contents in response to the erosion of the shell as result of hydrolysis reaction catalyzed by local change of pH associated with the corrosion process (Figure 7.29). The chemistry of these approaches resides on the use of PhPh as the pH indicator encapsulated in microcapsules based on the combination of film-forming monomer and pre-polymers such as urea–formaldehyde and melamine–formaldehyde with

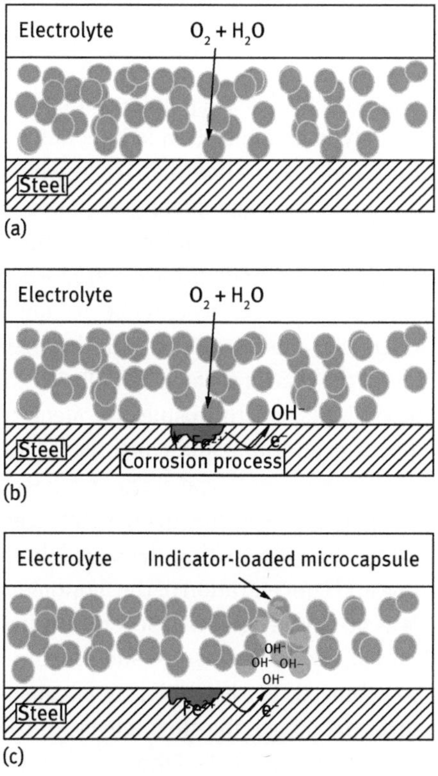

Fig. 7.29: Schematic representation of indicator-loaded microcapsuled coating. (a) Diffusion of water and oxygen, (b) onset corrosion, (c) response.

crosslinking agent such as pentaerythritol tetraki (3-mercaptopropionate) [60–62]. The microcapsule, being sensitive to an alkaline pH, breaks down and releases the indicator.

Maia et al. developed a sensing coating by encapsulating PhPh in silica nanocontainers. The authors discovered that PhPh, instead of being released in to the coating, reacts with the hydroxide anions diffused within the capsule and turns them from colourless to pink [119]. Later, the same authors encapsulated PhPh in polyurea microcapsules dispersed in polyurethane based coating [120]. The corrosion onset ability of the sensing coating was investigated by applying the formulation onto an aluminum sample and then exposing it to 3.5% of NaCl solution. The applied coating successfully detected the early stage corrosion through the formation of pink coloured spots scattered on all over the surface as a result of the reaction of the PhPh with the hydroxide ions generated in the cathodic areas (eq. 7.5). Another interesting aspect of this work is the combination of microcapsules of polyurea and polyurethane matrix. In fact, as both microcapsules and coating share a similar chemical nature, their adoption has allowed for enhancing the compatibility between the shell and matrix, obtaining uniform and homogeneous sensing coating and therefore, increasing the barrier properties of the coating itself.

7.5 Metal/liquid microcapsule composite coatings

Composite coatings can be prepared by mixing microcapsules in a paint formulation, or by electrodeposition (or co-deposition). Electrolytic co-deposition is a technique (Figure 7.15) used to embed small particles into a metal or alloy matrix, to which they impart special properties, in order to produce composite coating. Composite coatings produced by electrolytic co-deposition are broadly used in the automotive and aerospace industries due to their better wear, friction and corrosion properties [66, 67]. In addition to the co-deposition of small particles, it has recently been shown that by embedding encapsulated liquids products into the metallic matrix they can be released gradually, forming a protective film on the matrix surface [66, 68, 70–77] (Figure 7.30b). One of the first attempts to embed liquid microcapsules on metallic matrix dates back to 1994 and was performed by Alexandridou and co-workers. The study focused on the influence of the synthesis parameters, e.g. type of encapsulated, size, distribution etc., of an oil liquid core and polyamide microcapsules obtained by interfacial polymerization as a self-lubricating metallic coating deposited from Watts nickel plating baths. Depending of the type and concentration of the monomers used in the deposition bath, a coverage as high as 40% was obtained. Several studies showed that, based on the type of liquid embedded, several properties such as abrasion, friction, and corrosion resistance can be achieved. For example, by encapsulating corrosion inhibitor liquid, upon local surface damage, the slow release of the liquid inhibitor can either partially repair the damage and/or protect the surface from

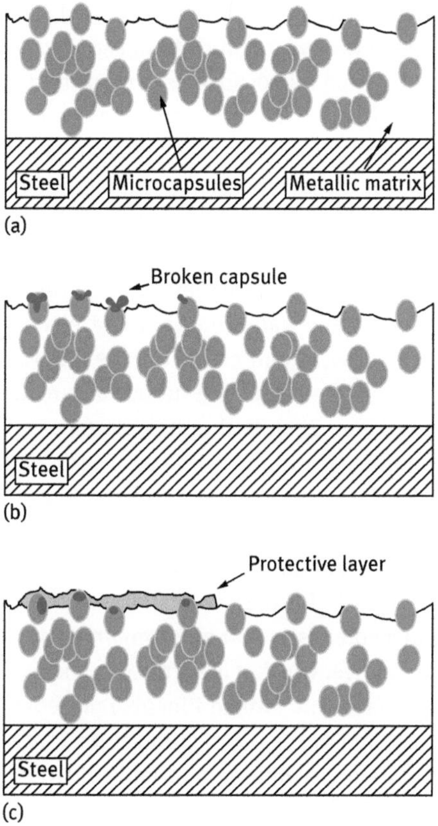

Fig. 7.30: Cracking of microcapsule with formation of the protective layer. (a) Intact coating, (b) broken capsules, (c) protective layer.

corrosion by the formation of passive layer on the metallic surface. Conversely, the release of encapsulated oil liquid products leads to the formation of a low surface energy hydrophobic layer on the metal surface enhancing the corrosion resistance of the coating [71, 75–77] (Figure 7.36d).

This new protective layer can provide long-term protection under a broad range of corrosive conditions either due to chemical stability or excellent water-repellent properties. The literature also reports others' successful attempts to co-deposite encapsulated liquid core microcapsules on others metallic matrix, e.g. nickel [66, 67, 69], tin [73], etc. Masayuki et al. successfully prepared embedded microcapsules containing abietic acid as a liquid core, (i.e. welding agent, incorporated with a tin matrix in order to improve the weldability of the final composite coating [73]) and lube oil-containing microcapsules in Watts nickel plating baths and co-deposited on nickel metallic [69]. Copper is a metal widely used in the chemical and microelectronics industries due to its high electrical conductivities and lower cost than gold or silver as protective layer. However, one of the major disadvantages of copper coating, as we have seen from the Pourbaix diagram, is that in aerated acid solutions copper is prone to corrode. In recent decades, different methods were used to try to increase the corrosion resistance of plated copper coatings, such as surface passivation [68], self-assembly technology [72], and different coatings were used for protection, e.g. by creating hydrophobic surfaces [71, 76]. At present, Cu/liquid microcapsules composite coatings, prepared by electrolytic co-deposition with hydrophobic surfaces, have attracted increasing attention [70–72, 74–77, 129]. In order to improve the corrosion resistance of plating copper coatings, Zhu and co-workers [71, 75–77] synthesized Cu/microcapsule composite coatings using BH-102 hydrophobic agent and 1H-Benzotriazole inhibitor as core and methyl cellulose as shell, by electrolytic co-deposition with metal ions in an acidic copper-plating bath (Figure 7.31). The hydrophobic protective layer and corrosion resistance of the composite was investigated by means of water contact angle, electrochemical and morphological techniques. Morphological and contact angle analysis showed that a thin hydrophobic film was formed on the composite surface as a result of the slow release of the encapsulated liquid (Figure 7.32). Hydrophilic surfaces are characterized by contact angles of 0°, while complete hydrophobicity surfaces display contact angles of 180°. For a good hydrophobic surface, the angle must be around 160°. Figure 7.32 depicts the contact angle at different exposure days; it is clearly visible that the angle increases with an increase of time, resulting in the formation of hydrophobic layers. Moreover, the findings displayed in Figure 7.34 for the immersion test performed in 0.1 M H_2SO_4 solution show that the copper composite coating, compared with a copper coating, exhibits excellent corrosion resistance, which it is ascribed to the hydrophobic nature of the film on the composite surface. The formed film also presented good stability over a long period of time, as suggested by the polarization curve for the composite coating stored in air at 25 °C and by the stable contact angle value of around 160° obtained after a long exposure time (100 days) (Figure 7.33).

Whatever the metal or the type of microcapsules used, the only requirement for the co-deposition of the microcapsules is that the shell must be thick enough to

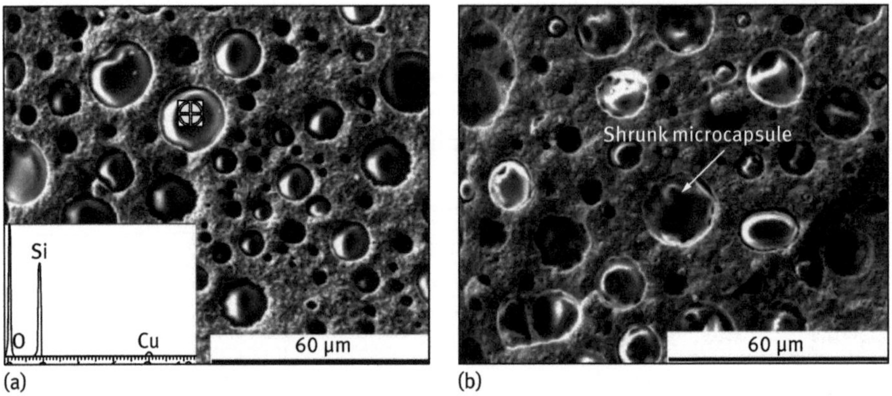

Fig. 7.31: Scanning electron microscope (SEM) images of (a) fresh prepared Cu/liquid microcapsule composite coating and (b) composite stored in air for 30 days at 25 °C (reprinted from [75] with permission from Elsevier).

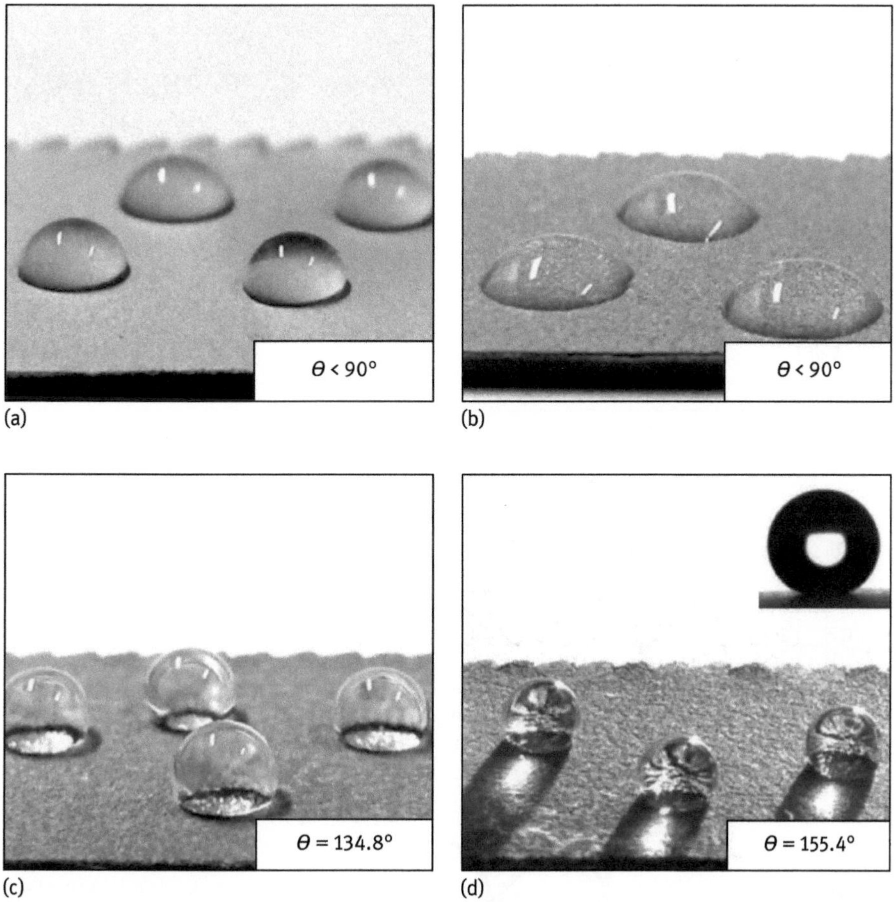

Fig. 7.32: Contact angle of water droplet on (a) pure copper surface, (b) copper composite surface, (c) composite stored in air for 7 days, (d) composite stored in air for 15 days (reprinted from [76] with permission from Elsevier).

withstand the forces generated during the application. They must be resistant to the chemicals agent of the plating bath and temperature. Moreover, by controlling the condition of plating, e.g. current applied, capsule size, composition of the plating bath, plating time and concentration of the microcapsules [67, 77], it is possible to tailor the amount of embedded microcapsules on the metal surface, thus tailoring the properties of the composite. Large microcapsules (>10 µm) are more likely to break during the process and are not easy to co-deposit with metal ions in the plating bath [77, 130]. Furthermore, it was found that composite coatings embedded with smaller microcapsules (2–8 µm) displayed better corrosion resistance compared with larger ones. In fact, large cracked embedded microcapsules act like superficial defects, compromising the stability of the coating with adverse effects from

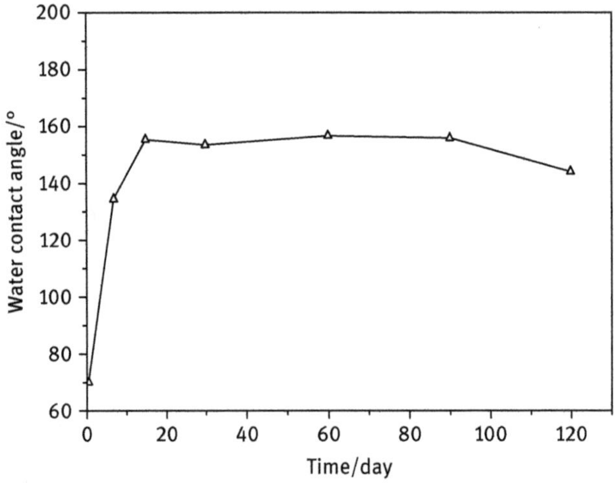

Fig. 7.33: Contact angle plot related to the stability of the hydrophobic protective layer (reprinted from [76] with permission from Elsevier).

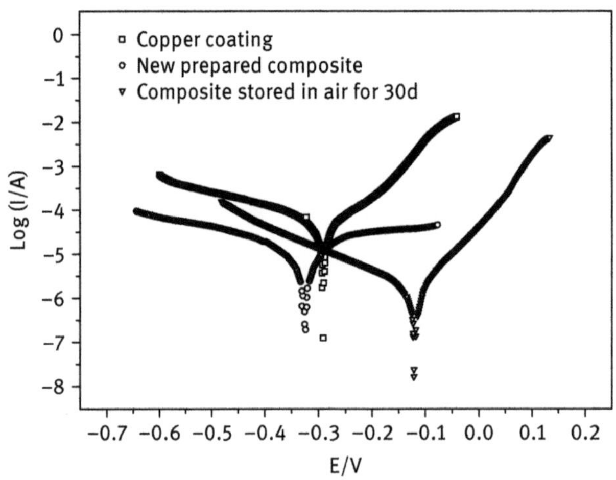

Fig. 7.34: Polarization curves of coating composite and copper coating (reprinted from [75] with permission from Elsevier).

a corrosion resistance standpoint (Figure 7.35) [130]. Moreover, it was found that different liquid-containing microcapsules have different deposition times under the same current density applied, suggesting a direct correlation between the current applied, the co-deposition time and type of microcapsules. As the microcapsules are chargeless particles, Zhu suggested that they may acquire the positive charge by chelation of their hydroxyl group and oxygen of the microcapsule shell (methyl cellulose) with copper ions (Cu^{2+}) and be forced to move towards the cathode by the

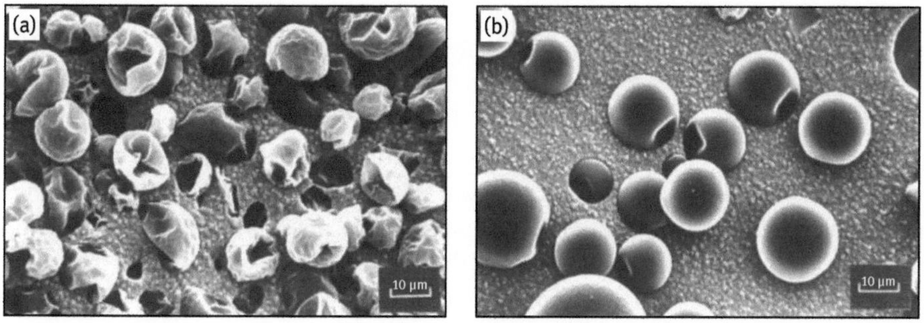

Fig. 7.35: (a) Cracked microcapsules embedded on the metallic surface, (b) large microcapsules partially embedded on the metal surface (reprinted from [66] with permission from Elsevier).

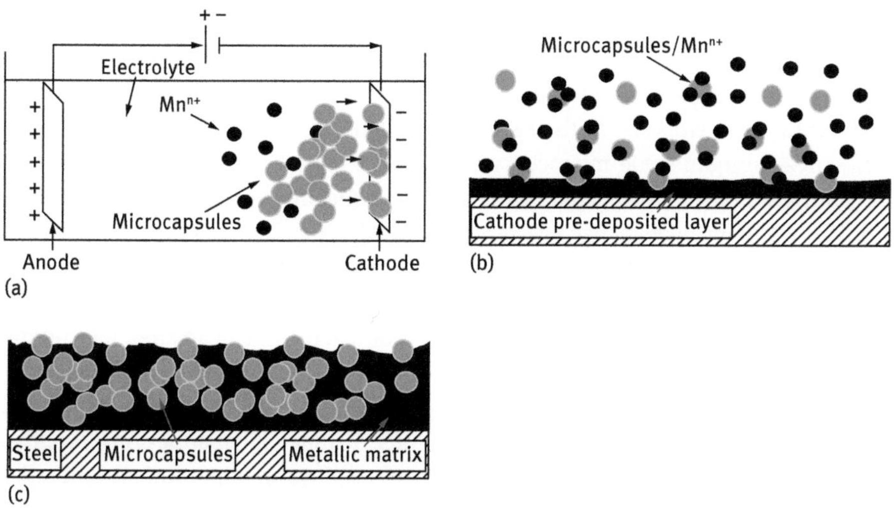

Fig. 7.36: (a) Codeposition of liquid-containing microcapsules on the metallic matrix, (b) ions/microcapsules move towards the cathode, (c) final composite coating.

electric field force and eventually co-deposited on the metal surface Figure 7.36 [77, 130].

7.6 Drawbacks

Whatever the type of microcapsule used, in order to perform their function properly a series of requirements must be fulfilled:

- remains intact during the application and storage;
- raptures readily when needed;
- does not compromise the properties of the coating, e.g. mechanical and adhesion;

- loads with sufficient amount of agent;
- ensures good compatibility with the polymeric matrix.

As during the application of the coasting the microcapsules have to withstand high compression forces, special attention must be paid to the mechanical stability of the microcapsules. If the wall of the microcapsules is not strong enough (thin wall), they may break during the application; conversely, microcapsules with thicker walls do not break easily, preventing the release of the healing agent. Furthermore, they must be stable enough to remain intact for many years in the coating.

Adhesion of the coating is another controversial issue that is far from simple. The microcapsules themselves should not provide any additional adhesive properties to the coating. Nevertheless, it was found that microcapsules negatively affect the adhesion between the metal and the coating. Microcapsules in contact with the metal surface act as interfacial defects, generating low adhesion spots, lowering the adhesion of the coating towards the substrata, defects that increase by increasing the concentration and the size of the microcapsules [83, 85, 90, 131].

Boura and co-workers, in their study of the optimization of process parameters of the urea–formaldehyde microcapsules, found out that the adhesion strength of the coating was strongly affected by the size and concentration of the microcapsules [90]. This was controversial because other authors [61, 62] claimed that the dispersion of the microcapsules into the coating did not affect the adhesion properties of the coating if the size was optimized within the range 0.2–20 μm). Figure 7.37 shows the adhesion strength as a function of the concentration of the microcapsules, also taking into account the size of the microcapsules. It is seems that increasing the concentration of the microcapsules leads to a decrease in the adhesion of the coating. Moreover, at the same concentration, smaller microcapsules display higher adhesion strengths. The amount of agent stored in the microcapsule also plays an important part. In order for the microcapsules to function properly, they must be loaded with enough amount healant. Lower amounts of healant means insufficient material available, either to entirely fill the damage in the case of the self-healing system or to be released in the polymeric matrix in the case of the corrosion inhibitors [132]. However, on one hand, increasing the size of the microcapsules allows an increase of the amount of healant agent, on the other hand, as was discussed above, larger microcapsules lead to a decrease of the adhesion performance of the coating. The amount of healing agent can also be incremented by increasing the concentration of the microcapsules themselves, but as the size, the adhesion strength of the coating also decreases.

Kumar et al. also studied the effect of microcapsules from an adhesion standpoint. In this work [83], they compared the adhesion strength of coatings obtained by applying the microcapsules using two different techniques. In the first one, the classic method, the coating was prepared by mixing the microcapsules directly with the primer and then applying it onto the metal (Figure 7.38). In the second technique, a layer of microcapsules mixed with the primer was sprayed on top of a first layer of

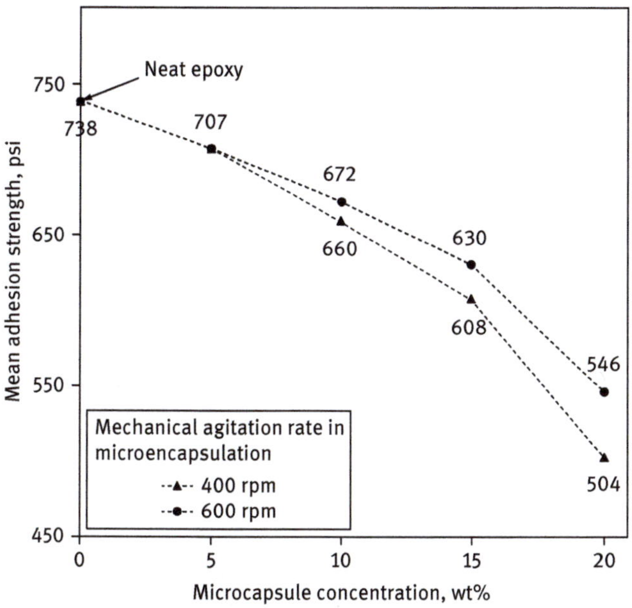

Fig. 7.37: Adhesion strength of neat epoxy coating and coating with embedded microcapsules obtained at different agitation rates (reprinted from [89] with permission from Elsevier).

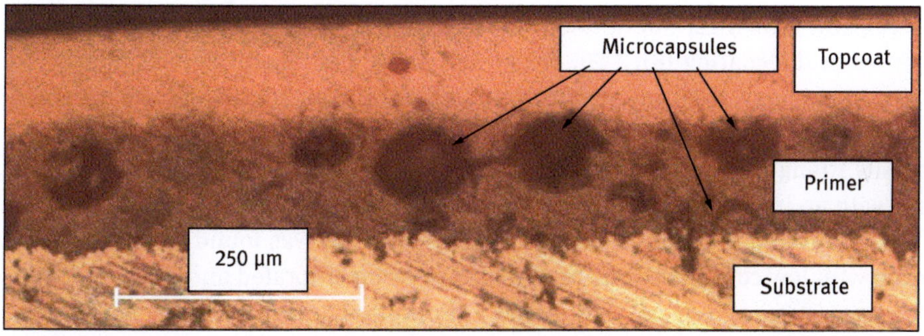

Fig. 7.38: Optical micrograph of cross section of coating prepared by mixing the microcapsule directly with the primer (reprinted from [83] with permission from Elsevier).

primer, previously applied, and then a second lining layer of primer was applied on top (sandwich) (Figure 7.39). The findings showed that the adhesion strength of the coatings prepared by sandwiching the microcapsules between two lining layers was remarkable higher than those prepared with first method. The authors suggested that the presence of the first lining layer of primer prevents the contact between the microcapsules and the metal surface, thereby leaving the primer free to adhere to the metal.

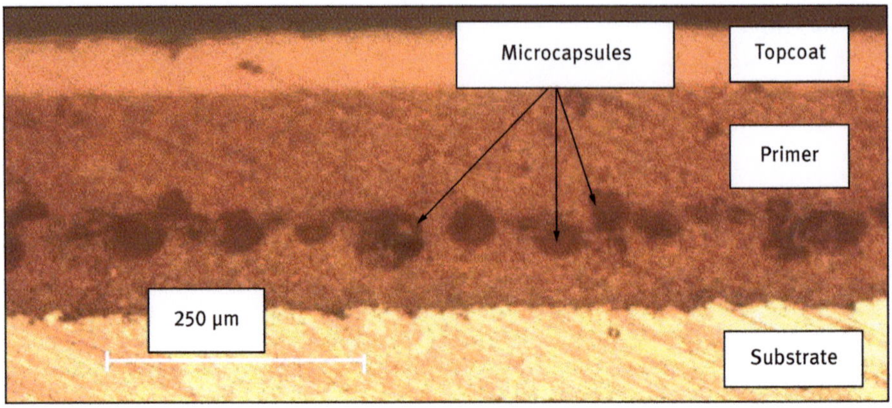

Fig. 7.39: Optical micrograph of a cross section of coating prepared by sandwiching a layer of microcapsules between two layers of primer (reprinted from [83] with permission from Elsevier).

Furthermore, as microcapsules are pigments, there is a limit to the amount of pigment that can be added (critical pigment volume concentration [CPVC]) to the coating. CPVC is a physical transition point at which the properties of an organic coating can change significantly at or near this point [24, 36, 38, 39]. When this value is exceeded, the appearance and behavior of the paint change significantly. Therefore, the addition of the microcapsules into the coating system can have a significant influence, not only on the adhesion but also on the mechanical properties of the coating [81, 133]. Tripathi et al. [134] studied the mechanical properties of epoxy resin matrix incorporated within microcapsules comprised of urea-formaldehyde and melamine-formaldehyde. According to the authors, both tensile strength, modulus, and impact resistance of the matrix was found to decrease with an increasing amount of microcapsule in the formulation. Blaiszik et al. [135] discovered that only a slight change in modulus was found from the control resin with low concentrations of microcapsules incorporated and a size of 1.5 µm. However, Rzeszutko et al. [136] noted a proportional decrease with the concentration with microcapsules of 180 µm. Nevertheless, all agreed that a significant improvement of the toughness of the final product was observed. As Toughness is defined as the ability to absorb energy and undergo extensive plastic deformation without rupturing [137]. Tougher coatings can withstand the impact of tools. The compatibility and the adhesion of the microcapsules with the polymeric matrix is also a crucial aspect that can determine the final quality of the coating and which coating scientists have to consider. Microcapsules should be matched to the coating matrix. A judicious choice of the chemistry of the microcapsules can remarkable increase this compatibility, leading to the formation of uniform and homogeneous continuous film. Tatya and co-workers [131] found out that by using microcapsule of polyurea dispersed in a polyurethane matrix, as the two materials shared similar chemistry, the adhesion of the microcapsules with the matrix was improved. FTIR analysis showed that chemical bonds were formed by the isocyanate group of the polyurethane matrix and the amine groups present on the wall of the microcapsules. Suryanarayana et al. [84] discovered that the surface of the microcapsules themselves could affect the compatibility between the two parts. It was observed that rough microcapsules enhanced the adhesion of the microcapsules with the matrix, allowing the formation of strong bonds with the coating. Furthermore, it was observed that a good adhesion helped breaking the microcapsules when stressed.

The use of a *one-step* self-healing approach, i.e. a self-healing system comprised of only healing agent that does not require the use of catalyst, e.g. linseed oil, tung oil, etc., seems an effective viable way of overcoming the above drawbacks. The presence of a second phase decreases the likelihood of having both phases present at the same time and in the required ratio at the region of the scratch; hindering the restoration of the barrier properties of the coating itself and reducing the concentration of the microcapsules into the system. Figures 7.19 and 7.22 depict the corrosion results obtained after exposing the specimens in an aggressive media by using different *one-step* self-healing approaches, i.e. tung oil [85] and POTS [89], respectively. The difference is mainly due to the type of action by which the healant repairs the damage, i.e. the first tends to reduce the growth of the damage by filling it with new material released from the microcapsule (Figure 7.17), while in the second the damage is isolated by the formation of a passive layer on the metal surface rather than repair. Both figures clearly display no sign of undercutting corrosion, indicating the effectiveness of the healing system used. However, based on the final use of the coating, the choice of correct methodology makes a difference. In fact, even though the second method showed a significant increase of the anticorrosion property of the healed coating, from an aesthetic stand point it is not suitable in a field where the decorative aspect is important, i.e. the automotive industry, on jewelry, etc.

As we can see, all those requirements are strictly connected to each other and can improve or degrade the quality of the coating, and they have to be takes into account in order to develop self-healing coatings through microcapsules with outstanding anticorrosion properties. These adverse effects are leading coating scientists to focus their attention on the development of coatings comprised of pH-sensitive inorganic nanocontainers as intelligent controlled release corrosion inhibitors [63–65, 93, 116–118, 138–140]. The use of smaller capsule will enhance this probability, ensuring higher homogeneous dispersion of the nanocontainers within the coating matrix [141]. Due to their high lateral surfaces, particles are able to delivery significant amount of either healing agents or corrosion inhibitors. In addition, due to their nanoscale size it is possible to distribute them uniformly in the matrix.

7.7 Conclusion

This work is intended to provide the reader with an appreciation of the complexities of this problem and some possible solutions. Some selected recent contributions are reported here and critically assessed.

7.8 References

[1] Borgnakke, C., Sonntag, R. E., Fundamentals of Thermodynamics, John Wiley & Sons, Inc., New York, NY, 2012.
[2] Koretsky, M. D., Engineering and Chemical Thermodynamics, John Wiley & Sons, Inc., New York, NY, 2013.
[3] Fontana, M. G., Corrosion Engineering, McGraw-Hill International, Singapore, 1987.
[4] Trethewey, K. R., Chamberlain, J., Corrosion for Science and Engineering, Longman Group Ltd, Essex, UK, 1995.
[5] Uhlig, H. H., Corrosion & Corrosion Control, John Wiley & Sons, Inc., New York, NY, 1971.
[6] Roberge, P. R., Handbook of Corrosion Engineering, McGraw-Hill Professional, New York, NY, 1999.
[7] Scully, J. C., The Fundamentals of Corrosion, Pergamon Press, Oxford, 1990.
[8] Tait, W. S., An Introduction to Electrochemical Corrosion Testing for Practicing Engineers and Scientists, PairODocs Publications, Racine, WI, 1994.
[9] Bockris, J. M., Reddy, A. K. N., Modern Electrochemistry, Plenum, New York, NY, 1970.
[10] Kelly, R. G., Scully, J. R., Electrochemical techniques in corrosion science and engineering, Marcel Dekker, Inc., New York, USA, 2003.
[11] Jones, D. A., Principles and Prevention of Corrosion, Prentice Hall, Upper Saddle River, USA, 1996.
[12] Pourbaix, M., Atlas of Electrochemical Equilibria in Aqueous Solutions, NACE International, Houston, TX, 1974.
[13] Friend, W. Z., Corrosion of Nickel and Nickel–Base Alloys, John Wiley & Sons, Inc., New York, NY, 1980.
[14] Bunshah, R. F., Handbook of Hard Coatings, Noyes Publications/William Andrew Publishing, LLC, Norwich, NY, 2001.
[15] NACE. Materials Performance, Special Issue, Houston, Texas, USA, July, Jointly with C. C. Technologies and FHWA, 2002.
[16] (FHWA) FHA Office of the Infrastructure and Development, Report FHWA-RD-01-156, 2001.
[17] Petrucci, R. H., Herring, F. G., Madura, J. D., Bissonnette, C., General Chemistry, Principles and Modern Applications, Pearson Canada Inc., Toronto, 2011.
[18] Sastri, V. S., Green Corrosion Inhibitors, Theory and Practice, John Wiley and sons, Inc., Hoboken, NJ, 2011.
[19] Leidheiser, H., Mechanism of corrosion inhibition with special attention to inhibitors in organic coatings, J Coat Tech **53** (1981) 29–39.
[20] Deslouis, C., Lafont, M. C., Pebere, N., You, D., Corrosion inhibition of pure iron in neutral solutions by electrochemical techniques, Corros Sci **34** (2004) 1567–1579.
[21] Schweitzer, P. A., Corrosion of Lining and Coating, CRC Press, Boca Raton, 2007.
[22] Mackhlouf, A. S. H., Handbook of smart coatings for materials selection, Woodhead Publishing, Cambridge, 2014.
[23] Marcus, P., Corrosion Mechanisms in Theory and Practice, Marcel Dekker, Inc., New York, NY, 2002.
[24] Forsfren, A., Corrosion Control Through Organic Coatings, CRC Press, Boca Raton, 2006.

[25] Ceres, L., Vargas, T., Herrera, L., Influence of pitting and iron oxide formation during corrosion of carbon steel in unbuffered NaCl solutions, Corros Sci **51** (2009) 971–978.
[26] Cheng, Y. F., Wilmott, M., Luo, J. L., The role of chloride ions in pitting of carbon steel studied by the statistical analysis of electrochemical noise, Appl Surf Sci **152** (1999) 161–168.
[27] Morcillo, M., Underfilm corrosion of steel induced by saline contaminants at the metal/paint nterface, J Oil Colour Chem As **73** (1990) 24–28.
[28] Evans, U. R., The Corrosion & Oxidation of Metals, Butler & Tanner, London, 1960.
[29] Hare, C. H., Barrier coatings, Journal of Protective Coatings & Linings **15** (1998) 17–37.
[30] Ali, M., Ali, B., Handbook of Industrial Chemistry–Organic Chemicals, McGraw-Hill Companies, Inc., New York, 2005.
[31] Mayne, J. E., O., The mechanism of the protective action of an unpigmented film of polystyrene, J Oil Colour Chem As **32** (1949) 481–487.
[32] Brunt, N. A., Blistering of paint layers as an effect of swelling by water, J Oil Colour Chem As (1964) 31–42.
[33] Gay, P. J., Blistering of paint films on metal, J Oil Colour Chem As **32** (1949) 488–498.
[34] Greenfield, D., Scantlebury, J. D., Blistering and delamination processes on coated steel, J Corros Sci Eng 2000, 2.
[35] Kobayashi, A., Kuroda, T., Kimura, H., Inoue, A., Effect of spraying condition on property of Zr-based metallic glass coating by gas tunnel type plasma spraying, Mater Sci Eng B **173** (2010) 122–125.
[36] Ulrich, D. A., Users's guide to powder coating, SME, Michigan, 1993.
[37] Utech, B. A., guide to high-performance powder coating, SME, Michigan, USA, 2002.
[38] Lambourne, R., Strivens, T. A., Paint and surface coatings, theory and practice, Woodhead Publishing Ltd, Cambridge, 1999.
[39] Wicks, Z. W., Jones, J. F. N., Pappas, S. P., Organic coatings, science and technology, Volume I, film formation, components and apperance, John Wiley & Sons, Inc., Hoboken, NJ, 1992.
[40] Hung, W., Chang, K. C., Chang, Y. H., Yeh, J. M., Advanced anticorrosive coatings prepared from polymer–clay nanocomposite materials, in, Dr. Boreddy Reddy (editor), Advances in Nanocomposites–Synthesis, Characterization and Industrial Applications, RijekaCroatia: InTech., DOI: 10.5772/15355, (accessed 31 October 2014 at: http://www.intechopen.com/books/advances-in-nanocomposites-synthesis-characterization-and-industrial-applications/advanced-anticorrosive-coatings prepared-from-polymer-clay-nanocomposite-materials).
[41] Dry, C. M., Sottos, N., Passive smart self-repair in polymer matrix composite materials, Smart Struct Mat **1916** (1993) 438–444.
[42] White, S. R., Sottos, N. R., Geubelle, P. H., Moore, S., Kessler, M. R., Sriam, S. R., et al., Autonomic healing of polymer composites, Nature **409** (2001) 794–797.
[43] Zhao, Y., Zhang, W., Liao, L-p., Wang, S-j., Li, W-j., Self-healing coatings containing microcapsule, Appl Surf Sci **258** (2012) 1915–1918.
[44] García, S. J., Fischer, H. R., White, P. A., et al., Self-healing anticorrosive organic coating based on an encapsulated water reactive silyl ester, synthesis and proof of concept, Prog Org Coat **70** (2011) 142–149.
[45] Aguilar, M. R., Román, J. S., Smart Polymers and their Applications, Woodhead Publishing Limited, Cambridge, 2014.
[46] van der Zwaag, S., Self Healing Materials, An Alternative Approach to 20 Centuries of Material Science, Springer, Dordrecht, The Netherlands, 2007.
[47] Gosh, S. K., Functional Coatings, Wiley-VCH Verlag Gmbh & Co., Germany, 2006.
[48] 1st International Conference on Self-Healing Materials, Delft, Netherlands, 18–20 April 2007. Accessed 31 October 2014, at: http://tudelft.nl/en/current/latest-news/article/detail/eerste-internationale-conferentie-over-zelfherstellende-materialen-1/).

[49] 2nd International Conference on Self-Healing Materials, Chicago, Illinois USA June 2 – July 1 2009, Accessed 31 October 2014, at: http://www.springer.com/physics? SGWID=0–10100.-2–723509.-0).
[50] 3rd International Conference on Self-healing Materials, Bath, UK, 27–29 June 2011, Accessed 31 October 2014, at: http://www.bris.ac.uk/composites/events/2011/3icshm.html).
[51] 4th International Conference on Self-healing Materials, Ghent, Belgium, 16–20 June 2013, Accessed 31 October 2014 at: http://www.icshm2013.be/index.html.
[52] Shu, B., Yu, W., Zhao, Y., Liu, X., Study on microencapsulation of lycopene by spray-drying, J Food Eng **76** (2006) 664–669.
[53] Benita, S., Microencapsulation – Methods and Industrial Applications, CRC Press, Taylor & Francis Group, Boca Raton, FL, 2006.
[54] Bruschi, M. L., Cardoso, M. L .C., Lucchesi, M. B., Gremiao, M. P. D., Gelatin microparticles containing propolis obtained by spray-drying technique: preparation and characterization, Int J Pharm **264** (2003) 45–55.
[55] Brooman, E. W., Modifying organic coatings to provide corrosion resistance: Part II—Inorganic additives and inhibitors, Metal Finishing **100** (2002) 42–53.
[56] Sinko, J., Challenges of chromate inhibitor pigments replacement in organic coatings, Prog Org Coat **42** (2001) 267–282.
[57] Howard, R. L., Zin, I. M, Scantlebury, J. D., Lyon, S. B., Inhibition of cut edge corrosion of coil-coated architectural cladding, Prog Org Coat **37** (1999) 83–90.
[58] Kuang, F., Shi, T. H., Wang, J., Jia, F., Microencapsulation technology for thiourea corrosion inhibitor, J Solid State Electrochem **13** (2009) 1729–1735.
[59] Mansfeld, F., Hsu, C., Sun, Z., Ornek, D., Wood, T., Technical Note: Ennoblement—a common phenomenon? Corrosion **58** (2002) 187–91.
[60] Li, W., Calle, L. M., Micro-encapsulation for corrosion detection and control, Proceedings of the First International Conference on Self Healing Materials, 18–20 April 2007, Noordwijk aan Zee, The Netherlands. Accessed 31 October 2014, at: http://extras.springer.com/2007/978–1.-4020–6250.-6/documents/68.pdf
[61] Calle, L. M., Li, W., Coatings and methods for corrosion detection and/or reduction, 2010, USB, 2010.US 7,790,225 B1.
[62] Li, W., Calle, L. M., Controlled release microcapsules for smart coatings, in, NACE Corrosion 2007, Paper 07228, Nashville, TN; March 2007.
[63] Chen, T., Fu, J., An intelligent anticorrosion coating based on pH-responsive supramolecular nanocontainers, Nanotechnology **23** (2012) 505–705.
[64] Falcón, J. M., Batista, F. F., Aoki, I. V., Encapsulation of dodecylamine corrosion inhibitor on silica nanoparticles, Electrochim Acta **124** (2014) 109–118.
[65] Snihirova, D., Lamaka, S. V., Montemor, M. F., "SMART" protective ability of water based epoxy coatings loaded with CaCO3 microbeads impregnated with corrosion inhibitors applied on AA2024 substrates, Electrochim Acta **83** (2012) 439–447.
[66] Alexandridou, S., Kiparissides, C., Fransaer, J., Celis, J. P., On the synthesis of oil-containing microcapsules and their electrolytic codeposition, Surf Coat Tech **71** (1995) 267–276.
[67] Kentepozidou, A., Kiparissides, C., Kotzia, F., Kollia, C., Spyrellis, N., Nickel/microcapsule composite electrocoatings; The synthesis of water-containing microcapsules and preparation of the coatings, J Mater Sci **31** (1996) 1175–1181.
[68] Burke, L. D., Murphy, M. A., Multilayer hydrous oxide growth on copper in base and its correlation with previously reported electrocatalytic data for this electrode system, J Solid State Electrochem **5** (2001) 43–49.

[69] Itagaki, M., Shitanda, I., Watanabe, K., Koyano, H., Oil-containing microcapsule composite nickel coating—preparation of microcapsules and confirmation of co-deposition, J Surface Finishing Soc Jap **54** (2003) 230–234.
[70] Liqun, Z., Wei, Z., Feng, L., He, Y., Electrodeposition of composite copper/liquid-containing microcapsule coatings, J Mater Sci **39** (2004) 495–499.
[71] Liu, T., Yin, Y., Chen, S., Chang, X., Cheng, S., Super-hydrophobic surfaces improve corrosion resistance of copper in seawater. Electrochim Acta **52** (2007) 3709–3713.
[72] Ma, H., Chen, S., Niu, L., et al., Studies on electrochemical behavior of copper in aerated NaBr solutions with schiff base, N,N′o-phenylen-bis(3-methoxysalicylidenimine). J Electrochem Soc **148** (2001) B208-B16.
[73] Masayuki, I., Isao, S., Kunihiro, W, Composite coating of tin and microcapsule containing solder flux, Surf Technol **54** (2003) 599–604.
[74] Wang, J., Chen, L., He, Y., Preparation of environmental friendly coatings based on natural shellac modified by diamine and its applications for copper protection, Prog Org Coat **62** (2008) 307–312.
[75] Xu, X., Liu, H., Li, W., Zhu, L., A novel corrosion self-protective copper/liquid microcapsule composite coating, Mater Lett **65** (2011) 698–701.
[76] Xu, X., Zhu, L., Li, W., Liu, H., A variable hydrophobic surface improves corrosion resistance of electroplating copper coating, Appl Surf Sci **257** (2011) 5524–5528.
[77] Xu, X-q., Zhu, L-q., Li, W-p., Liu, H-c., Microstructure and deposition mechanism of electrodeposited Cu/liquid microcapsule composite, T Nonferr Metal Soc **21** (2011) 2210–2215.
[78] Cho, S. H., Andersson, H. M., White, S. R., Sottos, N. R., Braun, P. V., Polydimethylsiloxane-based self-healing materials, Adv Mater **18** (2006) 997–1000.
[79] Cho, S. H., White, S. R., Braun, P. V., Self-healing polymer coatings, Adv Mater **21** (2009) 645–649.
[80] Rule, J. D., Brown, E. N., Sottos, N. R., White, S. R., Moore, J. S., Wax-protected catalyst microspheres for efficient self-healing materials, Adv Mater **17** (2005) 205–208.
[81] Brown, E. N., White, S. R., Sottos, N. R., Microcapsule induced toughening in a self-healing polymer composite, J Mater Sci **39** (2004) 1703–1710.
[82] Blaiszik, B. J., Caruso, M. M., McIlroy, D. A., Moore, J. S., White, S. R., Sottos, N. R., Microcapsules filled with reactive solutions for self-healing materials, Polymer **50** (2009) 990–997.
[83] Kumar, A., Stephenson, L. D., Murray, J. N., Self-healing coatings for steel, Prog Org Coat **55** (2006) 244–253.
[84] Suryanarayana, C., Rao, K. C., Kumar, D., Preparation and characterization of microcapsules containing linseed oil and its use in self-healing coatings, Prog Org Coat **63** (2008) 72–78.
[85] Samadzadeh, M., Boura, S. H., Peikari, M., Ashrafi, A., Kasiriha, M., Tung oil: an autonomous repairing agent for self-healing epoxy coatings, Prog Org Coat **70** (2011) 383–387.
[86] Huang, M., Yang, J., Facile microencapsulation of HDI for self-healing anticorrosion coatings, J Mater Chem **21** (2011) 11123–11130.
[87] Huang, M., Yang, J., Salt spray and EIS studies on HDI microcapsule-based self-healing anticorrosive coatings, Prog Org Coat **77** (2014) 168–175.
[88] Latnikova, A., Grigoriev, D. O., Hartmann, J., Mohwald, H., Shchukin, D. G., Polyfunctional active coatings with damage-triggered water-repelling effect, Soft Matter **7** (2011) 369–372.
[89] Huang, M., Zhang, H., Yang, J., Synthesis of organic silane microcapsules for self-healing corrosion resistant polymer coatings, Corros Sci **65** (2012) 561–566.
[90] Boura, S. H., Peikari, M., Ashrafi, A., Samadzadeh, M., Self-healing ability and adhesion strength of capsule embedded coating—Micro and nano sized capsules containing linseed oil, Prog Org Coat **75** (2012) 292–300.

[91] van Ooij, W. J., Zhu, D., Stacy, M., Seth, A., Mugada, T., Gandhi, J., et al., Corrosion protection properties of organofunctional silanes—an overview. Tsinghua Sci Technol **10** (2005) 639–664.
[92] Nesterova, T., Dam-Johansen, K., Kiil, S., Synthesis of durable microcapsules for self-healing anticorrosive coatings: A comparison of selected methods, Prog Org Coat **70** (2011) 342–352.
[93] Selvakumar, N., Jeyasubramanian, K., Sharmila, R., Smart coating for corrosion protection by adopting nano particles, Prog Org Coat **74** (2012) 461–469.
[94] Zhao, J., Xia, L., Sehgal, A., Lu, D., McCreery, R. L., Frankel, G. S., Effects of chromate and chromate conversion coatings on corrosion of aluminum alloy 2024-T3, Surf Coat Tech **140** (2001) 51–57.
[95] Lu, G. J., Ada, E. T., Zangari, G., Investigations of the effect of chromate conversion coatings on-the corrosion resistance of Ni-based alloys, Electrochim Acta **49** (2004) 1461–1473.
[96] Xia, L., McCreery, R. L., Chemistry of a chromate conversion coating on aluminum alloy AA2024 T3 probed by vibrational spectroscopy, J Electrochem Soc **145** (1998) 3083–3089.
[97] Cramer, D. S., Jr. ASM Handbook Volume 13 A: Corrosion, Fundamentals, Testing and Protection, Materials Park OHASM International, 2003.
[98] Breslin, C. B., Treacy, G., Carroll, W. M., Studies on the passivation of aluminum in chromate and molybdate solutions, Corros Sci **36** (1994) 1143–1154.
[99] Park, R. M., Bena, J. F., Stayner, L. T., Smith, R. J., Gibb, H. J., Lees, P. S., Hexavalent chromium and lung cancer in the chromate industry: a quantitative risk assessment, Risk Anal **24** (2004) 1099–108.
[100] Xianglin, S., Dalal, N. S., Generation of hydroxyl radical by chromate in biologically relevant systems: role of Cr(V) complexes versus tetraperoxochromate(V), Environ Health Perspect **102** (1994) 231–236.
[101] Register, E. F., National emission standards for hazardous air pollutants for source categories: aerospace manufacturing and rework facilities, **60** (1995) 45947.
[102] Directive 2000/53/EC of the European Parliament and the council of 18 September 2000 on end-life vehicles, Official Journal of Eur Communities 2000, L269:34.
[103] Yang, H., van Ooij, W. J., Plasma deposition of polymeric thin films on organic corrosion-inhibiting paint pigments: a novel method to achieve slow release, Plasmas Polymers **8** (2003) 297–323.
[104] Khramov, A. N., Voevodin, N. N., Balbyshev, V. N., Donley, M. S., Hybrid organo-ceramic corrosion protection coatings with encapsulated organic corrosion inhibitors, Thin Solid Films **447** (2004) 549–557.
[105] Khramov, A. N., Voevodin, N. N., Balbyshev, V. N., Mantz, R. A., Sol–gel-derived corrosion-protective coatings with controllable release of incorporated organic corrosion inhibitors, Thin Solid Films **483** (2005) 191–196.
[106] McIntyre, J. M., Pham, H. Q., Electrochemical impedance spectroscopy, a tool for organic coatings optimizations, Prog Org Coat **27** (1996) 201–207.
[107] Carbonini, P., Monetta, T., Nicodemo, L., Mastronardi, P., Scatteia, B., Bellucci, F., Electrochemical characterisation of multilayer organic coatings, Prog Org Coat **29** (1996) 13–20.
[108] Vreijling, M. P. W., Electrochemical characterisation of metallic thermally sprayed coatings, The Hague: TNO Institute of Industrial Technology, 1998.
[109] Lin, Y. S., Chang, C. H., Huang, T. J., Enhanced corrosion protection of cold rolled steel by low-temperature plasma cleaning and plasma polymerization, Surf Coat Tech **200** (2006) 4929–4938.
[110] Okubo, M., Tahara, M., Saeki, N., Yamamoto, T., Surface modification of fluorocarbon polymer films for improved adhesion using atmospheric-pressure nonthermal plasma graft-polymerization, Thin Solid Films **516** (2008) 6592–6597.
[111] Marathe, R. J., Chaudhari, A. B., Hedaoo, R. K., Sohn, D., Chaudhari, V. R., Gite, V. V., Urea formaldehyde (UF) microcapsules loaded with corrosion inhibitor for enhancing the anti-corrosive properties of acrylic-based multi-functional PU coatings, RSC Advances **5** (2015) 15539–15546.

[112] Ilyushchenko, M. A., Danilenko, A. V., Corrosion potentiometric sensors: properties, theory and origin mechanism of pseudo-Nernstian electrode response, Sensor Actuat B Chem **44** (1997) 542–550.
[113] Philip-Chandy, R., Scully, P. J., Thomas, D., A novel technique for on-line measurement of scaling using a multimode optical fibre sensor for industrial applications, Sensor Actuat B Chem **71** (2000) 19–23.
[114] Hilbert, L. R., Monitoring corrosion rates and localised corrosion in low conductivity water, Corros Sci **48** (2006) 3907–3923.
[115] Mak, D. K., Gauthier, J., Ultrasonic measurement of longitudinal and shear velocities of materials at elevated-temperatures, ultrasonics **31** (1993) 245–249.
[116] Shchukin, D. G., Mohwald, H., Smart nanocontainers as depot media for feedback active coatings, Chem Commun **47** (2011) 8730–8739.
[117] Shchukin, D. G., Zheludkevich, M., Yasakau, K., Lamaka, S., Ferreira, M. G. S., Mohwald, H., Layer-by-layer assembled nanocontainers for self-healing corrosion protection, Adv Mat **18** (2006) 1672.
[118] Shchukin, D. G., Zheludkevich, M., Moehwald, H., Feedback active coatings based on incorporated nanocontainers, J Mater Chem **16** (2006) 4561–4566.
[119] Maia, F., Tedim, J., Bastos, A. C., Ferreira, M. G., Zheludkevich, M. L., Nanocontainer-based corrosion sensing coating, Nanotechnology **24** (2013) 415502.
[120] Maia, F., Tedim, J., Bastos, A. C., Ferreira, M. G. S., Zheludkevich, M. L., Active sensing coating for early detection of corrosion processes, RSC Advances **4** (2014) 17780–17786.
[121] El-Nahhal, I. M., Zourab, S. M., El-Ashgar, N. M., Encapsulation of phenolphthalein pH-indicator into a sol-gel matrix, J Disper Sci Technol **22** (2001) 583–590.
[122] Zhang, J., Frankel, G. S., Corrosion-sensing behavior of an acrylic-based coating system, Corrosion **55** (1999) 957–967.
[123] Isaacs, H. S., Adzic, G., Jeffcoate, C. S., 2000 W. R. Whitney award lecture: visualizing corrosion, Corrosion **56** (2000) 971–978.
[124] Li, S-m., Zhang, H-r., Liu, J-h., Preparation and performance of fluorescent sensing coating for monitoring corrosion of A1 alloy 2024, T Nonferr Metal Soc **16** (2006) 159–164.
[125] Gharaibeh, B., Omar, M., Salazar, A. J., Saito, K., Fluorescence emission sensing in coatings: Method for defects detection in coated surfaces of structural elements, Prog Org Coat **58** (2007) 282–289.
[126] Augustyniak, A., Ming, W., Early detection of aluminum corrosion via "turn-on" fluorescence in smart coatings, Prog Org Coat **71** (2011) 406–412.
[127] Engelmann, G., Jobmann, M., Rafler, G., Dextran carbamates—materials for microencapsulation, Ind Crop Prod **20** (2004) 37–48.
[128] Lv, W., Feng, J., Yan, W., Faul., C. F. J., Self-assembly and pH response of electroactive liquid core-tetra(aniline) shell microcapsules, J Mater Chem B **2** (2014) 4720–4725.
[129] Xu, X-q., Guo, Y-h., Li, W-p., Zhu, L-q., Electrochemical behavior of different shelled microcapsule composite copper coatings, Int J Min Metall Mater **18** (2011) 377–384.
[130] Ghosh, S. K., Functional Coatings and Microencapsulation: A General Perspective, in, Functional Coatings, Weinheim: Wiley-VCH Verlag GmbH & Co. KGaA, 2006.
[131] Tatiya, P. D., Hedaoo, R. K., Mahulikar, P. P., Gite, V. V., Novel polyurea microcapsules using dendritic functional monomer: synthesis, characterization, and Its use in self-healing and anticorrosive polyurethane coatings, Ind Eng Chem Res **52** (2012) 1562–1570.
[132] Rule, J. D., Sottos, N. R., White, S. R., Effect of microcapsule size on the performance of self-healing polymers, Polymer **48** (2007) 3520–3529.
[133] Yin, T., Rong, M. Z., Zhang, M. Q., Yang, G. C., Self-healing epoxy composites – preparation and effect of the healant consisting of microencapsulated epoxy and latent curing agent, Compos Sci Technol **67** (2007) 201–212.

[134] Tripathi, M., Rahamtullah, Kumar, D., Rajagopal, C., Kumar Roy, P., Influence of microcapsule shell material on the mechanical behavior of epoxy composites for self-healing applications, J Appl Polym Sci 2014;131:n/a-n/a.40572.
[135] Blaiszik B. J., Sottos, N. R., White S. R., Nanocapsules for self-healing materials, Compos Sci Technol **68** (2008) 978–986.
[136] Rzesutko, A. A., Brown, E. N., Sottos, N. R., Tensile properties of self-healing epoxy, TAM Technical Report (2003) 1041–1055.
[137] Ebewele, R. O., Polymer science and technology, CRC press, New York, NY, 2000.
[138] Zheludkevich, M. L., Serra, R., Montemor, M. F., Yasakau, K. A., Salvado, I. M. M., Ferreira, M. G. S., Nanostructured sol–gel coatings doped with cerium nitrate as pre-treatments for AA2024-T3 corrosion protection performance, Electrochim Acta **51** (2005) 208–217.
[139] Choi, H., Song, Y. K., Kim, K. Y., Park, J. M., Encapsulation of triethanolamine as organic corrosion inhibitor into nanoparticles and its active corrosion protection for steel sheets, Surf Coat Tech **206** (2012) 2354–2362.
[140] Jafari, A. H., Hosseini, S. M., Jamalizadeh E. investigation of smart nanocapsules containing inhibitors for corrosion protection of copper, Electrochim Acta **55** (2010) 9004–9009.
[141] Kouhi, M., Mohebbi, A., Mirzaei, M., Evaluation of the corrosion inhibition effect of micro/nano-capsulated polymeric coatings: a comparative study by use of EIS and Tafel experiments and the area under the Bode plot, Res Chem Intermediat **39** (2013) 2049–2062.

Karolina Wieszczycka and Katarzyna Staszak

8 Microcapsules in extraction technology

8.1 Introduction

Contamination of surface and groundwater is an important worldwide problem. The sources of the pollutants are mainly human activities, through the wide use of various types of chemical compounds. Because these impurities affect human health and environment, it is very important to look for effective methods of removing them. Several methods already exist for the separation and removal of metals from aqueous solutions, but the most commonly applied are liquid-liquid extraction [1–2], subsequently ion exchange [3–4], and membrane techniques such as reverse osmosis [7–9]. The extraction process is very selective and efficient, but its main disadvantage is the need to use organic diluents, which generate important secondary wastes. Moreover, the relatively high operation cost and complicated operation units significantly restrict its use. The application of ion exchange resins is limited due to their lower selectivity in comparison to the extraction, and as consequence, several problems exist, such as calcium or iron fouling and adsorption of organic, bacterial, and chlorine contaminations. On the other hand, this technique is characterized by easy operating conditions and simple methods of metal recovery as well as resin regeneration. The main problem with the relevance of membrane process is a well-known problem with membrane fouling and its low chemical resistance. Nowadays, many attempts have been made to develop new methods in which not only technological and economic but also environmental constraints are taken into consideration. It is very important to find such technologies that do not generate secondary waste and in which only easy-to-regenerate and recycle materials are used. Some of these new proposals are imprinted resins [5], solvent impregnated polymers [6], and microcapsules, including extractants, which are mostly packed in a column, while the most promising technique is microencapsulation. This solution is a combination of stationary solvent extraction – similar to ion-exchange column and with the use of selective extractant immobilized in solid polymer matrix. Their application is supported by a large interfacial area and, as a consequence, high mass transfer rates, as well as high selectivity. Moreover, production of microcapsules does not require the use of large amounts of organic solvents, and the procedure of metal-loaded microcapsule recovery is relatively simple [7]. The selection of extractant is a key issue from the practical point of view. The property of extractant entrapped in polymer core depends mainly on the separation ability and efficiency of the microcapsules used. Therefore, knowledge about liquid-liquid extraction systems is helpful in choosing the appropriate extractant, as well as in explaining the mechanism of separation.

https://doi.org/10.1515/9783110642070-008

For example, Cyanex 302 (*bis*(2,4,4-trimethylpentyl) monothiophosphinic acid) is recommended for noble metal extraction [8, 9] because of its strong affinity. Thus, it is also proposed to remove palladium [10, 11] or silver [12] using alginate microcapsules with immobilized Cyanex 302. Moreover, this extractant is proposed to remove cadmium(II) ions by liquid-liquid extraction [13] and using microcapsules [14]. Another example is commercial extractant: PC-88A (2-ethylhexyl phosphonic acid mono-2-ethylhexyl ester) applied for recovery and separation of nickel and cobalt in liquid-liquid extraction systems [15], as well as in column separation using microencapsulated extractant [16].

Depending on the extractant and the method of final usage of microcapsules, various microencapsulation techniques and shells have been proposed in the literature, and the most interesting examples (solutions) with their applications in metal separation are presented in the review.

8.2 Extractant encapsulated in polymer shell

The immobilization of compounds having extraction properties in a polymeric matrix has been intensively studied over the last 20 years [17]. This conception was an answer to the classical issue that reactive solvent extraction is ineffective for diluted aqueous solutions. However, the novel material should be more selective than most of the used ion-exchange resins. The literature reported that only two methods were proposed to retain the extractant in a solid polymeric phase without changes in extractant structure. These methods are impregnation (solvent impregnation resins [SIRs]) and microencapsulation. The SIRs can be obtained by incorporation of a liquid extractant into a porous polymer resin by dry or wet impregnation methods [18, 19]. None of the methods used for impregnation provides permanent incorporation of extractant inside pores of the polymer. Therefore, a procedure to prevent the extractant from leaking into the aqueous phase is required, e.g. stabilization by coating or chemical cross-linking [20–21]. Although impregnated resin provides an important opportunity to be immobile without blocking the complexing center, a too small sorption capacity limits the applicability of this type material to a small range of aqueous solutions. Encapsulation of the extracting agent inside a polymer particle is a more promising technique. It is assumed that the use of an efficient and selective extractant encapsulated into the solid polymer matrix enabled combining the advantages of both extractant and the solid support based. At the beginning of the studies, the most important problem to solve was developing a synthesis procedure that allows obtaining microspheres, to select components from which the polymeric shell does not act as a barrier between the extractant and the solution in which metal is dissolved.

8.2.1 Synthesis

In general, macroporous polymer particles are produced by heterogeneous polymerizations using the immiscibility of two or more liquids. Suspension, dispersion, precipitation, multistage, microchannel emulsification, and microfluidic polymerizations are the main techniques to form porous particles [22, 23]. However, in case of extractant containing macroporous copolymer networks, the suspension polymerization technique has been readily used to form porous particles. Using this technique, drops of an organic phase containing monomer, cross-linkers, initiator, and porogen are dispersed in a continuous liquid phase with dissolved surfactant, polymeric stabilizer.

As illustrated in Figure 8.1, the initiation and propagation take place in the monomer droplets. Thus, both monomer solubility in the continuous phase and the interfacial activity of stabilizers are of utmost importance during the polymerization process and can cause delay in capsule formation.

The initiator is thermally decomposed by homolytic cleavage, generating free radicals that initiate polymerization of the vinyl monomers. At the polymerization stage, the extractant is enclosed in the polymer particle, while the diluent or porogen leaves the formed particle acting as a preforming agent. Toluene, xylene, or heptane can be used as a porogen, which influences the polymer bead formation, surface morphology, and swelling ratio [24, 25]. During the polymerization process, organic phase droplets are produced and maintained in suspension, thanks to vigorous mixing during the reaction and the use of a slurry stabilizer, mainly soluble in water. As the reaction progresses, the monomer conversion increases with the viscosity of the organic droplets to give a characteristic solid particle. As the polymer particle formation continues, droplets are subjected to disintegration under the influence of intense mixing and adhesion of partially polymerized particles interacting

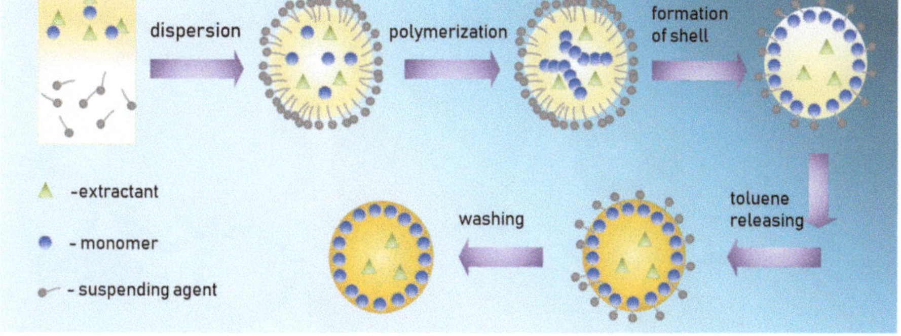

Fig. 8.1: Scheme of encapsulation at suspension polymerization.

with each other. The encapsulation is followed by evaporation of the organic swelling solvent, during which the hydrophobic microsphere shrinks, leaving the extractant trapped within the microsphere. The most important issue in the bead production is the control of the final particle size distribution, which, according to the equilibrium between drop rates and coalescence, depends mainly on the agitator systems, reaction time and temperature, dispersion to continental phase ratio, density and viscosity of both phases, and type and concentration of suspending agent [26–27]. It was also shown that high-speed stirring and, in some cases, ultrasonic dispersion favor the formation of small and relatively uniform-sized beads; however, too high-speed stirring may make the formation of beads impossible or cause the formation of cracked microspheres [28].

As mentioned above, the production of polymer capsules is affected by various parameters (Figure 8.2) [29], but in the case of the polymer beads with entrapped extractant, the effect of organic compounds should be also considered. Besides lipophilic character, the extractant should be soluble in monomers and cross-linkers that guarantee entrapping into a polymer shell. Moreover, this compound should be stable at processing conditions and its boiling point must be high enough to avoid its vaporization.

Also important is the continuous phase composition, in which suspending agents or stabilizers are dissolved. The function of suspending agents or stabilizers consists of the stabilization of the monomer droplets during the polymerization reaction, reducing their coalescence, and thus making possible the synthesis of microcapsules with an adequate particle size distribution and a spherical morphology (Table 8.1).

The enclosing of extractant in the copolymer of styrene and divinylbenzene has been the most frequently considered system. Moreover, commercial sorbent

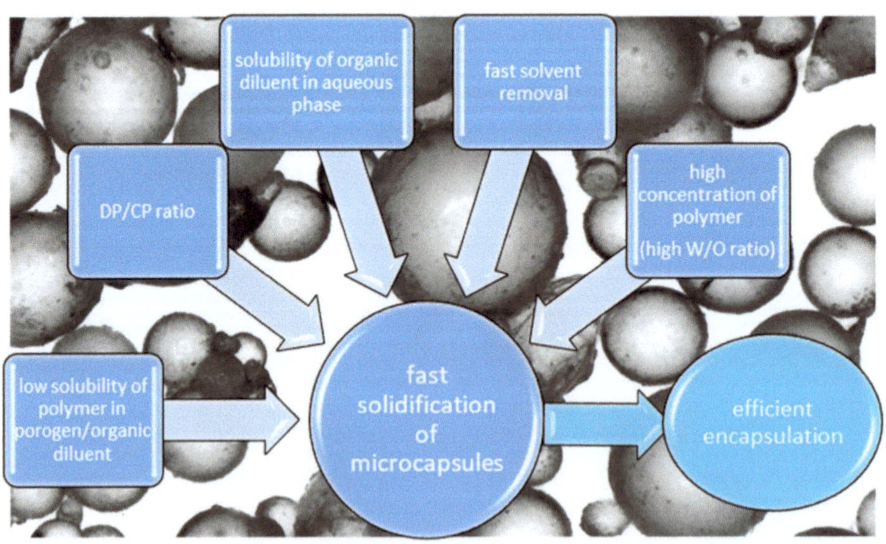

Fig. 8.2: Factors affecting encapsulation efficiency.

Table 8.1: Synthesis conditions and properties of microcapsules formed during suspension polymerization using styrene and divinylbenzene as monomers.

Extractant	Conditions	Properties	Ref.
Without extractant	**Continuous phase:** HEC, water and NaCl (2%). **Discontinuous phase:** ST and DVB in varying ratios AIBN 1% in relation to ST and DVB. Porogen: toluene and mixture of toluene with cyclohexanol and heptane. Phase volume ratio 3:1. Polymerization temperature 70 °C dispersion classical, using high-speed stirring, and using ultrasonic.	Toluene: Surface area: 52.2 m^2/g Pore diameter: 15.6 nm Fraction of adsorption: 25.4% Toluene: cyclohexanol 1:2 (v/v) Surface area: 179.7 m^2/g Pore diameter: 68.7 nm Fraction of adsorption: 117.9% Toluene: heptane 1:2 (v/v) Surface area: 137 m^2/g Pore diameter: 55.2 nm Fraction of adsorption: 68.2% High-speed stirring Percentage (<10 µm), % D_1: 32 µm D_2: 43 µm Ultrasonic-assisted dispersion Percentage (<10 µm), % D_1: 25 µm D_2: 35 µm	[34]
TOA	**Continuous phase:** 0.5 wt% arabic gum with HCl for protonation of TOA **Discontinuous phase:** Benzoyl peroxide 0.9% TOA 4–15% DVB 63–80% ST 20–37% Porogen: toluene Agitation: 600 rpm Polymerization temperature 75 °C for 3 h under a nitrogen atmosphere	Microcapsules are circular and amorphous with tendency to agglomeration (at excess of ST) Composition: TOA 7%, DVB 63.1%, ST 36.9% rnmol TOA/g MC: 0.43 mg TOA/g MC: 152.9 %TOA retained: 55.9 TOA 6.6%, DVB 80%, ST 20% rnmol TOA/g MC: 0.06 mg TOA/g MC: 20.4 %TOA retained: 20.3	[32]
TOA	**Continuous phase:** 2 wt% arabic gum **Discontinuous phase:** DVB: 1.6 mol/L ADVN: n.d. TOA: 2.29 mol/L Porogen: toluene or TOA Agitation: 400 rpm Polymerization temperature 70 °C under a nitrogen atmosphere	Porogen: TOA Yield 97% Average diameter 187 µm Porogen: toluene + TOA Yield 100% Average diameter 159 µm (increases with the increase in DVB concentration in continuous phase). Porous particles on presence toluene.	[33]

Tab. 8.1 *(continued)*

Extractant	Conditions	Properties	Ref.
D2EHPA	**Continuous phase:** 0.5 wt% arabic gum **Discontinuous phase:** DVB: 0.28 mol/L ST: 1.72 mol/L AIBN: 0.02 mol/L D2EHPA: 0.2 mol/L Porogen: xylene Phase volume ratio 5:1 Agitation: 200 rpm Polymerization temperature 70 °C for 5 h under a nitrogen atmosphere	Microcapsules of ca. 1–2 μm with tendency to agglomeration Wider size distribution ca. 1–500 μm	[36]
TOMAC	**Continuous phase:** 0.5 wt.% arabic gum **Discontinuous phase:** DVB: 0.51 mol/L ST: 3.19 mol/L AIBN: 0.04 mol/L TOMAC: 0.2 mol/L Porogen: xylene Phase volume ratio 5:1 Agitation: 200 rpm Polymerization temperature 70 °C for 5 h under a nitrogen atmosphere	Spherical microcapsules of ca. 50 μm diameter with a narrow size distribution – ca. 30–150 μm	[34]
TOMAC	**Continuous phase:** 2 wt% arabic gum and PVA (1:1) **Discontinuous phase:** DVB: 50 wt% ST: 50 wt% BOX: 2.13 g TOMAC: 30 wt% Porogen: toluene Phase volume ratio 20:1. Agitation: 400 rpm Polymerization temperature 80 °C for 5 h under a nitrogen atmosphere	$dpn_{0.5}$ μm: 5.80±0.24 $dpv_{0.5}$ μm: 9.63±0.06 % TOMAC retained: 38%	[34]
TBP TOPO TOA	**Continuous phase:** 2 wt% arabic gum **Discontinuous phase:** ST: 7.64 g	Average $dpn_{0.5}$ (μm) and $dpv_{0.5}$ (μm) TBP: 642 ± 4 and 794 ± 9 TOPO: 572 ± 26 and 678 ± 9 TOA: 685 ± 27 and 1,040 ± 26	[35]

Tab. 8.1 *(continued)*

Extractant	Conditions	Properties	Ref.
TOMAC	DVB: 1.77 g Extractant: 30 wt% BOX: 2.13 g Porogen: heptane Phase volume ratio 5:1 Agitation: 400 rpm Polymerization temperature 80 °C for 5 h under a nitrogen atmosphere	TOMAC: 19 ± 2 and 758 ± 2	
EHPNA	**Continuous phase:** 2 wt% arabic gum **Discontinuous phase:** DVB, ADVN, EHPNA Porogen: toluene Agitation: 300 rpm Polymerization temperature 70 °C for 6 h under a nitrogen atmosphere	Amount of EHPNA: 1.6 mmol/g MC Average radius: 0 7.5 × 10^{-5} m Porosity: 5.2 × 10^{-1} Density: 9.9 × 105 g/m^3 MC Surface area: 0.054 m^2/g MC	[36,37]

ST, styrene; DVB, divinylbenzene; AIBN, 2;2′-azobis(2-methylpropionitrile); ADVN, *N;N′*-azobis(2;4-dimethyl-valeronitrile); PVA, poly(vinylalcohol); BOX, benzoyl peroxide; HEC, hydroxyethylcellulose; EHPNA, 2-ethylhexylphosphonic acid mono-2-ethylhexyl ester; MC, microcapsules.

contained encapsulated extractants (Levextrel series: *Lewatit* TP 272, Lewatit OC 1023, and *Lewatit* VP OC 1026) are prepared by adding an extractant to the mixture of styrene and divinylbenzene during the suspension polymerization [30]. In this combination of monomer and cross-linker, the discontinuous phase also consists of an initiator such as 2,2′-azobis(2-methylpropionitrile) (AIBN) or benzoyl peroxide (BOX), and toluene as porogen, whereas the continuous phase can be an aqueous solution of poly(vinyl alcohol), gum arabic, hydroxyethylcellulose, or gelatine. Using these compounds, micro- and meso-pores are predominant, resulting in beads with high surface area values but low pore volumes [31]. Moreover, the used polymerization conditions cause a single bead formation by an agglomeration of multiple microbeads with the void space between them filled with the extractant (e.g. Lewatit OC 1023 revealed a porosity of 63%, surface area of 345 m^2/g, **and** average pore diameter of 180 A).

The extractant can be successfully encapsulated not only by copolymer of styrene with divinylbenzene polymer cover, but also using as cross-linker ethylene glycol dimethacrylate [38, 39]. The microcapsules obtained were spherical and had a relatively high extractant content. This type of shell improves the characteristics of microcapsules (Table 8.2). Further studies on extractant encapsulation have also considered polysulfone, polyvinyl chloride, and hydrophilic polyvinyl alcohol as polymer

Table 8.2: Synthesis conditions and properties of microcapsules formed during suspension polymerization using as monomers styrene and ethylene glycol dimethacrylate.

Extractant	Conditions	Properties	Ref.
PC-88A	**Continuous phase:** 2 wt% arabic gum **Discontinuous phase:** EGDMA: 1.84 g ST: 2.20 g BOX: 0.9 g PC-88A: 2.43 g Porogen: toluene Agitation: 600 rpm Polymerization temperature 70 °C for 3 h under a nitrogen atmosphere	Microcapsule weight 6.2 g (yield 84.5%) Retained extractant 92.7% (mmol PC-88A/g MC) 1.197	[46]
Aliquat 336	**Continuous phase:** 1 wt% arabic gum **Discontinuous phase:** EGDMA: 7 g ST: 1 g BOX: 2 g Aliquat 336: 6 g Porogen: toluene 14.5 g Agitation: 500 rpm Polymerization temperature 70 °C for 3 h under a nitrogen atmosphere	*S*BET: 18 m^2/g Pore diameter: 7.2 nm Pore volume: 0.0327 cm^3/g Mean particle size: 20 µm	[47]
D2EHPA	**Continuous phase:** 1 wt% PVP **Discontinuous phase:** EGDMA: 0.1 wt% ST: 78.32 g BOX: 0.9 g DEHPA: 27.04 g Porogen: – Agitation: 850 rpm Polymerization temperature 98 °C for 5 h under a nitrogen atmosphere	Spherical microcapsules with average particle size of 353 µm DEHPA content: 28.2 wt%	[44]
Cyanex 923	**Continuous phase:** 1 wt% arabic gum **Discontinuous phase:** PSU: 2.0 g Cyanex 923: 2.0 g Porogen: CH2Cl2 Agitation: 600 rpm Polymerization temperature 70 °C for 3 h under a nitrogen atmosphere	Spherical microcapsules with a diameter of 273 µm; retained extractant: 68%	[45]

Tab. 8.2 *(continued)*

Extractant	Conditions	Properties	Ref.
D2EHPA	**Continuous phase:** 5 wt% of gelatine **Discontinuous phase:** PSU: 2 g D2EHPA: 2 g additive: CH_2Cl_2 Agitation: 600 rpm Temperature 25 °C for 24 h under a nitrogen atmosphere	Spherical microcapsules with an average diameter of 30 µm. Encapsulation efficiency was 45%.	[46]
D2EHPA	**Continuous phase:** 1 wt% PVA or NMP **Discontinuous phase:** PSU: 12% D2EHPA: 3 mol/L (at 5:1 ratio) Incorporation temperature: 30 °C Time: 24 h	PSU/D2EHPA/NMP beads are spherical with uniformly distributed pores (1 µm). PSU/D2EHPA/PVA are coconut type structure with inside placed extractant PSU/D2EHPA/PVA beads are more porous than PSU/D2EHPA/NMP	[47]

EGDMA, ethylene glycol dimethacrylate; PVP, polyvinylpyrrolidone; PSU, polysulfone; NMP, 1-methyl-2-pyrolidone.

shell [40, 41]. Especially polysulfone, an environment-friendly matrix material widely used in water purification membranes has been proposed as a suitable base matrix for the preparation of extractant contained microcapsules (Table 8.2). Very interesting is fact, that in the preparation of polysulfone microcapsules containing extractant required other methodology than used for polystyrene copolymers. The procedure consisted of two steps, preparation of empty microcapsules and loading of extractant by phase inversion method [42, 43].

The styrene-divinylbenzene microcapsules have shown high selectivity for the desired solute but relatively slow kinetics and low capacity compared to commercial ion-exchange resins, making sorbents less attractive. The solution has been found to improve physical and mechanical properties, which may influence positively on the application potential of the sorbent, which can be achieved not only by entirely changing the polymer shell but also by modifying well-known polystyrene structure. Incorporation of hydrophilic sulfonic acid groups into polymers improves their permeability for selected ions. This is very useful in the case of microcapsules containing cationic extractant agents such as Di-(2-ethylhexyl)phosphoric acid (D_2EHPA), mono-2-ethylhexyl ester (EHPNA), and dialkyl phosphinic acid (Cyanex 272). For the functionalized microcapsule, the negatively charged sulfonate groups attract the metal cations to the surface of the microcapsule. Once they reach the surface, the cations enter into the polymer shell, where the functional groups affect their mobility, promoting transport to the extractant, which is the core of the capsule [48–49]. Even though the favored mass and kinetic transfer of the metal cations from the aqueous phase is much higher than observed for styrene-divinylbenzene microcapsules, the production of the sulfonated microcapsules required another optimization. Classical methods of synthesis of the polymeric beads with sodium styrene sulfonate are employed the most; however, the monomer works as a cross-linker or source of coating layer [50, 51].

8.2.2 Extractant encapsulated in metal removal

Material sorption containing an encapsulated extractant was developed to bridge the gap between solvent extraction and ion exchange, but only a few examples have been found to be efficient adsorbents of desired metals species. Commercial extractants containing microcapsules being cross-linked to polystyrene and extractant such as TBP, D2EHPA, and Cyanex 272 have been evaluated mainly for processes strictly related to complexing properties of the extractant (Table 8.3); e.g. in the case of TBP, recovery of uranium from a plutonium-uranium solutions was proposed, while for Cyanex 272, the selective Co(II) extraction over Ni(II) from sulfate media was developed.

In the case of noncommercial materials, the results have also shown that the microcapsules have abilities to recover metal ions from dilute solutions and sufficient stabilities for repeated processing. Moreover, the microcapsules have revealed

Table 8.3: Commercial resins being cross-linked to polystyrene containing extractant.

Product name	Matrix	Extractant	Application	Ref.
Lewatit OC 1023	P(ST-DVB)	TBP	U(VI) recovery from a plutonium-uranium solution (5M HNO_3-0.1 M $Fe(SO_3NH_2)_2$); sorption capacity = 45 g U(VI)/L (53% if TBP capacity); desorption using 0.1 M $Fe(SO_3NH_2)_2$	[52]
Lewatit VP OC 1026	P(ST-DVB)	D2EHPA	Selective Zn(II) recovery from a nickel electrolyte; selectivity over Ni and Co(II) at pH below 3; sorption capacity 17 mg Zn(II)/g; desorption using 1 M H_2SO_4	[53]
			Zn(II) separation from electroplating effluent containing Cr(VI); selectivity at pH of 6.2; sorption capacity 58 mg Zn(II)/g; desorption using 5% H_2SO_4.	[54]
Lewatit TP 272	P(ST-DVB)	Cyanex 272	Co(II) removal over Ni(II) from sulfate media; electivity at pH range of 4.8–6.0; sorption capacity 18 mg/g; desorption using diluted H_2SO_4	[55]

a high selectivity for the metal ions, similar to those of the pure extractant agent. However, sufficient mass transfer kinetic, as well as loading capacity, is characteristic for selected sorption systems (extractant-polymer shell). An example of ineffective material is microcapsules consisting of a polysulfone shell. For instance, polysulfone beads prepared by immobilizing two extractants, D2EHPA and TBP, remove Sr(II) from the aqueous solution, but the capacities are very low, 1.71 mg/g and 2.41 mg/g, respectively [56, 57]. Yadav et al. obtained much better results testing polysulfone capsules with PC88A as extractant for the removal of yttrium ions from chloride solutions [58]. It was shown that sorption of Y(III) onto the PC88A encapsulated polymeric composite beads occurs with a q_{max} value of 82.82 mg/g. Another promising material in which polymer shell are polyvinyl chloride indicated very low capacity, 0.33 mg Cu(II)/g [59]. Moreover, the sulfonation of *poly(styrene*-co-*divinylbenzene)* beads with the encapsulated D2EHPA did not improve the sorption properties of the microcapsules [58]. Although not all considered materials can compete with an ion-exchange resin, many promising materials have been developed, which led to the proposal of new technological solutions. The most interesting examples are listed in Table 8.4.

8.3 Encapsulation of extractants in biopolymers

Biopolymeric microcapsules are an interesting solution, especially in times in which the possibility of replacing petroleum-based compounds with compounds of natural origin is still being sought. The most popular biopolymer used for the preparation of microcapsules is alginate, an anionic polysaccharide. As natural polymers, alginates can be found in brown seaweeds and bacterium, and their compositions are not regular and depend on the source from which they are isolated. Alginates are linear polymers, with two monomeric forms of uronic acid in their structure: β-D-mannuronic acid (M) and α-L-guluronic acid (G). These two uronic acids form the irregular structure of the polymer, with variable proportions of MM, MG, and GG blocks [65]. From a practical point of view, the most important is sodium salts of alginic acid – sodium alginate, which, due to the presence of free carboxyl groups, can form a complex with calcium ions. The main reason for alginate application is easy capsulation procedure [66], relatively low cost [67], high acid resistance [68], and, the most important, ion-exchange properties for metals ions. These microcapsules are usually applied for the immobilization of biologically valuable materials, such as living cells, peptides, and enzymes [69, 70]. Their utility in separation technology is a rather new proposition.

It should be noted that other biopolymers are often used in the production of polymeric microcapsules. For example, polylactide finds application in medicine as an ultrasound contrast agent catalyst [71], poly(lactic-co-glycolic acids) are employed in therapeutics and antigen microencapsulation [72], and chitosan microcapsules are

Table 8.4: Noncommercial microcapsules containing extractant.

Extractant	Matrix	Application	Ref
TOA	P(ST-DVB)	Removal of Zn(II) and Cu(II) from chloride solutions. The best results obtained for microcapsules prepared using the ratio ST:DVB (6:4) and 36.2% TOA; sorption capacities: 0.279 mmol Zn(II)/g and = 0.283 mmol Cu(II)/g	[60]
TOMAC	P(ST-DVB)	Removal of Hg(II) from neutral and acidic aqueous chloride solutions. The uptake of Hg was favored at acid pH and low chloride concentration; sorption capacity: 1 eq Hg/L	[61]
PC-88A	P(ST-co-EGDMA)	Removal of Cd(II) and Cu(II) from acidic aqueous solutions; k and q_e higher for Cd(II) than for Cu(II). Cd(II) k= 0.48 mmol/gs, q_e = 0.25 mmol/g Cu(II) k= 0.28 mmol/gs, q_e = 0.22 mmol/g	[46]
Aliquat 336	P(ST-co-EGDMA)	Efficient sorbent of Cr(VI) from aqueous solutions with pH = 2 (99% removal); maximum sorption capacity between 0.86 and 0.97 mmol Cr(VI)/g	[47]
Cyanex 923	PSU	Removal of Cr(VI) from acidic aqueous solution; high selectivity toward Cr(III), Ni(II), Pb(II), Cu(II), and Co(II); sorption capacity = 0.430 mmol/g	[54]
D2EHPA	PSU	La(III), Sm(III), and Y(III) removal from acidic solutions (0.3–4 N HCl). The best results obtained for Y(III) removal from 0.3 M HCl; stripping solutions: 20–40% H_2SO_4	[56]
		Sorption Ga(III) and In(III) at pH of 5; maximum sorption capacities: 0.61 mmol Ga(III)/g MC and 0.35 mmol In(III)/g MC; stripping solutions: Ga(III) 2 mol/L H_2SO_4, In(III) 5 mol/L H_2SO_4	[62]
[A336][P507]	PSU	La(III), Sm(III) and Er(III) removal from diluted aqueous solutions. Maximum extraction capacities (at pH of 5): of 58.4 mg La(III)/g 56.6 mg Sm(III)/g and 81.7 mg Er(III)/g: desorption with 0.5 M HNO_3	[63]
BenzoDODA	PSU	Plutonium recovery from nitric acid medium; maximum sorption capacity: 649 µg Pu(IV)/g MC; stripping solutions: 0.3 M Hydrazine + 0.5 M HNO_3	[64]

[A336][P507], [trialkylmethylammonium][di(2-ethylhelxyl) orthophosphinate]; BenzoDODA, *bis*(2-ethylhexyl) carbamoyl methoxy phenoxy-*bis*(2-ethylhexyl) acetamide.

used in cosmetics, food, and pharmaceuticals [73, 74]. However, these biopolymers have not yet found application in the encapsulation of extractants. The only example of application mentioned for biopolymers is using alginate microcapsules coated with chitosan or polyethyleneimine. Such microcapsules are very stable and can be used for the extraction of heavy metal ions [75].

A typical procedure for obtaining such microcapsules is different from that in the case of conventional polymers (e.g. interfacial or *in situ* polymerization). The most common technique here is gelation. The encapsulation procedure includes the formation of multinuclear microcapsules, with the following composition: polymeric granule and extractant microdroplets dispersed in their matrix. In the next step, kneaded sol (a mixture of extractant and biopolymer) is added dropwise to the inorganic salt solution (mainly as calcium salts) to gelify and form gel particles – microcapsules. In this procedure, the addition of divalent cation (mainly Ca(II)) to sodium alginate solutions causes a cross-linked reaction due to the exchange of sodium cations from the glucuronic acid with divalent cations. As a consequence, the formation of water-insoluble gel (calcium alginate) with a characteristic "egg-box" structure is observed [76]. Schematically, the entire procedure is shown in Figure 8.3. In another method, the extractant is mixed with a surfactant, as in the case of the procedure proposed by Mimura et al. [15], where Cyanex 302 and nonionic surfactant Span 20 (sorbitan laurate) were added and mixed with the sodium alginate solution to obtain a fully dispersed mixture. This method allows obtaining a larger microcapsule particle, while the diameters of the oil drops encapsulated are significantly lower.

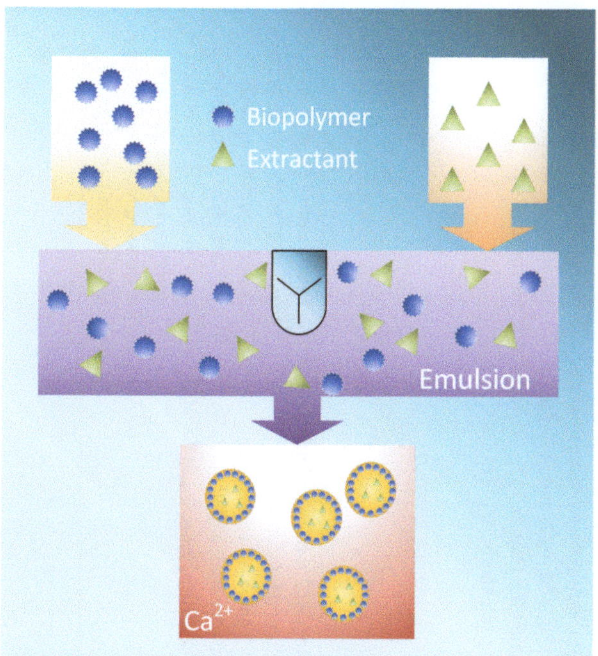

Fig. 8.3: Procedure for obtaining biopolymeric microcapsules with immobilized extractant.

Table 8.5: Microcapsule parameters.

Composition	Condition	Microcapsules parameters	Ref.
– Polymer (P): 1 wt.% sodium alginate – Extractant (E): Cyanex 302 – Nonionic surfactant (S): Span 20 – Gelling salt solution: 0.5 M, $CaCl_2$ or $BaCl_2$	– Stirring at room temp., 300 rpm for 10 min – Mixing ratio E/S (in g)/P (in cm^3) from 0.5/0/50 to 0.5/1.5/50 g/g/cm^3	– Diameters increase with increase of surfactant's content, from 0.79 to 1.01 mm for calcium alginate MCs, for 0 and 1.5 g Span 20, respectively. – Diameter of barium alginate MCs is higher in comparison to the calcium ones: 0.89 and 0.79 mm, respectively for mixing ratio 0.5/0/0.5.	[15]
– Polymer (P): 1 or 1.5 wt.% sodium alginate, with different viscosities of 100–150, 300–400, and 500–600 cP – Extractant (E): Cyanex 302, – Gelling salt solution: 0.5 M $CaCl_2$	– Mixing ratio E (in g)/P (in cm^3) from 0.1/50 to 3.0/50 g/cm^3 – 1 M HNO_3 for treatment of microcapsules	– Various shapes: spherical, columnar, and fibrous forms, depending on the method of dipping kneaded sol into $CaCl_2$ solution. – Viscosity of sodium alginate solution has no effect on the MC size. – Diameters of microcapsules increase with an increase of extractant content, i.e. 1.2 and 1.6 mm for mixing ratio of 1.5/50 and 3.0/50 g/cm^3, respectively; however, at higher mixing ratio above 5.0/50 g/cm^3, ooze of extractant through the wall of gel matrices is observed. – Contraction of about 8% in volume of MCs after acid treatment, caused by the replacement of cross-linking cations of Ca^{2+} with H^+ in the gel matrices.	[16]
– Polymer (P): 2.4 wt.% sodium alginate – Extractant (E): Cyanex 302, – Coating polymer: chitosan or polyethyleneimine (PEI) – Gelling salt solution: 0.1 M $Ca(NO_3)_2$	– Mixing ratio E (in g)/P (in cm^3) from 0.6/50 to 1.8/50 g/cm^3; – Coating procedure: fresh alginate MCs (i) were dipped in chitosan or (PEI) solution and then washed and dried or (II) were repeated coatings with chitosan and alginate.	– Diameter of uncoated MCs is lower (0.605 mm) than coated ones (0.62–0.66 mm). – Losing of extractant due to the multiple coating, as a consequence of physical stress during the coating; effect is not observed for PEI-coated MCs.	[85]

Tab. 8.5 *(continued)*

Composition	Condition	Microcapsules parameters	Ref.
– Polymer (P): 2 wt.% sodium alginate – Extractant (E): TBP (tributylphosphate) – Gelling salt solution: 0.1 or 0.3 M $Ca(NO_3)_2$, $Ba(NO_3)_2$ or 1 M HNO_3.	– Mixing ratio E (in g)/P (in cm^3) from 0.1/50 to 2.0/50 g/cm^3 – 1 M HNO_3 for treatment of microcapsules	– Shape: spherical and elastic granules for calcium and barium MCs or flaky and hard-to-form spherical particles for acid treatment MCs. – Calcium and barium MCs indicate similar surface morphology. – Diameters of MCs in the range 0.6 to 1.15 mm, depending on: – kind of the MCs – decreasing in the order barium>calcium, acid MCs; – content of TBP – increasing with the increase of extractant content (i.e. 0.65 and 0.75 mm for calcium MCs with mixing ratio 1.0/50 and 1.0/50 g/cm^3, respectively). – Higher mixing ratio over 3.0/50 g/cm^3 caused low MC stability – TBP ooze through the wall of gel matrices, the largest extractant content = 51.9 wt.%).	[90]
– Polymer (P): 1 or 2.5 wt.% sodium alginate, with different viscosities of 100–150 and 500–600 cP – Extractant (E): Cyanex 301 – Gelling salt solution: 0.5 M $Ca(NO_3)_2$, $Sr(NO_3)_2$, $Ba(NO_3)_2$, $Al(NO_3)_3$	– Mixing ratio E (in g)/P (in cm^3) from 0.5/50 to 5.0/50 g/cm^3 – 1 M HCl for treatment of microcapsules	– Various shapes: spherical, columnar, and fibrous forms, depending on the method of dipping kneaded sol. – Diameter of spherical MCs increase with increasing extractant content (0.8 and 0.9 mm for 0.5/50 and 2.0/50 g/cm^3, respectively). – Size is independent of the viscosity of sodium alginate solution. – Higher mixing ratio over 5.0/50 g/cm^3 caused low MCs stability – Cyanex 301 ooze through the wall of gel matrices.	[91]
– Polymer (P): sodium alginate – Extractant (E): 0.1, 0.2 or 0.3 M Aliquat 336 (trioctylammonium chloride) – Dispersing agent: carboxymethyl cellulose (CMC) – Gelling salt solution: $CaCl_2$	– Encapsulation procedure: solution of 1.5% $CaCl_2$ and 3% CMC were mixed with Aliquat 336 and added dropwise to sodium alginate.	– Shape: spherical – Diameter: 3 mm	[92]

MC, microcapsule.

Table 8.6: Exemplary microcapsules with biopolymers applied in separation processes.

Microcapsules composition	Effects	Ref.

Alginate MCs immobilizing Cyanex 302 ((*bis*(2,4,4-trimethylpentyl) monothiophosphinic acid); gelling salt solution: $CaCl_2$, $BaCl_2$	– Removal of Pd(II), Ru(III) (in the form of $RuNO^{3+}$). and Rh(III) ions.
	– Efficiency depends on the gelling salt solutions used:
	  – for calcium alginate MCs: Pd^{2+} up to 100% after 5 days,
	  – for barium alginate MCs: Pd^{2+}, $RuNO^{3+}$, and Rh^{3+} up to 100%, 60, and 50%, respectively, after 5 days; moreover, they exhibit high adsorbability for $RuNO^{3+}$ in comparison to the calcium MCs.
	– Barium MCs show a relatively large separation factor of Pd/Ru and Pd/Rh and can be used in the separation of palladium using the packed column.
		[15]

Alginate MCs immobilizing Cyanex 302; gelling salt solution: $CaCl_2$, $SrCl_2$, $BaCl_2$ or HCl	– Removal of Pd(II), Ru(III) (in the form of $RuNO^{3+}$), and Rh(III) ions.
	– Proposed mechanism: Pd(II) ion based on the extraction with Cyanex 302, according to the reaction:
	$$Pd^{2+}_{aq} + 2(HA)_{2,org} \leftrightarrow Pd(HA_2)_{2,org} + 2H^{+}_{aq}$$
	while Ru(III) and Rh(III) based on the ion-exchange reaction in gel matrix.
	– Efficiency depends on the gelling salt solutions used, as well as concentration of extractant:
	  – removal (uptake) increases with increasing extractant content;
	  – similar results for barium and strontium alginate MCs;
	  – higher values were obtained for palladium ion for all considered salt solutions.
	– There is the possibility of separating Pd from Ru and Rh ions using the packed column (separation factors (α) of Pd/Ru and Pd/Rh were 10^2–10^3 and 10^2–10^5, respectively).
		[16]

Alginate MCs immobilizing Cyanex 302; gelling salt solution $Ca(NO_3)_2$	– Removal of Ag(I) ion, up to 100%. – Proposed mechanism: both extraction with Cyanex 302 (HA) and ion exchange with calcium alginate, according to the reactions: $$nAg^{+}\frac{n}{2}(HA)_2 \leftrightarrow \overline{Ag_nA_n} + nH^{+}$$ $$2Ag^{+}_{aq} + Ca(ALG)_2 \leftrightarrow \overline{2AgALG} + Ca^{2+}_{aq}$$ where the overline depicts both organic and alginate phases of MCs;	[17]

Tab. 8.6 *(continued)*

Microcapsules composition	Effects	Ref.
	– Very selective process for Ag(I) ion (efficiency at least 90% from the solution containing both Ag(I) and Na(I) ions). – MCs showed high removal efficiency in a wide range of pH or acid concentrations. The lowering of efficiency to 94% at 3 M HNO_3 (after 24 h) is caused by the slow degradation of extractant in acidic solution. – Unfortunately, in the presence of coexisting ions, alginate significantly loses its ability to remove the metal ions from the solution.	
Coated alginate MCs immobilizing Cyanex 302; gelling salt solution $Ca(NO_3)_2$	– Removal of Ag(I) ion. – Efficiency about 70%, lower for coated MCs, in comparison to the uncoated one, due to the lower ability for adsorption of Ag(I) ions by coating layers. – Results indicate that stable extractive microcapsules could be applied for the selective recovery of heavy metal ions.	[85]

Alginate MCs immobilizing TBP; gelling salt solution: $Ca(NO_3)_2$, $Ba(NO_3)_2$ or HNO_3	– Removal of: Fe(III), Sr(II), Co(II), U(VI), and Pu(IV) ions.
	– Efficiency depends on the metal ions and acid concentration (HNO_3):
	  – 80% for Fe(III), Sr(II), and Co(II) ions, in the presence of 10^{-3} M HNO_3, and decrease with increasing acid concentration to 5 M; equilibrium time = 5 h;
	  – Below 60% for U(VI) and Pu(IV) ions; equilibrium time 1 and 3 h, respectively.
	– Proposed mechanism for Fe(III), Sr(II) and Co(II) ions – ion exchange:
	  – at low HNO_3 concentration – ion-exchange reactions with calcium alginate,
	  – at higher HNO_3 concentration (>1 M) – Ca^{2+} ions are eluted from MC matrices ion-exchange reaction between H+ ions of carboxylic acids and metal ions dominant.
	– Proposed mechanism for U(VI) and Pu(IV) ions – both ion exchange and extraction, according to the reactions:
	$UO^{2+}_{2,aq} + Ca - ALG_s \rightarrow UO_2^- ALG_s + Ca^{2+}_{aq}$
	$UO^{2+}_{2,aq} + H_2ALG_s \rightarrow UO_2^- ALG_s + 2H^+_{aq}$
	$UO^{2+}_{2,aq} + 2NO^-_{3,aq} + 2TBP_{org} \rightarrow UO_2(NO_3)_2 \cdot 2TBP_{org}$
		[100]

Tab. 8.6 *(continued)*

Microcapsules composition	Effects	Ref.

Alginate MCs immobilizing Cyanex 301 (*bis*(2,4,4-trimethylpentyl) dithiophosphinic acid); gelling salt solution: $Ca(NO_3)_2$, $Sr(NO_3)_2$, $Ba(NO_3)_2$, $Al(NO_3)_3$	– Removal of ^{241}Am and ^{152}Eu in the form of trivalent ions.
	– Efficiency depends on the metal ions and MCs used:
	  – above 97% for Am(III) ions, except for aluminum MCs (47%);
	  – above 93% for Eu(III) ions, except for aluminum MCs (30%), and treated with acid (67%).
	– Proposed mechanism for Am(III) and Eu(III) ions – both ion exchange and extraction, according to the reaction:
	$M^{3+}_{aq} + 3(HA)_{2,org} \leftrightarrow M(HA_2)_{3,org} + 3H^+_{aq}$
	– The separation of Am(III) from Eu(III) is possible using the column packed with MCs treated with acid.
		[101]

Alginate MCs immobilizing Aliquat 336; gelling salt solution: $CaCl_2$	– Removal of Au(III) ion in the presence Pt(IV) and Pd(II) ions. – MCs wit 0.2 M Aliquat-336 revealed the best Au(III) selectivity (above 97% purity of gold solution). – Proposed mechanism: electrostatic interaction between Aliquat 336 (R_4N^+) and metal chlorocomplexes, according to the reactions: $R_4N^+Cl^- + AuCl_4^- \rightarrow R_4N^+AuCl_4^- + Cl^-$ $\overline{2(R_4N^+Cl^-)} + PtCl_6^{2-} \rightarrow \overline{2(R_4N^+)}PtCl_6^{2-} + 2Cl^-$ $\overline{2(R_4N^+Cl^-)} + PdCl_4^{2-} \rightarrow \overline{2(R_4N^+)}PtCl_4^{2-} + 2Cl^-$	[102]

MC, microcapsule.

The result obtained also implied that the addition of surfactants reduced the aggregation of dispersed oil drops. A similar procedure was proposed by Zwar et al. [77]. They used two types of emulsifiers, Span 85 and Tween 80, for obtaining magnetic alginate capsules.

Depending on the used reagents as well as the reaction conditions, different microcapsules can be obtained. Several examples of this relationship are presented in Table 8.5, while their application in separation process is presented in Table 8.6.

There are several mechanisms of separation of metals ion using microcapsules with immobilized extractants proposed in the literature. Outokesh et al. [17], based on their results of silver uptake by multinuclear alginate microcapsules with Cyanex 302 as organophosphinic acid extractant, have indicated that the kinetic mechanism depends on the presence and concentration of coexisting ions in separation system and change from physical-chemical shielding to pore diffusion controlled kinetics. In the case of negligible amount or low concentration of coexisting ions, the slow step of the kinetics is adsorption of Ag(I) ions by a microdroplet of extractant; thus, alginate plays a role of ion carrier for microdroplets. A mixed mechanism, where alginate has two functions – ion carrier and adsorbent – occurs in the case when Ag(I) ion concentration is above the breakpoint of alginate and there are no coexisting ions in the solution. A higher concentration of coexisting ions causes the lowering of alginate's uptake ability. As a consequence, alginate matrix cannot longer perform the role of ion carrier for extractant and ions are transferred to microdroplets via the water trapped in the pore of microcapsules. In this case, kinetics is controlled by the pore diffusion mechanism.

It should be noted that the characteristic structure of sodium alginate, with a carbonyl group in the molecule, results in alginate being able to bind multivalent cations, in the following order: Pb(II)>Cu(II)>Cd(II)>Ba(II)>Sr(II)>Ca(II)>Co(II)>Ni(II)>Zn(II)>Mn(II) [78]. This fact can explain the ability to adsorb metal ions in the separation process by alginate matrix. Moreover, as was mentioned by Deze et al. [79], alginate plays a role of chelator with a unique sorption capacity for metal ions.

8.3.1 Microcapsules with magnetic nanoparticles

A very interesting approach proposed in the literature is the addition of magnetic materials and nanoparticles into microcapsules. Incorporating magnetic materials with a microcapsule structure is very convenient in isolation or recovery process due to controllable motion in magnetic fields. This method finds application also in the removal of contaminations from aqueous solutions. MnO_2-impregnated alginate beads are successfully used for the removal of arsenic and cadmium [80], while magnetic alginate beads based on maghemite nanoparticles (γ-Fe_2O_3) are proposed for lead [81, 82] or copper [83] recovery. Several authors proposed the addition of cross-linking ions to improve the stability and reusability of magnetic alginate. Zirconium(IV) ions were proposed to enhance magnetic alginate beads based on maghemite nanoparticles

Table 8.7: Magnetic microcapsule parameters.

Composition	Condition	Microcapsules parameters	Ref.
– Polymer (P): sodium alginate – Extractant (E): Cyanex 272 – Magnetic nanoparticle (M): γ-Fe_2O_3 – Gelling salt solution: 0.5 M $CaCl_2$	– Mixing ratio E (in g)/M/ (in mol/P (in g) 4/0.095/ 5 g/mol/g	– Shape: spherical and red-brown due to the presence of maghemite particles – Diameter: 0.95 mm – Iron content in MCs: close to the initial amount of magnetic nanoparticles – Sodium amount: smaller than calcium, as a consequence of ion-exchange reaction during the gelation	[98]
– Polymer (P): 0.02 wt.% sodium alginate, – Extractant (E): P507, – Magnetic nanoparticle (M): Fe_3O_4, – Gelling salt solution: 0.5 M $CaCl_2$.	– Mixing ratio E/M (in g)/P (in cm^3) 1/0/50, 0/0.25/50, 1/0.25/50 and 2/0.25/50 g/g/cm^3	– Diameters depend of the initial composition: the highest value for MCs without extractant (0.92 mm), the lowest one for MCs without Fe_3O_4 (0.8 mm); only small difference for various contents of P507 (0.82–0.83 mm).	[99]

MC, microcapsule.

(γ-Fe_2O_3) in the lead recovery method [84]. Furthermore, magnetic alginate microcapsules based on Fe_3O_4 particles stabilized by cerium(III) ions were applied for efficient chromium(VI) removal [85]. Periyasamy et al. [86] proposed the modification of magnetic (Fe_3O_4) alginate microcapsules by the addition of hydroxyapatite. The reason for fabrication of such hybrid material was the fact that hydroxyapatites are already proposed for removing metals ions from aqueous solution and, similar to extractant, can increase the efficiency of the separation process, as confirmed in chromium(VI) removal studies. The results obtained have shown that the used magnetic alginate is a promising material to solve the problem of chromium contaminated groundwater. Other examples of magnetic alginate microcapsules are described in another work [87]. The authors showed that it is possible to recover Cu(II) and U(VI) ions by encapsulated nano-Fe_3O_4 alginate-chitosan hydrogel beads.

As in the case of "classical" microcapsules, it is possible to introduce an extractant into magnetic alginate microcapsules. Such an option was proposed at first by Ngomsik et al [88]. In their work, both extractant (Cyanex 272) and magnetic nanoparticles (γ-Fe_2O_3) were immobilized into alginate microcapsules. After the process, a simple and quick method separation of microcapsules by an external magnet was applied. Moreover, as it was presented in [89], the presence of magnetic nanoparticles

Table 8.8: Exemplary magnetic microcapsules with biopolymers applied in separation processes.

Microcapsules composition	Effects	Ref.
Alginate MCs containing an extractant Cyanex 272 (*bis*(2,4,4-trimethylpentyl) phosphinic acid) and magnetic nanoparticles (γ-Fe_2O_3).	– Removal of nickel (II) from aqueous solutions. – Efficiency: max 70% (after 8 hours), depends on the pH (nickel removal increases with increase of pH). – The presence of magnetic particles in the MCs helps in the isolation of the beads from the aqueous solutions after the sorption process; no effect on the efficiency of metal removal.	[98]
Alginate MCs containing an extractant 2-ethylhexyl phosphonic acid mono-2-ethylhexyl ester (P507) and magnetic nanoparticles (Fe_3O_4)	– Removal of Nd(III) ion in the presence of Pb(II), Cu(II), Zn(II), Co(II), and Ni(II) ions. – Efficiency for Nd(III) ion: from 60% to almost 100%, equilibrium time from 5 to 11 h, depending on the used MCs; higher for the higher extractant content and the presence of Fe_3O_4. – Proposed mechanism for Nd(III) ions – both ion exchange and extraction, according to the reactions: $2Nd^{3+} + 3Ca(ALG)_2 \leftrightarrow 2Nd(ALG)_3 + 3Ca^{2+}$ $Nd^{3+} + 3H_2L_{2,org} \rightarrow NdL_3(HL)_{3,org} + 3H^+$ – The adsorption capacity changes in the order Pb(II)>Nd(III)>Cu(II)>Zn(II)>Co(II)>Ni(II). – Nd(III) could be selectively separated from the mixed Nd(III)/Co(II) and Nd(III)/Zn(II) solutions.	[99]

MC, microcapsules.

helps to increase the efficiency of the metal recovery. It was explained by the increasing surface and internal size of the magnetic microcapsules due to the addition of magnetic nanoparticles (in this case Fe_3O_4).

The method of preparation of magnetic microcapsules and their properties, as well as the obtained results of separation, are presented in Tables 8.7 and 8.8.

8.4 Conclusions

This work highlights the unique features of microcapsules based on extractant encapsulated in polymer shell in the field of metal ion recovery from aqueous solutions. Recent results have been reported concerning the use of two types of polymers: artificial, such as styrene and divinylbenzene, and biopolymers, mainly alginate. The enhancement of such techniques by incorporation inside polymer capsules significantly improves the efficiency and selectivity of the separation process.

8.5 References

[1] Wieszczycka, K., Tylkowski, B., Staszak, K., Metals in Wastes. Walter de Gruyter GmbH, Berlin/ Boston, 2018.

[2] Staszak, K., Regel-Rosocka, M., Wieszczycka, K., Burmistrzak, P., Copper(II) sulphate solutions treatment by solvent extraction with Na-Cyanex 272, Sep Purif Technol **85** (2012) 183–192.

[3] Wołowicz, A., Staszak, K., Hubicki, Z., Static sorption of heavy metal ions on ion exchanger in the presence of sodium dodecylbenzenesulfonate, Adsorption 2019, in press. https://doi.org/10.1007/s10450-019-00014-8

[4] Azimi, A., Azari, A., Rezakazemi, M., Ansarpour, M., Removal of heavy metals from industrial wastewaters: a review, ChemBioEng Rev **4** (2017) 37–59.

[5] Ashraf, S., Cluley, A., Mercado, C., Mueller, A., Imprinted polymers for the removal of heavy metal ions from water, Water Sci Technol **64** (2011) 1325–1332.

[6] Alexandratos, S. D., Smith, S. D., High stability solvent impregnates resins: Metal ion complexation as a function of time, Solv Extr Ion Exch 22 (2004) 713–720.

[7] Borreguero, A. M., Leura, A., Rodríguez, J. F., Vaselli, O., Nisi, B., Higueras, P. L., Carmona, M., Modelling the mercury removal from polluted waters by using TOMAC microcapsules considering the metal speciation, Chem Eng J **341** (2018) 308–316.

[8] Laki, S., Shamsabadi, A. A., Kargari, A., Comparative solvent extraction study of silver(I) by MEHPA and Cyanex 302 as acidic extractants in a new industrial diluent (MIPS), Hydrometallurgy **160** (2016) 38–46.

[9] Sarkar, S. G., Dhadke, P. M. Solvent extraction separation of gold with Cyanex 302 as extractant, J Chin Chem Soc 47 (2000) 869–873.

[10] Mimura, H., Ohta, H., Akiba, K., Wakui, Y., Onodera, Y. Uptake and recovery of platinum group metals ions by alginate microcapsules immobilizing Cyanex 302 emulsions, J Nucl Sci Technol **39** (2002) 1008–1012.

[11] Mimura, H., Ohta, H., Hoshi, H., Akiba, K., Onodera, Y., Uptake properties of palladium for biopolymer microcapsules, enclosing Cyanex 302 extractant, Sep Sci Technol **36** (2001) 31–44.

[12] Outokesh, M., Mimura, H., Niibori, Y., Tanaka, K., Equilibrium and kinetics of silver uptake by multinuclear alginate microcapsules comprising an ion exchanger matrix and Cyanex 302 organophosphinic acid extractant, Ind Eng Chem Res **45** (2006) 3633–3643.

[13] Staszak, K., Wieszczycka, K., Burmistrzak, P., Removal of cadmium(II) Ions from Chloride solutions by Cyanex 301 and Cyanex 302, Sep Sci Technol **46** (2011) 1495–1502.

[14] Mimura, H., Outokesh, M., Niibori, Y., Tanaka, K., Preparation of biopolymer microcapsules and their uptake properties for Cd2+ ions, Wit Trans Ecol Envir **79** (2004) 99–109.

[15] Luo, L., Wei, J., Wu, G., Toyohisa, F., Atsushi, S., Extraction studies of cobalt (II) and nickel (II) from chloride solution using PC88A, Trans Nonferrous Met Soc China **16** (2006) 687–692.

[16] Kondo, K., Hirao, S., Matsumoto, M., Column separation of nickel and cobalt using a microencapsulated extractant. Hydrometallurgy **149** (2014) 87–96.

[17] Wieszczycka, K., Staszak, K., Polymers in separation processes, in Polymer Engineering, (Tylkowski, B., Wieszczycka, K., Jastrzab, R., (Eds)), Walter de Gruyter, GmbH, Berlin, Germany (2017) pp. 235–76.

[18] Cortina, J. L., Warshawsky, A., Developments in solid-liquid extraction by solvent-impregnated resins in ion exchange and solvent extraction (Marinsky, J. A., Marcus, Y., (Eds.)), Marcel Dekker, NY **13** (1997) p. 195.

[19] Kabay, N., Cortina, J. L., Trochimczuk, A., Streat, M., Solvent-impregnated resins (SIRs) – methods of preparation and their applications, React Funct Polym **70** (2010) 484–96.

[20] Trochimczuk, A. W., Kabay, N., Arda, M., Streat, M., Stabilization of solvent impregnated resins (SIRs) by coating with water soluble polymers and chemical crosslinking, React Funct Polym **59** (2004) 1–7.

[21] Bao, S., Chen, B., Zhang, Y., Tang, Y., Synthesis of coated solvent impregnated resins by PVA cross-linked with vapor-phase glutaraldehyde for adsorption of vanadium (IV), Reac Funct Polym **128** (2018) 58–66.
[22] Gokmen, M. T., Du Prez, F. E., Porous polymer particles – a comprehensive guide to synthesis, characterization, functionalization and applications, Progress Pol Sci **37** (2012) 365–405.
[23] Brooks, B. W., Suspension polymerization processes, Chem Eng Technol **33** (2010) 1737–44.
[24] Zhang, R., Cherdhirankorn, T., Graf, K., Koynov, K., Berger, R., Swelling of cross-linked polystyrene beads in toluene, Microelectronic Eng **85** (2008) 1261–4.
[25] Kangwansupamonkon, W., Damronglerd, S., Kiatkamjornwong, S., Effects of the crosslinking agent and diluents on bead properties of styrene-divinylbenzene copolymers, J Appl Polym Sci **85** (2002) 654–69.
[26] Hao, D. X., Gong, F. L., Hu, G. H., Lei, J. D., Ma, G. H., Su, Z. G., The relationship between heterogeneous structures and phase separation in synthesis of uniform PolyDVB microspheres, Polymer **50** (2009) 3188–95.
[27] Loiseau, E., Niedermair, F., Albrecht, G., Frey, M., Hauser, A., Rühs, P. A., Studart, A. R., Strong microcapsules with permeable porous shells made through phase separation in double emulsions, Langmuir **33** (2017) 2402–10.
[28] Nishihama, S., Nishimura, G., Hirai, T., Komasawa, I., Separation and recovery of Cr(VI) from simulated plating waste using microcapsules containing quaternary ammonium salt, extractant and phosphoric acid extractant, Ind Eng Chem Res **43** (2004) 751–7.
[29] Jyothi, N. V. N., Prasanna, P. M., Sakarkar, S. N., Prabha, K. S., Ramaiah, P. S., Srawan, G. Y., Microencapsulation techniques, factors influencing encapsulation efficiency, J Microencapsul **27** (2010) 187–97.
[30] US Patent 3960762A *Production of absorbents.*
[31] Kauczor, H. W., Meyer, A., Structure and properties of Levextrel resins, Hydrometallurgy **3** (1978) 65–73.
[32] Fonseca, C., Araneda, C., Vargas, F., Yazdani-Pedram, M., Valenzuela, F., Icroencapsulation of trioctilamine in a polymeric matrix, for metal uptake from aqueous solutions. Xii Entmme / Vii Mshmt – Ouro Preto-Mg, (2007) pp. 145–51.
[33] Shiomori, K., Yoshizawa, H., Fujikubo, K., Kawano, Y., Hatate, Y., Kitamura, Y., Extraction equilibrium of precious metals from aqueous acidic solutions using divinylbenzene homopolymeric microcapsules encapsulating ternary amine as a core material, Sep Sci Technol **38** (2003) 4057–77.
[34] Alcazar, A., Borreguero, A. M., de Lucas, A., Rodriguez, J. F., Carmona, M., Microencapsulation of TOMAC by suspension polymerisation: process optimisation, Chem Eng Res Design **117** (2017) 1–10.
[35] Alcázar, Á., Garrido, I., García, E. M., De Lucas, A., Carmona, M., Rodriguez, J. F., New type of highly selective microcapsules for the removal of mercury from surface polluted waters, Sep Purif Technol **154** (2015) 255–62.
[36] Kamio, E., Matsumoto, M., Kondo, K., Extraction mechanism of rare metals with microcapsules containing organophosphorus compounds. J Chem Eng Jpn **35** (2002) 178–85.
[37] Kamio, E., Fujiwara, Y., Matsumoto, M., Valenzuela, F., Kondo, K., Investigation on extraction rate of lanthanides with extractant-impregnated microcapsule, Chem Eng J **139** (2008) 93–105.
[38] Araneda, C., Fonseca, C., Sapag, J., Basualto, C., Yazdani-Pedram, M., Kondo, K., Kamio, E., Valenzuela F. Removal of metal ions from aqueous solutions by sorption onto microcapsules prepared by copolymerization of ethylene glycol dimethacrylate with styrene, Sep Purif Techn **63** (2008) 517–23.
[39] Barassi, G., Valdes, A., Araneda, C., Basualto, C., Sapag, J., Tapia, C., Valenzuela, F., Cr(VI) sorption behavior from aqueous solutions onto polymeric microcapsules containing a

long-chain quaternary ammonium salt: Kinetics and thermodynamics analysis, J Hazard Mater **172** (2009) 262–8.
[40] Ochoa, N. A., Illanes, C., Marchese, J., Basualto, C., Valenzuela, F., Preparation and characterization of polymeric microspheres for Cr(VI) extraction, Sep Purif Technol **52** (2006) 39–45.
[41] Kobayashi, T., Yoshimoto, M., Nakao, K., Preparation and characterization of immobilized chelate extractant in PVA gel beads for an efficient recovery of copper (II) in aqueous solution, Ind Eng Chem Res **49** (2010) 11652–60.
[42] Panisello, C., Peña, B., Gumí, T., Garcia-Valls, R., Polysulfone microcapsules with different wall morphology, J Appl Polym Sci **129** (2013) 1625–36.
[43] Garcia-Valls, R., Panisello, C., Microencapsulation by Phase Inversion Precipitation, Handbook of Encapsulation and Controlled Release, (2015) 347–57.
[44] Alcázar, A., Perez, A., de Lucas, A., Carmona, M., Rodríguez, J. F., Sulfonation of microcapsules containing selective extractant agents J Chem Sci Technol **2** (2013) 25–38.
[45] Ozcan, S., Tor, A., Aydin, M. E., Removal of Cr(VI) from aqueous solution by polysulfone microcapsules containing Cyanex 923 as extraction reagent, Desalination **259** (2010) 179–86.
[46] Yang, W. W., Luo, G. S., Wu, F. Y., Chen, F., Gong X. C., Di-2-ethylhexyl phosphoric acid immobilization with polysulfone microcapsules, React Funct Polym **61** (2004) 91–9.
[47] Yadav, K. K., Singh, D. K., Anitha, M., Varshney, L., Singh, H., Studies on separation of rare earths from aqueous media by polyethersulfone beads containing D2EHPA as extractant, Sep Purif Techn **118** (2013) 350–8.
[48] Toro, C. A., Rodrigo, R., Cuellar, J., Sulfonation of macroporous poly(styrene-co-divinylbenzene) beads: Effect of the proportion of isomers on their cation exchange capacity, React Funct Polym **68** (2008) 1325–36.
[49] Alcázar, A., Gutiérrez, C., Lucas, A. D., Carmona, M., Rodríguez, J. F., Equilibrium treatment for highly selective sulfonated microcapsules containing di(2-ethylhexyl)phosphoric acid, Ind Eng Chem Res **55** (2016) 1033–42.
[50] Zhang, F., Cao, L., Yang, W., Preparation of monodisperse and anion-charged polystyrene microspheres stabilized with polymerizable sodium styrene sulfonate by dispersion polymerization, Macromol Chem Phys **211** (2010) 744–51.
[51] Khopade, A. J., Caruso, F., Stepwise self-assembled poly(amidoamine) dendrimer and poly(styrenesulfonate) microcapsules as sustained delivery vehicles, Biomacromolecules **3** (2002) 1154–62.
[52] Alford, C. E., Navratil, J. D., Tributyl-phosphate-impregnated sorbents for plutonium-uranium separations Rockwell International Corp., Golden, CO (USA), Rocky Flats Plant.
[53] Cortina, J. L., Miralles, N., Aguilar, M., Sastre, A. M., Extraction Studies of Zn(II), Cu(II) and Cd(II) with Impregnated and Levextrel resins containing di(2-ethylhexyl)phosphoric Acid, Hydrometallurgy **36** (1994) 131–42.
[54] Jha, M. K., Kumar, V., Lee Jae-chun., *Processing of electroplating effluent for the recovery of zinc and chromium using ion exchange technique.* In: Proceedings of International Symposium of Resource Recycling, 2007, resource recycling. Eds Davis, B., Free, M., TMS (The Minerals, Metals and Materials Society, (2007) pp. 77–86.
[55] Vaughan, J., Dieters, C., Fu, W., Byrne, K., Properties of Lewatit® TP272, a commercial solvent impregnated cation exchange resin for cobalt recovery, Miner Eng **88** (2016) 2–8.
[56] Lee, M. G., Kam, S. K., Suh, J. H., Lee, C. H., Removal of strontium ion from aqueous solution by PS-D2EHPA beads prepared by immobilizing di-2-ethylhexyl phosphoric acid (D2EHPA) with polysulfone (PS), Des Water Treat **124** (2018) 184–92.
[57] Kam, S. K., Jeon, J. W., Lee, M. G., Removal characteristics of Sr(II) by solid-phase extractant prepared by immobilizing di-(2-ethylhexyl)phosphoric acid (D2EHPA) and tri-butyl-phosphate (TBP) in polysulfone, J Environ Sci Inter **24** (2015) 267–74.

[58] Yadav, K. K., Singh, D. K., Varshney, L., Kain, V., Role of additives in the matrix of solvent encapsulated beads on its yttrium sorption from aqueous streams, Sep Sci Techn **52** (2017) 2274–83.
[59] Kam, S. K., Lee, S. W., Lee, M. G., Removal characteristics of Cu(II) ion in aqueous solution by solid-phase extractant immobilized D2EHPA and TBP in PVC, J Environ Sci Inter **24** (2015) 47–53.
[60] Fonseca, C., Araneda, C., Yazdani-Pedram, M., Borrmann, T., Basualto, C., Sapag, J., Valenzuela, F., Microencapsulation of trioctylamine in polymeric matrices for removing Zn (II) and Cu (II) from chloride aqueous solutions, J Chil Chem Soc **55** (2010) 408–14.
[61] Alcázar, A., Garrido, I., García, E. M., de Lucas, A., Carmona, M., Rodriguez, J. F., New type of highly selective microcapsules for the removal of mercury from surface polluted waters, Sep Purif Techn **154** (2015) 255–62.
[62] Kamio, E., Matsumoto, M., Valenzuela, F., Kondo, K. Sorption behavior of Ga(III) and In(III) into a microcapsule containing long-chain alkylphosphonic acid monoester, Ind Eng Chem Res **44** (2005) 2266–72.
[63] Wang, Y., Wang, Y., Jing, Y., Chen, J., Liu, Y., Microcapsules containing ionic liquid [A336][P507] for $La^{3+}/Sm^{3+}/Er^{3+}$ recovery from dilute aqueous solution, J Rare Earths **34** (2016) 1260–1268.
[64] Singh, K. K., Panja, S., Ruhela, R., Kumar, M., Tripathi, S. C., Singh, A. K., Chakravartty, J. K., Bajaj, P. N., Studies on BenzoDODA encapsulated polymeric beads for separation of plutonium from acidic solution, Sep Purif Technol **154** (2015) 186–92.
[65] Rhein-Knudsen, N., Ale, M. T., Meyer, A. S., Seaweed hydrocolloid production: an update on enzyme assisted extraction and modification technologies, Mar Drugs **13** (2015) 3340–59.
[66] Simó, G., Fernández-Fernández, E., Vila-Crespo, J., Ruipérez, V., Rodríguez-Nogales, J. M., Research progress in coating techniques of alginate gel polymer for cell encapsulation, Carbohydr Polym **170** (2017) 1–14.
[67] Leong, J. Y., Lam, W. H., Ho, K. W., Voo, W. P., Lee M. F. X., Lim, H. P., Lim, S. L., Tey, B. T., Poncelet, D., Chan, E. S., Advances in fabricating spherical alginate hydrogels with controlled particle designs by ionotropic gelation as encapsulation systems, Particuology **24** (2016) 44–60.
[68] Zhou, H. B., Chen, J., Li, S., Zhang, J., Zhu, C. E., Ran, H., Luo, M., Pan, X., Hu, H., Wu, C., Preparation of acid-resistant microcapsules with shell-matrix structure to enhance stability of Streptococcus Thermophilus IFFI 6038, J Food Sci **82** (2017) 1978–84.
[69] Benson, E. E., Harding, K., Ryan, M., Petrenko, A., Petrenko, Y., Fuller, B., alginate encapsulation to enhance biopreservation scope and success: a multidisciplinary review of current ideas and applications in cryopreservation and non-freezing storage, Cryoletters **39** (2018) 14–38.
[70] Orive, G., Tam, S. K., Pedraz, J. L., Hallé, J. P., Biocompatibility of alginate–poly-l-lysine microcapsules for cell therapy, Biomaterials **27** (2006) 3691–700.
[71] Sawalha, H., Schroën, K., Boom, R., Biodegradable polymeric microcapsules: Preparation and properties, Chem Eng J **169** (2011) 1–10.
[72] Freitas, S., Merkle, H. P., Gander, B., Microencapsulation by solvent extraction/evaporation: reviewing the state of the art of microsphere preparation process technology, J Control Release **102** (2005) 313–32.
[73] Souza, J. M., Caldas, A. L., Tohidi, S. D., Molina, J., Souto, A. P., Fangueiro, R., Zille, A., Properties and controlled release of chitosan microencapsulated limonene oil, Rev Bras Farmacogn **24** (2014) 691–8.
[74] Javid, A., Raza, Z. A., Hussain, T., Rehman, A., Chitosan microencapsulation of various essential oils to enhance the functional properties of cotton fabric, J Microencapsul **31** (2014) 461–8.
[75] Outokesh, M., Mimura, H., Niibori, Y., Tanaka, K., Preparation of stable alginate microcapsules coated with chitosan or polyethyleneimine for extraction of heavy metal ions, J Microencapsul **23** (2006) 291–301.

[76] Wang, B., Wan, Y., Zheng, Y., Lee, X., Liu, T., Yu, Z., Huang, J., Ok, Y. S., Chen, J., Gao, B., Alginate-based composites for environmental applications: a critical review, Crit Rev Environ Sci Technol **49** (2019) 318–56.

[77] Zwar, E., Kemna, A., Richter, L., Degen, P., Rehage, H., Production, deformation and mechanical investigation of magnetic alginate capsules, J Phys Condens Matter **30** (2018) 085101.

[78] Russo, R., Malinconico, M., Santagata, G., Effect of cross-linking with calcium ions on the physical properties of alginate films, Biomacromolecules **8** (2007) 3193–7.

[79] Deze, E. G., Papageorgiou, S. K., Favvas, E. P., Katsaros, F. K., Porous alginate aerogel beads for effective and rapid heavy metal sorption from aqueous solutions: effect of porosity in Cu^{2+} and Cd^{2+} ion sorption, Chem Eng J **209** (2012) 537–46.

[80] Shim, J., Kumar, M., Mukherjee, S., Goswami, R., Sustainable removal of pernicious arsenic and cadmium by a novel composite of MnO2 impregnated alginate beads: a cost-effective approach for wastewater treatment, J Environ Manage **234** (2019) Pages 8–20.

[81] Idris, A., Ismail, N. S. M., Hassan, N., Misran, E., Ngomsik, A. F., Synthesis of magnetic alginate beads based on maghemite nanoparticles for Pb(II) removal in aqueous solution, J Ind Eng Chem **18** (2012) 1582–9.

[82] Bée, A., Talbot, D., Abramson, S., Dupuis, V., Magnetic alginate beads for Pb(II) ions removal from wastewater, J Colloid Interface Sci **362** (2011) 486–92.

[83] Germanos, G., Youssef, S., Abboud, M., Farah, W., Lescop, B., Rioual, S., Diffusion and agglomeration of iron oxide nanoparticles in magnetic calcium alginate beads initiated by copper sorption, J Environ Chem Eng **5** (2017) 3727–33.

[84] Li, X., Qi, Y., Li, Y., Zhang, Y., He, X., Wang, Y., Novel magnetic beads based on sodium alginate gel crosslinked by zirconium(IV) and their effective removal for Pb^{2+} in aqueous solutions by using a batch and continuous systems, Bioresour Technol **142** (2013) 611–19.

[85] Gopalakannan, V., Viswanathan, N., Synthesis of magnetic alginate hybrid beads for efficient chromium (VI) removal, Int J Biol Macromol **72** (2015) 862–7.

[86] Periyasamy, S., Gopalakannan, V., Viswanathan, N., Hydrothermal assisted magnetic nano-hydroxyapatite encapsulated alginate beads for efficient Cr(VI) uptake from water, J Environ Chem Eng **6** (2018) 1443–54.

[87] Yi, X., He, J., Guo, Y., Han, Z., Yang, M., Jin, J., Gu, J., Ou, M., Xu, X., Encapsulating Fe_3O_4 into calcium alginate coated chitosan hydrochloride hydrogel beads for removal of Cu(II) and U(VI) from aqueous solutions, Ecotoxicol Environ Saf **147** (2018) 699–707.

[88] Ngomsik, A. F., Bee, A., Siaugue, J. M., Cabuil, V., Cote, G., Nickel adsorption by magnetic alginate microcapsules containing an extractant, Water Res **40** (2006) 1848–56.

[89] Zhang, L., Wu, D., Zhu, B., Yang, Y., Wang, L., Adsorption and selective separation of neodymium with magnetic alginate microcapsules containing the extractant 2-ethylhexylphosphonic acid mono-2-ethylhexyl ester, J Chem Eng Data **56** (2011) 2280–9.

[90] Wu, Y., Outokesh, M., Mimura, H., Niibori, Y., Selective uptake properties of metal ions by hybrid microcapsules enclosed with TBP, Prog Nuc Energ **50** (2008) 487–93.

[91] Mimura, H., Hoshi, H., Akiba, K., Onodera, Y., Separation of americium from europium by biopolymer microcapsules enclosing Cyanex 301 extractant, J Radioanal Nucl Chem **247** (2001) 375–9.

[92] Wei, Reddy, D. H. K., Bediako, J. K., Yun, Y. S., Aliquat-336-impregnated alginate capsule as a green sorbent for selective recovery of gold from metal mixtures, Chem Eng J **289** (2016) 413–22.

Renata Jastrząb

9 Micro and nanocapsules as supports for Surface-Enhanced Raman Spectroscopy (SERS)

9.1 Introduction

Raman spectroscopy (RS) is a relatively weak optical process that provides information about the unique vibrational modes of molecules. This technique is effective and essential, e.g. for solid materials or analysis in solution, however its sensitivity is poor. As a result of many attempts at improving the sensitivity of Raman spectroscopy, a successful solution was a combination of surface enhancement scattering and RS, which proved a highly selective and sensitive method referred to as surface-enhanced Raman scattering/spectroscopy (SERS). SERS is an extremely effective qualitative technique, which works on standard equipment. Moreover, this method permits the analyses of small amounts of samples, e.g. a drastically diluted solution (even 10^{-16} mol/dm^3). Enhancement in SERS is about 100 times that of standard Raman scattering. The most important point is that SERS signals are effective for a wide range of molecules and could be applied for detection, e.g. of cancer, anthrax [1, 2], chemical warfare-stimulants [3], explosive-agents [4, 5]; for environmental monitoring [6] or for the monitoring of heterogeneous catalytic reactions [7] as well as *in vitro* [8–10] and *in vivo* [11] glucose sensing. The best SERS-active materials are nanoparticles of silver and gold, but other metals such as copper are also effective. It is worth highlighting that the enhancement in resolution depends not only on the kind of material but particularly on the shape of the nanoparticles. The most common nanoparticles are colloids of the metals. In recent years, thousands of papers have been published about SERS, which illustrates the fast development of this method and its powerful possibilities.

9.2 History

Inelastic scattering of light was first presented by an Austrian theoretical physicist, Adolf Smekal in 1923 [12]. This type of scattering was observed by Indian physicists Chandrasekhara Venkata Raman and his student Kariamanikkam Srinivasa Krishnan on the 28 February 1928. A week earlier, Russian physicists Grigory Samuilovich Landsberg and Leonid Mandelstam discovered the same effect as Raman and Krishnan. Chandrasekhara Raman published his results before Landsberg, in an article entitled "A New Type of Secondary Radiation", and that is why the effect carries his name. In Russia this effect is still called "combinatorial scattering of light" [13–16]. In 1930, Sir Ch. V. Raman won the Nobel Prize in Physics "for his work on the scattering of light and for the discovery of the effect named in his honor". The limitation of

https://doi.org/10.1515/9783110642070-009

Raman spectroscopy in the early years was correlated with the preparation of samples (concentration above 1 mol/L and volumes of a minimum 5 mL). The milestone for simplified Raman spectrometers and improved sensitivity of this technique [17] was the use of the laser beam. The first functioning laser was constructed by Theodore H. Maiman in 1960 (a ruby crystal used to produced a red light laser of 694 nm wavelength) and later in 1960 the first gas laser was discovered by Ali Javan, William Bennett and Donald Herriott (helium and neon were used to produce a red light laser of 632.8 nm wavelength).

In 1973, Martin Fleischmann, Patrick J. Hendra and A. James McQuillan observed a much-enhanced Raman signal of pyridine adsorbed on a roughened silver surface [18]. After that, a lot of independent scientists confirmed this extraordinary effect, and a mechanism for the observed enhancement was proposed. Theoretical fundamentals of the SERS were laid in 1977 by two independent groups of scientists. The first research group headed by David L. Jeanmaire and Richard P. Van Duyne proposed the electromagnetic effect as a fundamental of SERS, while the second group of scientists, Albrecht M. Grant and Alan J. Creighton, proposed the charge-transfer effect to play this role [19, 20]. In 1997, a milestone was the recording of a SERS spectrum of a single-molecule, which made the grounds for elimination of the limitations of RS. It has resulted in the appearance of a new research field and development of SERS as an effective analytical technique [5, 21, 22].

9.3 Fundamentals

Classical RS is a technique used to detect vibrational and rotational oscillations of a molecule upon incidence of monochromatic light, as they are functions of frequency of the scattered light. The scattering of light can be elastic (Rayleigh scattering, the frequency of scattered light is the same as that of the initial photon $hv_0 = hv_0$) or

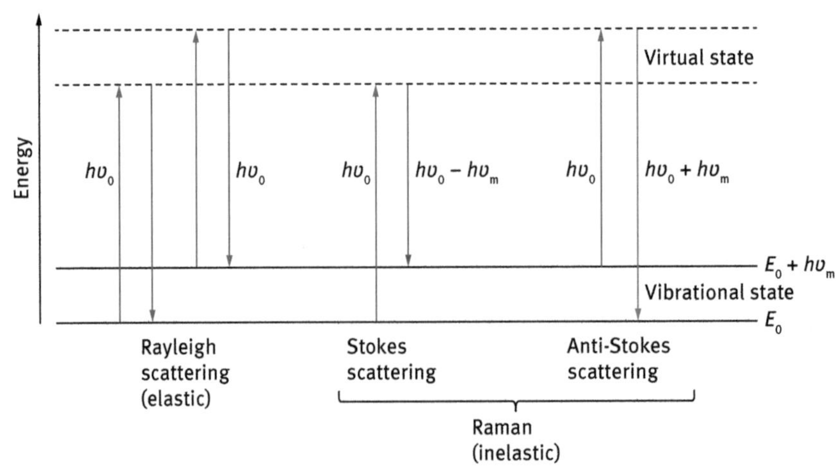

Fig. 9.1: Different forms of scattering.

inelastic (Raman scattering). Inelastic scattering can be divided into Stokes and anti-Stokes scattering. The laser light interacts with a molecule to endow it with higher energy and move it to a higher or lower level ($h\nu_0 = h\nu_0 - h\nu_m$ or $h\nu_0 = h\nu_0 + h\nu_m$) (Figure 9.1).

Stokes and anti-Stokes scattering are relatively weak and their emission intensities are up to several orders of magnitude lower when compared to those of elastic scattering bands, which could be overlapped by Rayleigh bands. Rayleigh and Stokes processes are more important in Raman spectroscopy than the weakest anti-Stokes (Figure 9.2), but the recording of these signals is essential for higher sensitivity detection. The SERS becomes an extremely effective technique as it permits enhancement of the Raman signal intensities by several orders of magnitude.

The surface enhancement factor of Raman signal $EF_{SERS} = 10^6$ or more could be explained as a result of two contributing effects: (i) the electromagnetic enhancement (EM) [23] and (ii) chemical (charge-transfer) enhancement (EC) [24, 25]. High levels of enhancement are always related to specific morphologies of nanostructures [26–30]. The molecule analyzed must be adsorbed on a SERS-active surface and irradiated by monochromatic radiation, usually from a laser and detected via a Raman spectrometer (Figure 9.3).

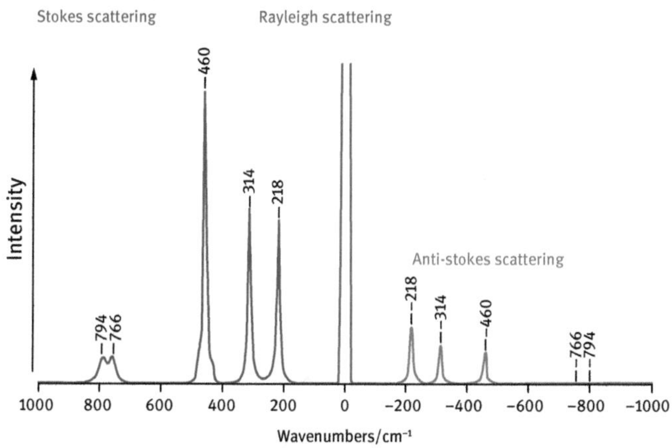

Fig. 9.2: Raman spectra of CCl_4 recorded using a laser line of 488 nm.

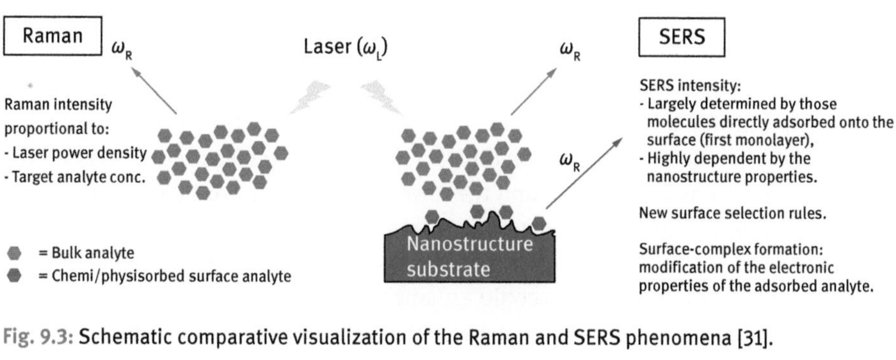

Fig. 9.3: Schematic comparative visualization of the Raman and SERS phenomena [31].

9.4 The electromagnetic enhancement (EM)

Electromagnetic enhancement (EM) involves interactions between surface plasmon of the metallic nanosized cluster of Ag, Au or Cu and the molecule located near the plasmon surface [32, 33]. Conversely, EM is related to resonances of the optical fields with surface plasmons [34]. Plasmon is a collective excitation of the electron cloud, while surface plasmons is an excitation confined to the nearest surface region (Figure 9.4).

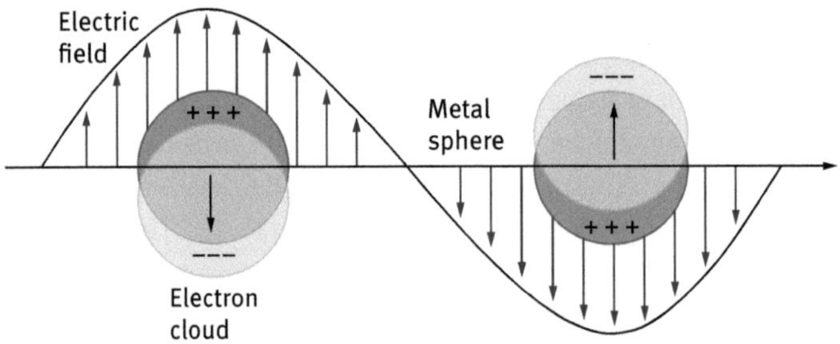

Fig. 9.4: Illustration of the localized surface plasmon resonance effect.

Moreover, nanostructures can be considered as nanoantennae for transmission and enhancement of Raman scattered light. EM depends on the resonance process between the plasmons, excitations and scattered fields. The enhancement is strong when excitation and scattered fields are in resonance with the surface plasmons. The shifts in frequencies between excitation and scattered light are small when compared to the width of the plasmon resonance [35].

9.5 The chemical enhancement

Chemical (charge-transfer) enhancement (CE) is a result of bonding between the analyzed molecule and metal nanoparticles surface [32, 33]. The exciting radiation interacts with the metal to form an electron-hole pair. The energy is transferred to the analyte through metal to the bonds of the molecule. The charge-transfer complex formed considerably increases the molecular polarizability of the molecule due to interaction with the metal electrons, which generates new, shifted or broadened Raman bands [28, 36–41]. The Raman process occurs on the analyte, and the energy is transferred back into the metal to be scattered [42]. The charge-transfer occurs between the conduction electrons of the metal and the lowest unoccupied orbital of the molecule chemisorbed to its surface. CE occurs only from the molecules directly attached to the surface and increases only up to monolayer coverage. It is possible that electromagnetic and/or chemical effects could enhance a Raman signal by up to $\times 10^{14}$. For the systems in which both mechanisms are simultaneously operative, the effects are

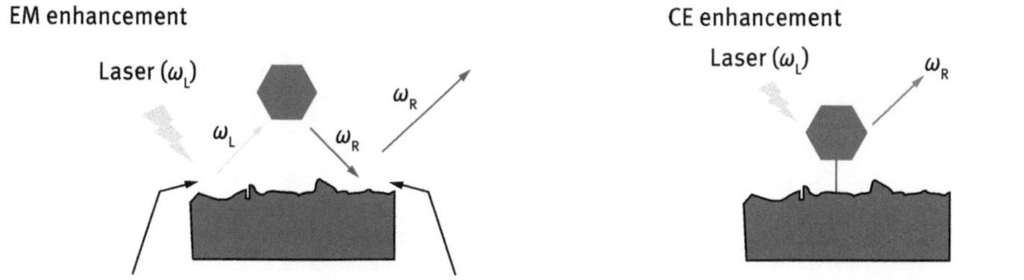

Fig. 9.5: Schematic outline of the electromagnetic and chemical enhancements in SERS [31].

multiplied. It has been almost impossible to separate these two effects [24]. Moreover, it has been proved that the electromagnetic enhancement based on plasmon resonances gives larger effects than charge-transfer enhancement [43] (Figure 9.5).

9.5.1 Effective SERS materials

As mentioned above, Raman scattering is a relatively weak optical phenomenon. SERS is the method allowing enhancement of the weak signal it gives on specially prepared materials. Current efforts in SERS probe development are aimed at reproducible preparation of highly sensitive SERS-active nanostructures with a narrow distribution of their enhancement factor (EF) values [44]. SERS-active substrates, providing nanoscale and atomic-scale roughness, include evaporated island films [45]. The best active materials are based on gold, silver or copper particles. The most common types of SERS-active substrates are clusters of colloidal silver or gold particles in the 10–100 nm size range, used in colloidal solution or "dry" on a surface. These nanoparticles are made by chemical reduction processes using, for example, citric acid, sodium borohydride or glucose [46–49]. Despite considerable success in synthesizing silver nanoparticles with different particle size distributions, many of the reported methods have certain limitations in terms of their control over shape, size and stability in the dispersion system [50, 51] (Figures 9.6 and Figure 9.7).

Achieving nanoparticles below 10 nm with high monodispersity and stability [52–55] is not easy, but the use of an excess of a strong reducing agent, for example sodium borohydride, permits a synthesis of monodisperse small uniform-sized nanoparticles. Moreover, it is difficult to obtain larger-sized nanoparticles via chemical reduction [47, 56, 57].

On the other hand, aggregated nanoparticles increase the intensity of Raman signals much more than separated particles, because of the enhanced field around the near-field coupling species [58, 59]. Confinement of nanoparticles in a limited space creates the so-called "hot spots", which increase even more the intensity of RS of molecules

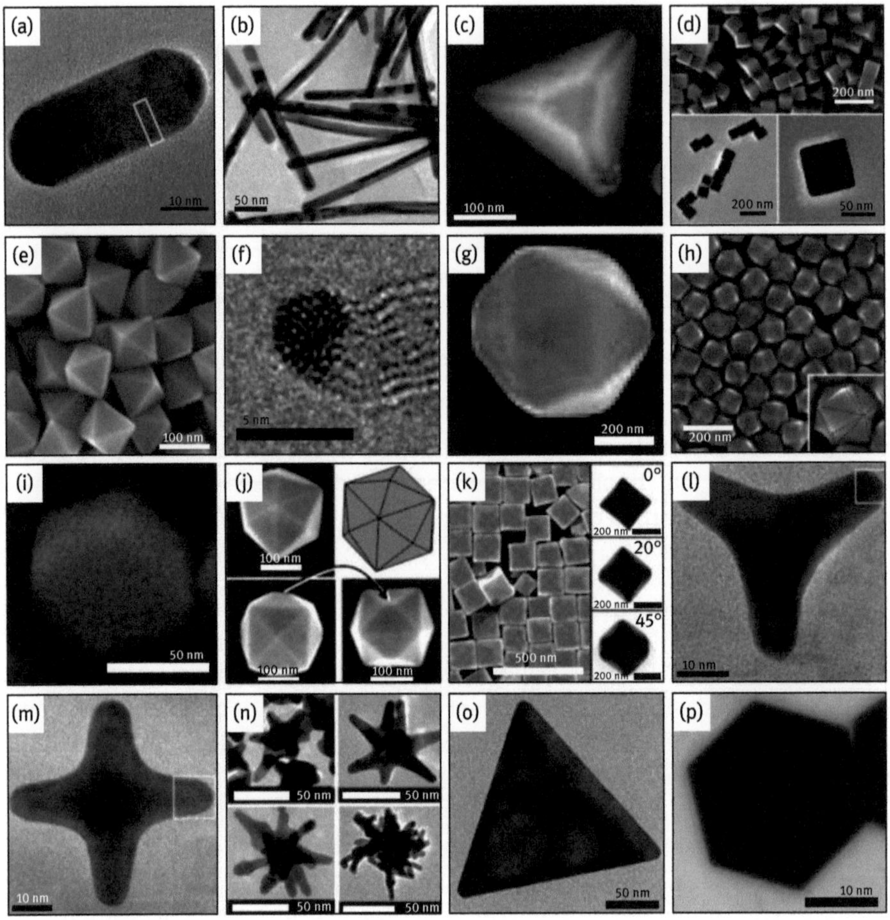

Fig. 9.6: The major classes of noble metal nanoparticle shapes seen through transmission electron microscopy (TEM) and/or scanning electron microscope (SEM): (a) Au octagonal single-crystal rod, (b) Au pentagonally twinned rods, (c) Au tetrahedron NP, (d) Pd hexahedron (i.e. cube) NPs, (e) Au octahedron NPs, (f) decahedron, (g) Au icosahedron NP, (h) Au trisoctahedron NPs, (i) Au rhombic dodecahedron NP, (j) Pt tetrahexahedron NPs, (k) Au concave hexahedron NPs, (l) Au tripod NP, (m) Au tetrapod NP, (n) Au star NPs, (o) Au triangular plate/prism NP, and (p) Au hexagonal plate/prism NP [60].

located in this area. Many reports have been published on different attempts to produce active hot spots to achieve high SERS signal by assembly of nanoparticles on the surface of various supports, such as polystyrene [61], silica [62] or agarose [63] microbeads, but also by use of agarose gels [64] or films of poly(diallyldimethylammonium chloride) and poly(acrylic acid) [65]. The most commonly used preparation method is the LbL assembly technique [63, 65]. Moreover, encapsulation of ready-to-use hot spots is the best method of obtaining active and long-life time materials.

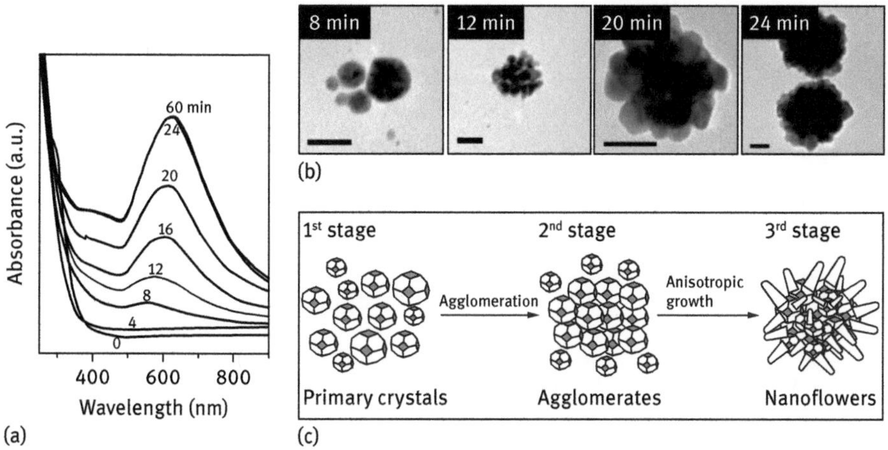

Fig. 9.7: (a) UV-vis spectra as a function of time of reaction between aqueous $AuCl_4^-$ solution (0.5 mM) and HEPES (10mM). (b) Representative transmission electron microscopy (TEM) images of the products harvested at 8, 12, 20 and 24 min into the reaction. All scale bars are 20nm. (c) Schematic illustration of the proposed mechanism for Au nanoflower formation in HEPES solution [52].

Capsules in general are spherical membranes that separate their inside from the outside environment. They can have different morphology, depending on the material used to their preparation. Furthermore, the preparation technique has a huge impact on the final outcome [66]. "Microcapsule" may be defined as a circular cross-section shaped particle with certain free volume inside, where a core material can be allocated. Their diameter size varies in the range 1–1000 µm (nanocapsules below 1 µm and macrocapsules above 1000 µm) [67, 68]. Depending on their structure they can be characterized as a continuous core/shell microcapsule, polycore capsule, continuous core capsule with more than one layer of shell material and matrix type, where the encapsulated agent is incorporated within the shell material [69].

An example of microcapsulation is a thiolated block copolymer consisting of a pH-responsive PMAA ploymethacrylic acid segment and an amphiphilic polyethylene glycol PEG segment for encapsulating gold nanocrystals. In materials of this type, SERS signals can be switched on and off by molecular conformational changes (Figure 9.8). It has been shown that neutralized PMAA molecules are able to interact with amphiphilic PEG chains, leading to highly compact and intermingled copolymer structures on the surface of nanoparticles. This type of molecular conformational change provides a new strategy for controling the distances and plasmonic interactions between two or more gold nanoparticles. This opens a possibility of using SERS nanoparticle tags for biomolecular binding and enzymatic cleavage studies [70].

The development of biocompatible nanoparticles for *in vivo* molecular imaging and targeted therapy is an area of considerable current interest across a number

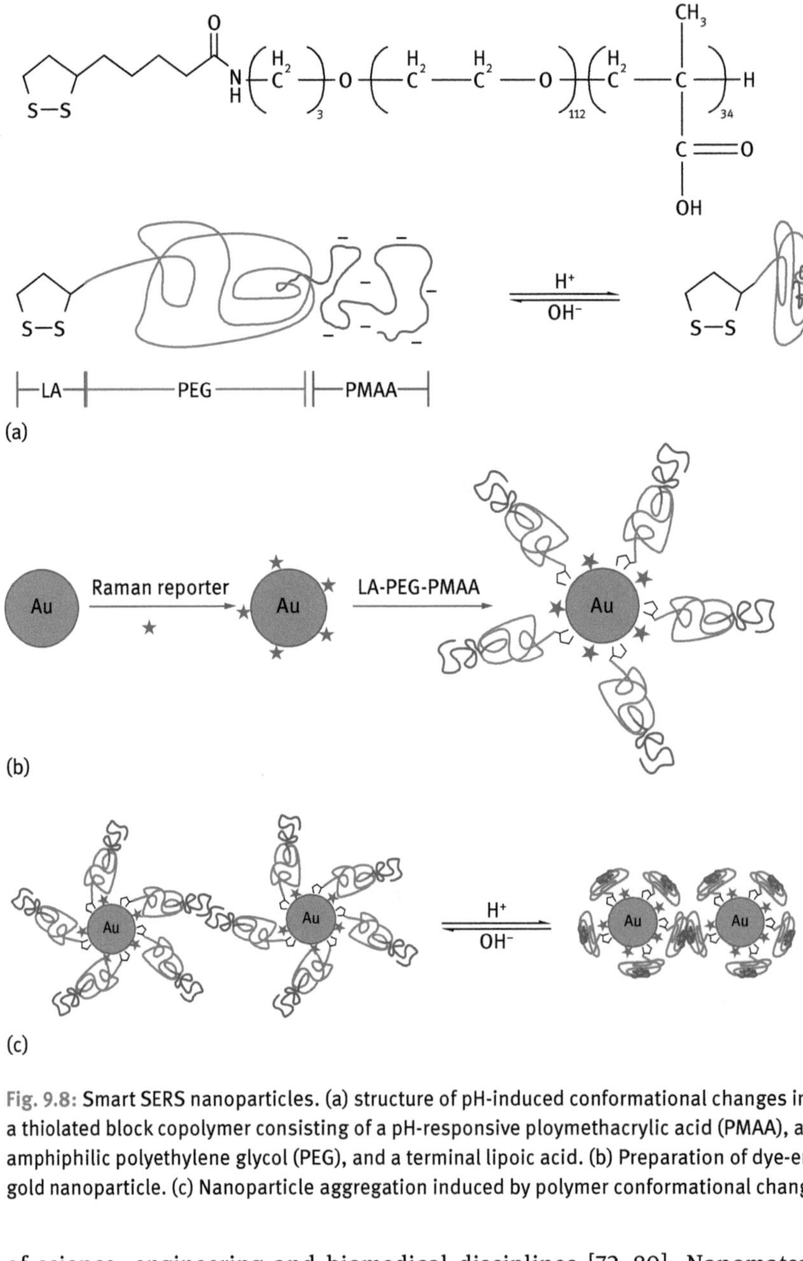

Fig. 9.8: Smart SERS nanoparticles. (a) structure of pH-induced conformational changes in a thiolated block copolymer consisting of a pH-responsive ploymethacrylic acid (PMAA), an amphiphilic polyethylene glycol (PEG), and a terminal lipoic acid. (b) Preparation of dye-encoded gold nanoparticle. (c) Nanoparticle aggregation induced by polymer conformational changes [71].

of science, engineering and biomedical disciplines [72–80]. Nanomaterials conjugated with bioligands such as monoclonal antibodies, peptides or small molecules can be used to target malignant tumors with high specificity and affinity [81–84]. Another advantage of nanoparticles is their large surface areas available for conjugation to multiple diagnostic (e.g. optical, radioisotopic or magnetic) and therapeutic (e.g. anticancer) agents. Recent advances have led to the development of biodegradable nontoxic nanostructures for drug delivery *in vivo*, for tumor targeting and spectroscopic detection [85–89] (Figure 9.9)

Moreover, colloidal gold nanoparticles have been safely used to treat rheumatoid arthritis for half a century [90, 91]. However, recent work indicates the pegylated gold

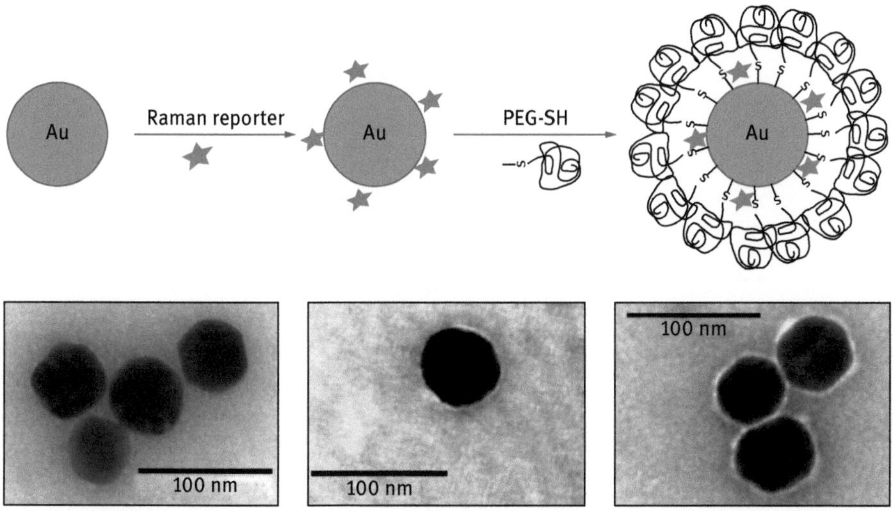

Fig. 9.9: Preparation and schematic structures of the gold pegylated colloid nanoparticles, and transmission electron microscopy (TEM) of each colloid particles [97].

nanoparticles (colloidal gold coated with a protective layer of polyethylene glycol or PEG) exhibit excellent *in vivo* biodistribution and pharmacokinetic properties upon systemic injection [92–94]. In contrast to quantum dots, containing cadmium and other toxic or immunogenic nanoparticles, gold colloids have almost no long-term toxicity [95–96].

Another very important feature of this technique is not only the compatibility in size and geometry of the sample of SERS-active substrates (Figure 9.10); but also "chemical" compatibility to a "biological environment". In general, due to its chemical inactivity, gold should be more suitable for incorporation inside biophysical systems. It has been shown that gold colloidal clusters have SERS enhancement factors (SERS EFs) comparable to those of silver clusters, when Near infrared (NIR) excitation is applied [45].

9.6 Application of SERS

The SERS has a very wide range of possible applications: SERS microscopy [98], biosensing [99], diagnostics [100], imaging [101], and clinical translation [102]. This spectroscopy is used to identify explosive materials [103], therapeutic agents [104],

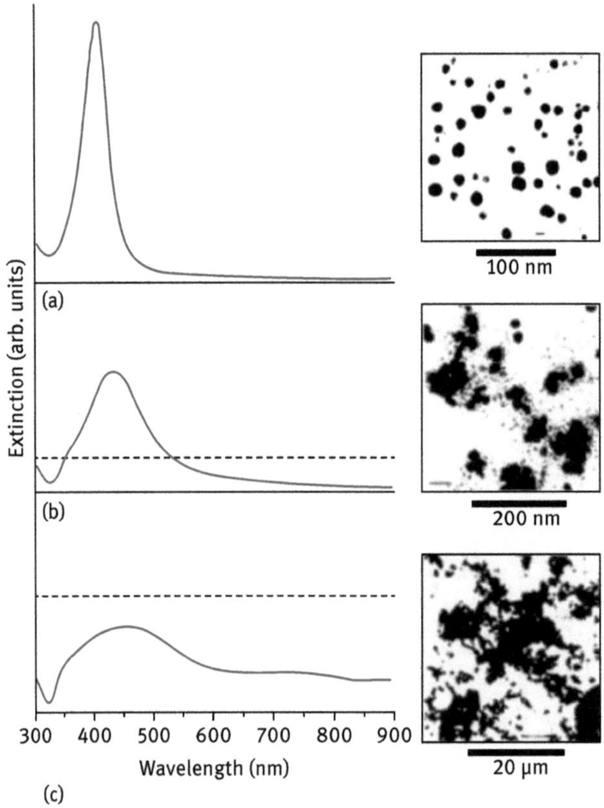

Fig. 9.10: SERS-active colloidal silver particles in different aggregation stages, demonstrating the fractal nature of these structures together with the appropriate extinction curves [96].

drugs of abuse [105], food additives [106], cells and spores and DNA sequence analysis [27, 107–115]. SERS allows detection of very small quantities of molecules, even a single-molecule and its identification moreover could be useful for detecting a known molecule and for monitoring its distribution [116].

Several reviews have described the preparation of SERS substrates (e.g. nanosphere lithography [5], nanofabrication techniques [31], surface functionalization [117], and fibre sensors [118]) and their performance in specific applications as well as *in vitro* cancer detection and diagnostics and for cancer imaging [1, 2].

SERS has been used to investigate a wide variety of problems in science. Recently Natan's group has published an interesting series of papers on the development of novel SERS substrates based on the self-assembly gold colloids [119]. They studied the compatibility of biomolecules with gold colloids coated with silver as a potential substances useful in SERS. The versatility of this approach is a promising development for analytical applications. Weaver's group has found a way to extend the SERS technique to transition metal surfaces by electrodepositing the metal of interest on a suitably prepared enhancing gold substrate. Although attempts at realization of this idea had been made before, the inability to make pinhole-free in the layer prevented the technique from being generally useful. High quality films were prepared by slow deposition at a constant cathode current rather than by a more rapid constant potential deposition method used formerly. These films must be thick enough to behave chemically as the bulk metal of interest and at the same time thin enough to support the electromagnetic enhancement of the underlying substrate. Recently, promising applications of Weaver's *in situ* method to study gas phase heterogeneous catalytic reactions and electrocatalytic processes have been reported. The future of this area looks very promising [120].

Gold nanoparticles, because of their biocompatibility, were investigated in cell biology. The nanoparticles were used directly as a probe of the chemical composition of endosomes of different stages and for the detection of specific cellular molecules, such as adenosine monophosphate (AMP) [121]. Gold as well as silver nanoparticles can be exported as labels that highlight cellular structures based on the enhanced Raman surface [122–124]. The Raman spectra recorded have fingerprint of molecules linked to the surface material. Other biomolecules such as as amino acids, purine or pyrimidine bases and "large" molecules such as proteins, DNA and RNA "intrinsically colored" biomolecules such as chlorophylls, hem-containing proteins and other pigments were studied by SERS. Gold nanoparticles were capped with a biofunctional molecule capable of forming a covalent link with the aromatic residues of the protein moiety, antithrombin as a sensitive recognition element [125].

Moreover SERS can be used to monitor transport through membranes and the results show that this technique can discriminate between the movement of different molecules across a membrane and to observe different interfacial arrival times and concentration growth rates in the receiving (colloidal silver) solution [126].

An extraordinary challenge is to apply SERS in living systems for real medical problems [127, 128]. The detection, identification and quantification of neurotransmitters in the brain fluid is an important question in neurochemistry [129]. SERS represents an interesting approach for studying charge-transfer processes, for instance in cytochrome c, which has been investigated on bare and coated silver electrodes [130].

The first SERS spectra of DNA adsorbed on a silver surface were reported in 1981 [131] and thereafter SERS studies of nucleic acids and their components quickly developed [110, 132–135]. In most of the SERS studies, the target nucleic acids were applied in relatively high concentrations (10^{-5}–10^{-8} mol/L), but the most interesting aspects of SERS on nucleic acids might be attributed to the ultrasensitive and single-molecule capabilities of the method.

A first *in vivo* application of SERS was demonstrated for quantitative glucose measurements in an animal model [11]. The result of the study indicated that glucose binds reversibly to the SERS-active surface and that changes in concentration as rapid as in 30 s can be measured. Glucose sensing has great medical value but the SERS spectrum of glucose is generally of relatively low intensity. Van Duyne and co-workers created a modified surface designed to adsorb glucose effectively on a solid state substrate made by depositing silver on a layer of polystyrene beads and then removing the beads to reveal a SERS-active surface of closely spaced essentially triangular deposits of silver [136]. Reproducible surface and strong signals were possible to create a monitor for glucose detection based on SERS technology.

Gold nanoparticles or nanoaggregates can be used as nanosensors to probe small biological structures such as cells and bacteria [121, 122, 137–141]. Applications of SERS in biomedical sensing includes SERS labels based on highly selective surface-enhanced Raman spectrum of a reporter attached to an enhancing silver or gold nanostructure [77, 142, 143]. Au or Ag cores functionalized with Raman active molecules can also be encapsulated in a glass shell, which provides the SERS label with mechanical and chemical stability [128].

Entities such as mammalian cells and spores give broader, more complex spectra. The origin of these signals is likely to be the part of the cell or spore closest to the enhancing surface on the spectra, and various bands that are indicative of proteins can be identified. Remarkably, the spectra can be used for effective discrimination of different cell types. The data are usually analyzed by methods such as least squares and principal component analysis and this information can discriminate between genetically different species of intact bacillus spores [42, 108].

The progress in SERS techniques could be used in cancer diagnostics, including multiplexed detection and identification of new biomarkers, single-nucleotide polymorphisms, and circulating tumor cells. In these experiments, colloidal silver particles were incorporated inside the cells and SERS was applied to monitor the intracellular distribution of drugs in the whole cell and to study the antitumor drugs/nucleic acid complexes. SERS is also used as a non-invasive tool for cancer imaging with immunoSERS microscopy, histological analysis of biopsies, and *in vivo* detection of tumors [144].

The high sensitivity and multiplexing capabilities of SERS technologies were supported by their integration into molecular diagnostics for *in vitro* cancer detection. A common approach involves immunoassays that rely on the recognition of biomarkers (cell surface markers, membrane receptors) with antibodies that are conjugated to SERS substrates. For instance mucin protein (encoded as *MUC4* gene) might be used as a serum marker for early detection of pancreatic cancer using a quantitative SERS-based platform [145]. The possibility of monitoring more than one biomarker enhanced the accuracy of lung cancer diagnostics (Figure 9.11) [146, 147].

SERS technologies, because of their high sensitivity, are ideal for the development of diagnostic assays and imaging tools and have progressed towards the quantification of biomarkers in the form of cell surface markers, mutant genes, and alleles. Moreover the assays require small sample volumes (a few microliters) and have extremely low detection limits (up to femtomolar level). SERS optical imaging with its modality for mapping biomolecules in cancer tissue with single-cell resolution has multifarious capabilities. Nowadays SERS technologies guide intraoperative imaging for tumor resection and endoscope-based imaging and have a significant impact on the next generation of molecular imaging tools for cancer detection and therapeutics [144].

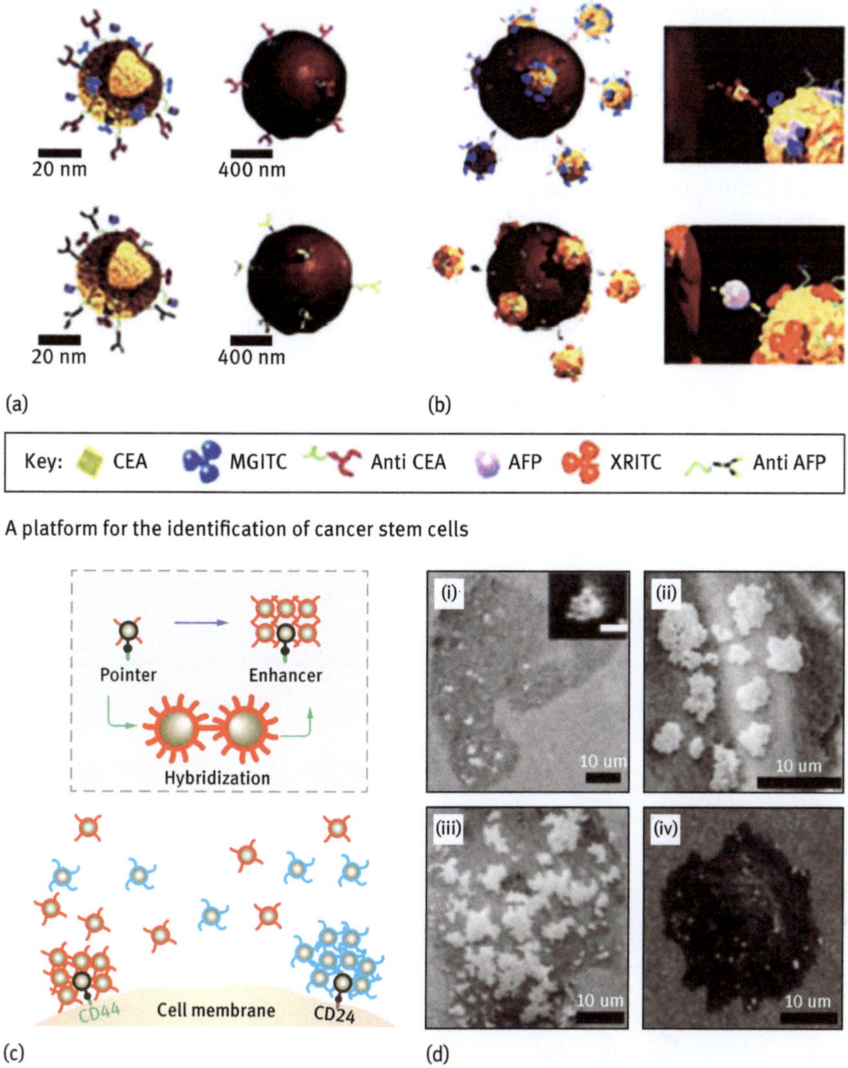

Fig. 9.11: Multiplex SERS immunoassays combining Au nanospheres and magnetic beads. (a) Conjugation of nanospheres and beads with reporters (malachite green isothiocyanate, MGITC; X-rhodamine isothiocyanate, XRITC), anti-carcinoembryonic antigen (CEA) and anti-α-fetoprotein (AFP). (b) Sandwich immunocomplexes after recognition of CEA and AFP [146]. (c, d) Platform for identification of cancer stem cells. (c) Schematic illustration, (d) backscattering SEM images [147].

Another biological application of SERS spectra is the enzyme immunoassay of the enzyme reaction product [148]. Antibodies immobilized on a solid substrate bind antigen, which binds to a second antibody labeled with peroxidase. If these immunocomplexes are subjected to the reaction with o-phenylenediamine, azoaniline is generated. The reaction product is adsorbed on colloidal silver particles, resulting in a strong SERS spectrum of azoaniline, whose signal strength is proportional to the concentration of the antigen. The method of silver colloidal SERS of the enzymatic product has been applied to direct detection of enzymes in cells. Moreover, the method has been successfully used to detect and to quantify prostaglandin H synthase-1 and 2 in normal human hepatocytes and human hepatocellular carcinoma cells [149].

Additionally, colloidal gold may be supported by the gold surface and that "SERS-active substrate" could bind immobilized antibodies and capture antigens from solution. Gold nanoparticles labeled with both specific antibodies and a specific reporter bind to the captured antigen. By immobilizing different antibodies and using different reporters, the presence of different antigens can be detected by the

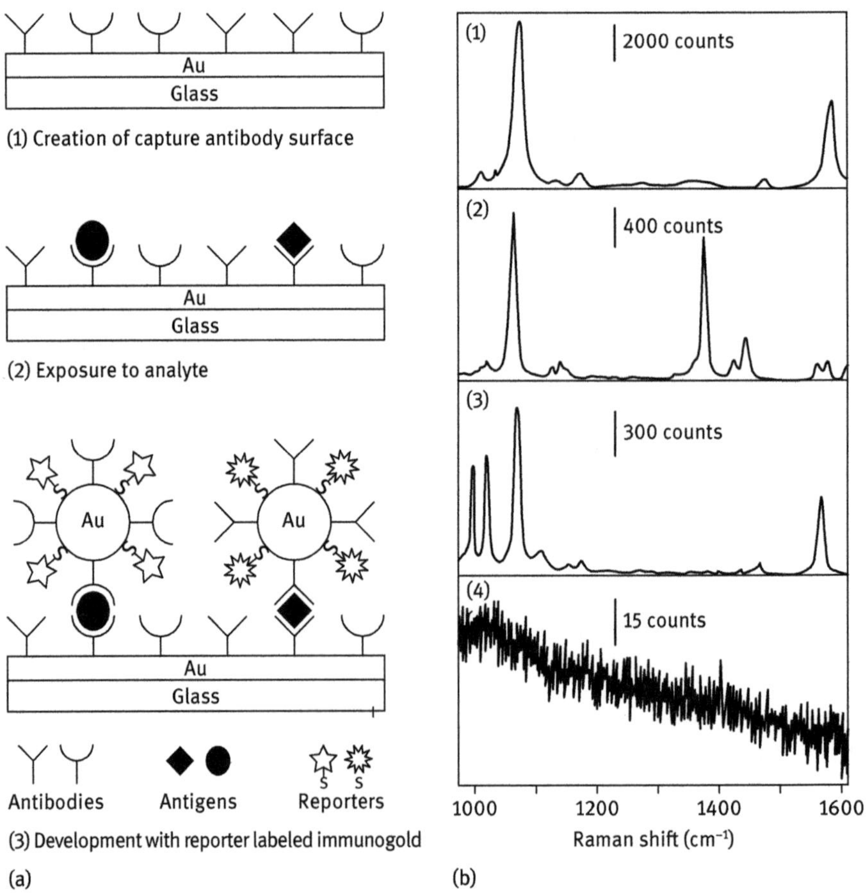

Fig. 9.12: (a) Scheme of an immunoassay system using two different SERS labels. (b) SERS signatures of three types of reporter-labeled immunogold colloid [150].

characteristic surface-enhanced Raman spectrum of the specific reporter molecules. There are several potential advantages of using SERS as a read-out method (Figure 9.12).

The SERS gene technique has been used for determination of the human immunodeficiency virus (HIV) gag gene sequence. The results of this study are potentially useful for HIV detection by SERS gene probes [151]. Progress in DNA and genome research results in quick development of SERS techniques for rapid characterization of DNA fragments [152]. A SERS-based method has been proposed for monitoring the concentration of double-stranded DNA amplified by polymerase chain reactions. Methods for labeling can employ radioactive or fluorescence reporters [153]. Recently, quantum dots and nanoparticles have been suggested as interesting new fluorescence labels for characterizing DNA fragments and for detecting specific nucleic acid sequences [154, 155].

Another example of using SERS spectroscopy is identification of artificial neural networks in aqueous solutions of different neurotransmitters [156]. Spectra of neurotransmitters have been measured on silver electrodes and on colloidal silver particles in water [114, 157–160]. The high sensitivity of SERS has been also applied for identification of relatively small amounts of bacteria. The first SERS studies of bacteria were reported in 1998 [161] and in that work, silver colloidal particles were produced selectively within the bacterium on its wall, forming there a rough silver coating. Considering the number and diversity of biomolecules in the bacterial wall, the SERS spectrum is a selective effect showing predominantly those molecules and functional groups that are in the immediate proximity of the silver colloid [161, 162].

Enkephalin, an endogenous substance in the human brain, was detected at the single-molecule level based on the surface-enhanced Raman signal of the ring breathing mode of phenylalanine, which is one building block of the molecule. The SERS

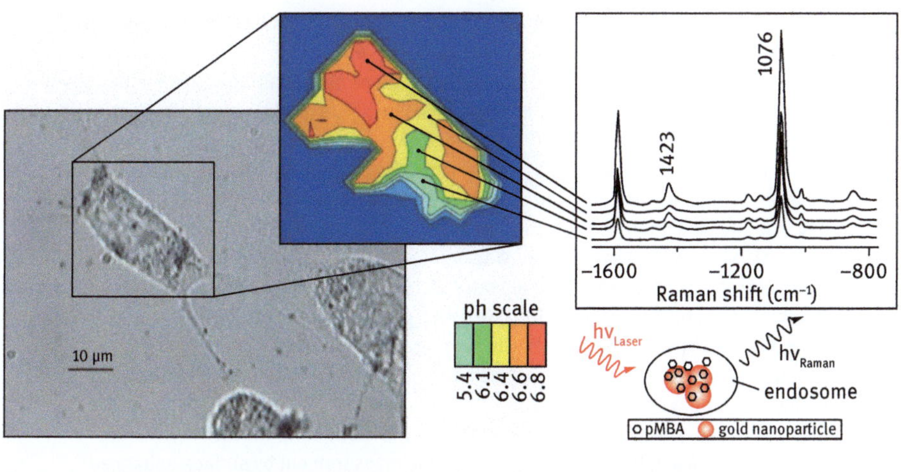

Fig. 9.13: Probing and imaging pH values in single live cells using a SERS nanosensor, which exploits the pH-sensitive SERS spectrum of 4-mercaptobenzoic acid (pMBA) [141].

signal of phenylalanine can be used as an intrinsic marker for detecting a single enkephalin molecule without the use of a specific label [116].

Moreover, gold or silver nanoparticles could be used as intracellular pH probe [124, 141, 163–165]. Determination and monitoring pH in cells and cellular compartments is of particular importance for a better understanding of a broad range of physiological and metabolic processes (Figure 9.13).

Conversely, SERS has been proposed for detection of a range of explosives and other trace materials. Some of the key explosives such as TNT, and components of plastic explosives such as RDX and PETN, have very low vapor pressures so the detection limits of any analytical method are required to be low. The SERS technique is effective if a gold substrate is treated with sodium hydroxide [103] and chemical derivatization of TNT produces a molecule that adheres strongly to silver surfaces, again giving good detection limits [166].

9.7 References

[1] Zhang, X., Young, M. A., Lyandres, O., Van Duyne, R. P., Rapid detection of an anthrax biomarker by surface-enhanced Raman spectroscopy, J Am Chem Soc **127** (2005) 4484–4489.

[2] Zhang, X., Zhao, J., Whitney, A., Elam, J., Van Duyne, R. P., Ultrastable substrates for surface-enhanced Raman spectroscopy fabricated by atomic layer deposition: improved anthrax biomarker detection, J Am Chem Soc **128** (2006) 10304–10309.

[3] Stuart, D. A., Biggs, K. B., Van Duyne, R. P., Surface-enhanced Raman spectroscopy of half-mustard agent, Analyst **131** (2006) 568–572.

[4] Sylvia, J. M., Janni, J. A., Klein, J. D., Spencer, K. M., Surface-enhanced Raman detection of 2,4-dinitrotoluene impurity vapour as a marker to locate landmines, Anal Chem **72** (2000) 5834–5840.

[5] Stiles, P. L., Dieringer, J. A., Shah, N. C., Van Duyne, R. P., Surface-enhanced Raman spectroscopy, Annu Rev Anal Chem **1** (2008) 601–626.

[6] Stokes, D. L., Alarie, J. P., Ananthanarayanan, V., Vo-Dinh, T., Fiber optic SERS sensor for environmental monitoring, Proc SPIE **3534** (1999) 647–654.

[7] Kundu, S., Mandal, M., Ghosh, S. K., Pal, T,. Photochemical deposition of SERS active silver nanoparticles on silica gel and their application as catalysts for the reduction of aromatic nitro compounds, J Colloid Interf Sci **272** (2004) 134–144.

[8] Stuart, D. A., Yonzon, C. R., Zhang, X. Y., Lyandres, O., Shah, N. C., Glucksberg, M. R., et al., Glucose sensing using near-infrared surface-enhanced Raman spectroscopy: gold surfaces, 10-day stability, and improved accuracy, Anal Chem **77** (2005) 4013–4019.

[9] Lyandres, O., Shah, N. C., Yonzon, C. R., Walsh, J. T., Glucksberg, M. R., Van Duyne, R. P., Real-time glucose sensing by surface-enhanced Raman spectroscopy in bovine plasma facilitated by a mixed decanethiol/mercaptohexanol partition layer, Anal Chem **77** (2005) 6134–6139.

[10] Shah, N. C., Lyandres, O., Walsh, J. T., Glucksberg, M. R., Van Duyne, R. P., Lactate and sequential lactate-glucose sensing using surface-enhanced Raman spectroscopy, Anal Chem **79** (2007) 6927–6932.

[11] Stuart, D. A., Yuen, J. M., Shah, N. C., Lyandres, O., Yonzon, C. R., Glucksberg, M. R., Walsh, J, T., Van Duyne, R. P., et al., In vivo glucose measurement by surface-enhanced Raman spectroscopy, Anal Chem **78** (2006) 7211–7215.

[12] Smekal, A., Zur Quantentheorie der dispersion, Naturwissnschaften **43** (1923) 873–875.
[13] Raman, C. V., Krishnan, K. S., A new type of secondary radiation, Nature **121** (1928) 501–502.
[14] Raman, C. V., Krishnan, K. S., The optical analogue of the compton effect, Nature **121** (1928) 377–378.
[15] Landsberg, G., Mandelstam, L., A novel effect of light scattering in crystals, Naturwissenschaften **16** (1928) 557–558.
[16] Raman, C. V., Krishnan, K. S., The production of new radiations by light scattering – part I, Proc R Soc Lond A **122** (1929) 23–25.
[17] Maiman, T. H., Stimulated optical radiation in ruby, Nature **187** (1960) 493–494.
[18] Fleischmann, M., Hendra, P. J., Mc Quillan, A. J., Raman spectra of pyridine adsorbed at a silver electrode, Chem Phys Lett **26** (1974) 163–166.
[19] Jeanmaire, D. L., Van Duyne, R. P., Surface Raman electrochemistry part I. Heterocyclic, aromatic and aliphatic amines adsorbed on the anodized silver electrode, J Electroanal Chem **84** (1977) 1–20.
[20] Grant, A. M., Creighton, A. J., Anomalously intense Raman spectra of pyridine at a silver electrode, J Am Chem Soc **99** (1977) 5215–5217.
[21] Nie, S. M., Emery, S. R., Probing single molecules and single nanoparticles by surface-enhanced Raman scattering, Science **275** (1997) 1102–1106.
[22] Kneipp, K., Wang, Y., Kneipp, H., Perelman, L.T., Itzkan, I., Dasari, R. R., et al., Single molecule detection using surface-enhanced Raman scattering (SERS), Phys Rev Lett **78** (1997) 1667–1670.
[23] Moskovits, M., Surface-enhanced spectroscopy, Rev Mod Phys **57** (1985) 783–826.
[24] Campion, A., Kambhampati, P., Surface-enhanced Raman scattering, Chem Soc Rev **27** (1998) 241–250.
[25] Otto, A., The 'chemical' (electronic) contribution to surface-enhanced Raman scattering, J Raman Spectrosc **36** (2005) 497–509.
[26] Kneipp, K., Kneipp, H., Deinum, G., Itzkan, I., Dasari, R. R., Feld, M. S., Single-molecule detection of a cyanine dye in silver colloidal solution using near-infrared surface-enhanced Raman scattering, Appl Spectrosc **52** (1998) 175–178.
[27] Kneipp, K., Kneipp, H., Kartha, V. B., Manoharan, R., Deinum, G., Itzkan, I., Dasari, R. R., Feld, M. S., Detection and identification of a single DNA-base molecule using surface-enhanced Raman scattering (SERS), Phys Rev E **57** (1998) R6281–6284.
[28] Michaels, A. M., Jiang, J., Brus, L., Ag nanocrystal junctions as the site for surface-enhanced Raman scattering of single rhodamine 6G molecules, J Phys Chem B **104** (2000) 11965–11971.
[29] Xu, H. X., Bjerneld, E. J., Kall, M., Borjesson, L., Spectroscopy of single hemoglobin molecules by surface enhanced Raman scattering, Phys Rev Lett **83** (1999) 4357–4360.
[30] Michaels, A. M., Nirmal, M., Brus, L. E., Surface enhanced Raman spectroscopy of individual rhodamine 6G molecules on large ag nanocrystals, J Am Chem Soc **121** (1999) 9932–9939.
[31] Guerrini, L., Graham, D., Molecularly-mediated assemblies of plasmonic nanoparticles for surface-enhanced Raman spectroscopy applications, Chem Soc Rev **41** (2012) 7085–7107.
[32] Smith, W. E., Rodger, C., Surface-enhanced Raman scattering (SERS), applications, in Lindon, J. C., (editor) Encyclopedia of Spectroscopy and Spectrometry. Elsevier Oxford, Oxford, UK, 1999, 2329–2334.
[33] Kneipp, K., Surface-enhanced raman scattering, Phys Today **60** (2007) 40–46.
[34] Wang, D. S., Kerker, M., Enhanced Raman scattering by molecules absorbed at the surface of colloidal spheroids, Phys Rev B **24** (1981) 1777–1790.
[35] Kneipp, J., Kneipp, H., Kneipp, K., SERS – a single-molecule and nanoscale tool for bioanalytics, Chem Soc Rev **37** (2008) 1052–1060.

[36] Otto, A., Surface-enhanced Raman scattering: "classical" and "chemical" origins, in Cardona, M., Guntherodt, G., (editors). Light Scattering in Solids IV. Electronic Scattering, Spin Effects, SERS and Morphic Effects. Springer-Verlag, Berlin, Germany, 1984, 289–418.
[37] Persson, B. N., On the theory of surface-enhanced Raman scattering, Chem Phys Lett **82** (1981) 561–565.
[38] Lombardi, J. R., Birke, R. L., Tianhong, L., Jia, X., Charge-transfer theory of surface enhanced Raman-spectroscopy – Herzberg-Teller contributions, J Chem Phys, **84** (1986) 4174–4180.
[39] Kambhampati, P., Child, C. M., Foster, M. C., Campion, A., On the chemical mechanism of surface enhanced Raman scattering: experiment and theory, J Chem Phys **108** (1998) 5013–5026.
[40] Persson, B. N., Zhao, K., Zhang, Z. Y., Chemical contribution to surface-enhanced Raman scattering, Phys Rev Lett **96** (2006).
[41] Lombardi, J. R., Birke, R. L., Time-dependent picture of the charge-transfer contributions to surface enhanced Raman spectroscopy, J Chem Phys **126** (2007) 244709–244712.
[42] McNay, G., Eustace, D., Smith, W. E., Faulds, K., Graham, D., Surface-enhanced raman scattering (SERS) and surface-enhanced resonance raman scattering (SERRS): a review of applications, Appl Spectrosc **65** (2011) 825–837.
[43] Smith, E., Dent, G., Moder Raman Spectroscopy – A Practical Approach. John Wiley & Sons Ltd, Hoboken, USA, 2005.
[44] Fang, Y., Seong, N. H., Dlott, D. D., Measurement of the distribution of site enhancements in surface-enhanced Raman scattering, Science **321** (2008) 388–392.
[45] Kneipp, K., Kneipp, H., Manoharan, R., Hanlon, E. B., Itzkan, I., Dasari, R. R., et al., Extremely large enhancement factors in surface-enhanced Raman scattering for molecules on colloidal gold clusters, Appl Spectrosc **52** (1998) 1493–1497.
[46] Lin, X. Z., Teng, X., Yang, H., Direct synthesis of narrowly dispersed silver nanoparticles using a single-source precursor, Langmuir **19** (2003) 10081– 10085.
[47] Creighton, J. A., Blatchford, C. G., Albrecht, M. G., Plasma resonance enhancement of Raman scattering by pyridine adsorbed on silver or gold sol particles of size comparable to the excitation wavelength, J Chem Soc Farad T2 **75** (1979) 790–798.
[48] Lee, P. C., Meisel, D., Adsorption and surface-enhanced Raman of dyes on silver and gold sols, J Phys Chem **86** (1982) 3391–3395.
[49] Osorio-Román, I. O., Ortega-Vásquez, V., Vargas, C. V., Aroca, R. F., Surface-enhanced spectra on D-gluconic acid coated silver nanoparticles, Appl Spectrosc **65** (2011) 18–23.
[50] Steinigeweg, D., Schlucker, S., Monodispersity and size control in the synthesis of 20–100 nm quasi-spherical silver nanoparticles by citrate and ascorbic acid reduction in glycerol–water mixtures, Chem Commun **48** (2012) 8682–8684.
[51] Lohse, S. E., Burrows, N. D., Scarabelli, L., Liz-Marzán, L. M., Murphy, C. J., Anisotropic noble metal nanocrystal growth: The role of halides, Chem Mater **26** (2014) 34–43.
[52] Xie, J., Zhang, O., Lee, J. Y., Wang, D. I. C., The synthesis of SERS-active gold nanoflower tags for in vivo applications, ACS Nano **2** (2008) 2473–2480.
[53] Shirtcliffe, N., Nickel, U., Schneider, S., Reproducible preparation of silver sols with small particle size using borohydride reduction: For use as nuclei for preparation of larger particles, J Colloid Interf Sci **211** (1999) 122–129.
[54] Caswell, K. K., Bender, C. M., Murphy, C. J., Seedless, surfactantless wet chemical synthesis of silver nanowires, Nano Lett **3** (2003) 667–669.
[55] Tan, Y., Dai, X., Li, Y., Zhu, D., Preparation of gold, platinum, palladium and silver nanoparticles by the reduction of their salts with a weak reductant—potassium bitartrate, Mater Chem **13** (2003) 1069–1075.

[56] Yang, J., Yin, H., Jia, J., Wei, Y., Facile synthesis of high-concentration, stable aqueous dispersions of uniform silver nanoparticles using aniline as a reductant, Langmuir **27** (2011) 5047–5053.
[57] Pyatenko, A., Yamaguchi, M., Suzuki, M., Synthesis of spherical silver nanoparticles with controllable sizes in aqueous solutions, J Phys Chem C **111** (2007) 7910–7917.
[58] Agnihotri, S., Mukherjiabc, S., Mukherji, S., Size-controlled silver nanoparticles synthesized over the range 5–100 nm using the same protocol and their antibacterial efficacy, RSC Adv **4** (2014) 3974–3983.
[59] Jain, P, K., El-Sayed, M. A., Plasmonic coupling in noble metal nanostructures, Chem Phys Lett **487** (2010) 153–164.
[60] Alvarez-Puebla, R., Liz-Marzán, L, M., García de Abajo, F. J., Light concentration at the nanometer scale, J Phys Chem Lett, **1** (2010) 2428–2434.
[61] Ahijado-Guzmán, R., Gómez-Puertas, P., Alvarez-Puebla, R. A., Rivas, G., Liz-Marzán, L. M., Surface-enhanced Raman scattering-based detection of the interactions between the essential cell division FtsZ protein and bacterial membrane elements, ACS Nano **6** (2012) 7514–7520.
[62] Tsoutsi, D., Montenegro, J, M., Dommershausen, F., Koert, U., Liz-Marzán, L. M., Parak, W. J., et al., Quantitative surface-enhanced Raman scattering ultradetection of atomic inorganic ions: the case of chloride, ACS Nano **5** (2011) 7539–7546.
[63] Abalde-Cela, S., Auguié, B., Fischlechner, M., Huck, W. T. S., Alvarez-Puebla R. A., Liz-Marzán, L. M., et al., Microdroplet fabrication of silver-agarose nanocomposite beads for SERS optical accumulation, Soft Matter **7** (2011) 1321–1325.
[64] Aldeanueva-Potel, P., Faoucher, E., Alvarez-Puebla, R. A., Liz-Marzán, L. M., Brust, M., Recyclable molecular trapping and SERS detection in silver-loaded agarose gels with dynamic hot spots, Anal Chem **81** (2009) 9233–9238.
[65] Abalde-Cela, S., Ho, S., Rodríguez-González, B., Correa-Duarte, M. A., Alvarez-Puebla, R. A., Liz-Marzán, L. M, et al., Loading of exponentially grown LBL films with silver nanoparticles and their application to generalized SERS detection, Angew Chem Int Edit **48** (2009) 5326–5329.
[66] Panisello, C., Peña, B., Gumí, T., Garcia-Valls, R., Polysulfone microcapsules with different wall morphology, J Appl Polym Sci **129** (2013) 1625–1636.
[67] Peña, B., Casals, M., Torras, C., Gumí, T., Garcia-Valls, R., Vanillin release from polysulfone macrocapsules, Ind Eng Chem Res **48** (2008) 1562–1565.
[68] Peña, B., Panisello, C., Aresté, G., Garcia-Valls, R., Gumí, T., Preparation and characterization of polysulfone microcapsules for perfume release, Chem Eng J **179** (2012) 394–403.
[69] Dubey, R., Shami, T. C., Bhasker Rao, K. U., Microencapsulation technology and applications, Defence Sci J **59** (2009) 82–95.
[70] Reinhard, B, M., Sheikholeslami, S., Mastroianna, A., Alivisatos, A. P., Liphardt, J., Use of plasmon coupling to reveal the dynamics of DNA bending and cleavage by single EcoRV restriction enzymes, J Proc Natl Acad Sci USA **104** (2007) 2667–2672.
[71] Qian, X., Li, J., Nie, S., Stimuli-responsive SERS nanoparticles: Conformational control of plasmonic coupling and surface Raman enhancement, J Am Chem Soc **131** (2009) 7540–7541.
[72] Alivisatos, P., The use of nanocrystals in biological detection. Nat Biotechnol **22** (2004) 47–52.
[73] Ferrari, M., Cancer nanotechnology: opportunities and challenges, Nat Rev Cancer **5** (2005) 161–171.
[74] Niemeyer, C. M., Nanoparticles, proteins, and nucleic acids: biotechnology meets materials science, Angew Chem Int Edit **40** (2001) 4128–58.
[75] Michalet, X., Pinaud, F. F., Bentolila, L. A., Tsay, J. M., Doose, S., Li, J. J., et al., Quantum dots for live cells, in vivo imaging, and diagnostics, Science **307** (2005) 538–544.
[76] Rosi, N, L., Mirkin, C. A., Nanostructures in biodiagnostics. Chem Rev **105** (2005) 1547– 1562.
[77] Cao, Y. C., Jin, R. C., Mirkin, C. A., Nanoparticles with Raman spectroscopic fingerprints for DNA and RNA detection, Science **297** (2002) 1536–1540.

[78] Gao, X., Yang, L., Petros, J. A., Marshall, F. F., Simons, J. W., Nie, S., In-vivo molecular and cellular imaging with quantum dots, Curr Opin Biotechnol **16** (2005) 63–72.
[79] Nie, S. M., Xing, Y., Kim, G. J., Simons, J. W., Nanotechnology applications in cancer, Annu Rev Biomed Eng **9** (2007) 257–288.
[80] Yezhelyev, M. V., Gao. X., Xing, Y., Al-Hajj, A., Nie, S., O'Regan, R. M., Emerging use of nanoparticles in diagnosis and treatment of breast cancer, Lancet Oncol **7** (2006) 657–667.
[81] Gao, X., Cui, Y. Y., Levenson, R. M., Chung, L. W. K., Nie, S. M., In vivo cancer targeting and imaging with semiconductor quantum dots, Nat Biotechnol **22** (2004) 969–976.
[82] Liu, Z., Cai, W., He. L., Nakayama, N., Chen, K., Sun, X., Chen, X., Dai, H., In vivo biodistribution and highly efficient tumour targeting of carbon nanotubes in mice, Nat Nanotechnol **2** (2007) 47–52.
[83] Weissleder, R., Kelly, K., Sun, E. Y., Shtatland, T., Josephson, L., Cell-specific targeting of nanoparticles by multivalent attachment of small molecules, Nat Biotechnol **23** (2005) 1418–1423.
[84] Lee, E. S., Na, K., Bae, Y. H., Polymeric micelle for tumor pH and folate-mediated targeting, J Control Release **91** (2003) 103–113.
[85] Torchilin, V. P., Micellar nanocarriers: pharmaceutical perspectives, Pharm Res **24** (2007) 1–16.
[86] Moghimi, S. M., Hunter, A. C., and Murray, J. C., Long-circulating and target-specific nanoparticles: theory to practice, Pharmacol Rev **53** (2001) 283–318.
[87] Couvreur, P., Vauthier, C., Nanotechnology: intelligent design to treat complex diseases, Pharm Res **23** (2006) 1417–1450.
[88] Duncan, R., Polymer conjugate as anticancer nanomedicines, Nat Rev Cancer **6** (2006) 688–701.
[89] Hood, J. D., Bednarski, M., Frausto, R., Guccione, S., Reisfeld, R. A., Xiang, R., et al., Tumor regression by targeted gene delivery to the neovasculature, Science **296** (2002) 2404–2407.
[90] Merchant, B., Gold, the noble metal and the paradoxes of its toxicology, Biologicals **26** (1998) 49–59.
[91] Root, S. W., Andrews, G. A., Kniseley, R. M., Tyor, M, P., The distribution and radiation effects of intravenously administered colloidal gold-198 in man, Cancer **7** (1954) 856–866.
[92] Paciotti, G. F., Kingston, D. G. I., Tamarkin, L., Colloidal gold nanoparticles: a novel nanoparticle platform for developing multifunctional tumor-targeted drug delivery vectors, Drug Dev Res **67** (2006) 47–54.
[93] Paciotti, G. F., Myer, L., Weinreich, D., Goia, D., Pavel, N., McLaughlin, R. E., Tamarkin, L., Colloidal gold: a novel nanoparticle vector for tumor directed drug delivery, Drug Deliv **11** (2004) 169–183.
[94] James, W. D., Hirsch, L. R., West, J. L., O'Neal, P. D., Payne, J. D., Application of INAA to the build-up and clearance of gold nanoshells in clinical studies in mice, J Radioanal Nucl Chem **271** (2007) 455–459.
[95] Connor, E. E., Mwamuka. J., Gole, A., Murphy, C. J., Wyatt, M. D., Gold nanoparticles are taken up by human cells but do not cause acute cytotoxicity, Small **1** (2005) 325–327.
[96] Kvítek, O., Siegel, J., Hnatowicz, V., Švorčík, V., Noble metal nanostructures influence of structure and environment on their optical properties, J Nanomater 2013. Article ID 743684, 15 pp.
[97] Qian, X., Peng, X. H., Ansari, D. O., Yin-Goen, Q., Chen, G. Z., Shin, D. M., Yang, L., Young, A. N., Wang, M. D., Nie, S., In vivo tumor targeting and spectroscopic detection with surface-enhanced Raman nanoparticle tags, Nat Biotechnola **26** (2008) 83–90
[98] Schlücker, S., SERS microscopy: nanoparticle probes and biomedical applications, Chem Phys Chem **10** (2009) 1344–1354.
[99] Bantz, K. C., Meyer, A. F., Wittenberg, N. J., Im, H., Kurtuluş, Ö., Lee, S. H., et al., Recent progress in SERS biosensing, Phys Chem Chem Phys **13** (2011) 11551–11567.
[100] Tu, Q., Chang, C., Diagnostic applications of Raman spectroscopy, Nanomedicine **8** (2012) 545–558.

[101] Kneipp, J., Kneipp, H., Wittig, B., Kneipp, K., Novel optical nanosensors for probing and imaging live cells, Nanomedicine **6** (2010) 214–226.
[102] Kho, K. W., Fu, C. Y., Dinish, U. S., Olivo, M., Clinical SERS: are we there yet? J Biophotonics **4** (2011) 667–684.
[103] Bertone, J. F., Spencer, K. M., Sylvia, J. M., Fingerprinting CBRNE materials using surface-enhanced Raman scattering, Proc SPIE (2008) 6954:69540J.
[104] Stokes, R. J., McBride, E., Wilson, C. G., Girkin, J. M., Smith, W. E., Graham, D., Surface-enhanced Raman scattering spectroscopy as a sensitive and selective technique for the detection of folic acid in water and human serum, Appl Spectrosc **62** (2008) 371–376.
[105] Faulds, K., Smith, W. E., Graham, D., Lacey, R. J., Assessment of silver and gold substrates for the detection of amphetamine sulfate by surface enhanced Raman scattering (SERS), Analyst **127** (2002) 282–286.
[106] Lin, M., He, L., Awika, J., Yang, L., Ledoux, D. R., Li, H., et al., Detection of melamine in gluten, chicken feed, and processed foods using surface enhanced Raman spectroscopy and HPLC, J Food Sci **73** (2008) T129–T134.
[107] Graham, D., Faulds, K., Quantitative SERRS for DNA sequence analysis, Chem Soc Rev **37** (2008) 1042–1051.
[108] Alexander, T. A., Le, D. M., Characterization of a commercialized SERS-active substrate and its application to the identification of intact *Bacillus* endospores, Appl Opt **46** (2007) 3878–3890.
[109] Reichert, W. M., Andrade, J. D., Surface Raman spectroscopy, in Andrade, J. D., editor Surface and Interfacial Aspects of Biomedical Polymers, New York Springer US, Plenum Press, New York, 1985, 421–442.
[110] Koglin, E., Sequaris, J. M., Surface enhanced Raman scattering of biomolecules, Top Curr Chem **134** (1986) 1–57.
[111] Schrader, B., (editor). Infrared and Raman spectroscopy: methods and applications, VCH Publishers Inc, New York, NY USA, 2008.
[112] Paisley, R. F., Morris, M. D., Surface enhanced Raman spectroscopy of small biological molecules. Prog Anal Atom Spect **11** (1988) 111–140.
[113] Cotton, T. M., Jae-Ho, K., Chumanov, G. D., Application of surface-enhanced Raman spectroscopy to biological systems, J Raman Spectrosc **22** (1991) 729–742.
[114] Garrell, R. L., Herne, T. M., Ahern, A. M., Sullenberger, E., Surface-enhanced Raman spectroscopy of peptides. Optical fibers in medicine V, Proc SPIE **1201** (1990) 451.
[115] Nabiev, I. R., Sokolov, K. V., Manfait, M., in Clark, R. J. H., Hester, R. E., (editors). Advances in Spectroscopy: Biomolecular Spectroscopy, Part A, Vol. 20. Hoboken, John Wiley and Sons, USA, 1993, 267–338.
[116] Kneipp, K., Kneipp, H., Abdali, S., Berg, R. W., Bohr, H., Single molecule Raman detection of enkephalin on silver colloidal particles, Spectros Int J **18** (2004) 433–440.
[117] Alvarez-Puebla, R. A., Liz-Marzan, L. M., Traps and cages for universal SERS detection, Chem Soc Rev **41** (2012) 43–51.
[118] Shi, C., Zhang, Y., Gu, C., Chen, B., Seballos, L., Olson, T., Zhang, J. Z., Molecular fiber sensors based on surface enhanced Raman scattering (SERS), J Nanosci Nanotechnol **9** (2009) 2234–2246.
[119] Grabar, K. C., Freeman, R. G., Hommer, M. B., Natan, M. J., Preparation and characterization of au colloid monolayers, Anal Chem **67** (1995) 735–743.
[120] Chan, H. Y. H., Takoudis, C. G., Weaver, M. J., High-pressure oxidation of ruthenium as probed by surface-enhanced Raman and X-Ray photoelectron spectroscopies, J Catal **172** (1997) 336–345.

[121] Kneipp, J., Kneipp, H., McLaughlin. M., Brown, D., Kneipp, K., In vivo Molecular probing of cellular compartments with gold nanoparticles and nanoaggregates, Nano Letters **6** (2006) 2225–2231.

[122] Kneipp, J., Kneipp, H., Rice, W. L., Kneipp, K., Optical probes for biological applications based on surface-enhanced Raman scattering from indocyanine green on gold nanoparticles, Anal Chem **77** (2005) 2381–2385.

[123] Wang, Y., Li, D., Li, P., Wang, W., Ren, W., Dong, S., et al., Surface enhanced Raman scattering of brilliant green on ag nanoparticles and applications in living cells as optical probes, J Phys Chem C **111** (2007) 16833–16839.

[124] Michota, A., Bukowska, J., Surface-enhanced Raman scattering (SERS) of 4-mercaptobenzoic acid on silver and gold substrates, J Raman Spectrosc **34** (2003) 21–25.

[125] Bizzarri, A. R., Cannistraro, S., SERS detection of thrombin by protein recognition using functionalized gold nanoparticles, Nanomed Namotech Bio Med **3** (2007) 306–310.

[126] Wood, E., Sutton, C., Beezer, A. E., Creighton, J. A., Davis, A. F., Mitchell, J. C., Surface enhanced Raman scattering (SERS) study of membrane transport processes, J Pharmaceutics **154** (1997) 115–118.

[127] Nabiev, I. R., Morjani, H., Manfait, M., Selective analysis of antitumor drug interaction with living cancer cells as probed by surface-enhanced Raman spectroscopy, Euro Biophys J **19** (1991) 311–316.

[128] Morjani, H., Riou, J. F., Nabiev, I., Lavelle, F., Manfait, M., Molecular and cellular interactions between intoplicine, DNA, and topoisomerase II studied by surface-enhanced Raman scattering spectroscopy, Cancer Res **53** (1993) 4784–4790.

[129] Westerink, B. H., Damsma. G., Rollema, H., de Vries, J. D., Horn, A. S., Scope and limitations of in vivo brain dialysis: A comparison of its application to various neurotransmitter systems, Life Sci **41** (1987) 1763–1766.

[130] Picorel, R., Chumanov, G., Torrado, E., Cotton, T. M., Seibert, M., Surface-enhanced resonance Raman scattering spectroscopy of plant photosystem II reaction centers excited on the red-edge of the qy band, J Phys Chem B **102** (1998) 2609–2613.

[131] Graham, D., Mallinder, B. J., Smith, W. E., Detection and identification of labeled DNA by surface enhanced resonance Raman scattering, Biopolymers **57** (2000) 85–91.

[132] Kneipp, K., Flemming, J., Surface enhanced Raman scattering (SERS) of nucleic acids adsorbed on colloidal silver particles, J Mol Struct **145** (1986) 173–179.

[133] Nabiev, I. R., Sokolov, K. V., Voloshin, O. N., Surface-enhanced Raman spectroscopy of biomolecules. Part III.—Determination of the local destabilization regions in the double helix, J Raman Spectrosc **21** (1990) 333–336.

[134] Kneipp, K., Pohle, W., Fabian, H., Surface enhanced Raman spectroscopy on nucleic acids and related compounds adsorbed on colloidal silver particles, J Mole Struct **244** (1991) 183–192.

[135] Koglin, E., Sequaris, J. M., Valenta, P., Hydrogen-deuterium exchange in adenosine 5′-monophosphate detected by surface enhanced Raman scattering (SERS), Zeitschrift für Naturforschung **360** (1981) 809–812.

[136] Shafer-Peltier, K. E., Haynes, C. L., Glucksberg, M. R., Van Duyne, R. P., Toward a glucose biosensor based on surface-enhanced Raman scattering, J Am Chem Soc **125** (2003) 588–593.

[137] Kneipp, J., Surface-enhanced Raman scattering: Physics and applications, Top Appl Phy **103** (2006) 335–349.

[138] Zeiri, L., Bronk, B. V., Shabtai, Y., Eichler, J., Efrima, S., Surface-enhanced Raman spectroscopy as a tool for probing specific biochemical components in bacteria, Appl Spectrosc **58** (2004) 33–40.

[139] Premasiri, W. R., Moir, D. T., Klempner, M. S., Krieger, N., Jones, G., Ziegler, L. D., Characterization of the surface enhanced raman scattering (SERS) of bacteria, J Phys Chem B **109** (2005) 312–320.

[140] Dijkstra, R. J., Scheenen, W., Dam, N., Roubos, E. W., ter Meulen, J. J., Monitoring neurotransmitter release using surface-enhanced Raman spectroscopy, J Neurosci Methods **159** (2007) 43–50.
[141] Kneipp, J., Kneipp, H., Wittig, B., Kneipp, K., One and two photon excited optical pH probing for cells using surface enhanced Raman and hyper Raman nanosensors, Nano Lett **103** (2007) 17149–17153.
[142] Allain, L. R., Vo-Dinh, T., Surface-enhanced Raman scattering detection of the breast cancer susceptibility gene BRCA1 using a silver -coated microarray platform, Anal Chim Acta **469** (2002) 149–154.
[143] Cao, Y. C., Jin, R. C., Nam, J. M., Thaxton, C. S., Mirkin, C. A., Raman dye-labeled nanoparticle probes for proteins, J Am Chem Soc **125** (2003) 14676–14677.
[144] Vendrell, M., Maiti, K. K., Dhaliwal, K., Chang, Y. T., Surface-enhanced Raman scattering in cancer detection and imaging, Trends Biotechnol **31** (2013) 249–257.
[145] Wang, G., Lipert, R. J., Jain, M., Kaur, S., Chakraboty, S., Torres, M. P., et al., Detection of the potential pancreatic cancer marker MUC4 in serum using surface-enhanced Raman scattering, Anal Chem **83** (2011) 2554–2561.
[146] Chon, H., Lee, S., Yoon, S. Y., Chang, S. I., Lim, D. W., Choo, J., Simultaneous immunoassay for the detection of two lung cancer markers using functionalized SERS nanoprobes, Chem Commun **47** (2011) 1251–1257.
[147] Lee, K., Drachev, V. P., Irudayaraj, J., DNA–gold nanoparticle reversible networks grown on cell surface marker sites: application in diagnostics, ACS Nano **5** (2011) 2109–2117.
[148] Dou, X., Takama, T., Yamaguchi, Y., Yamamoto, H., Ozaka, Y., Enzyme immunoassay utilizing surface-enhanced Raman scattering of the enzyme reaction product, Anal Chem **69** (1997) 1492–1495.
[149] Hawi, S. R., Rochanakij, S., Adar, F., Campbell, W. B., Nithipatikom, K., Detection of membrane-bound enzymes in cells using immunoassay and Raman microspectroscopy, Anal Biochem **259** (1998) 212–217.
[150] Ni, J., Lipert, R. J., Dawson, G. B., Porter, M. D., Immunoassay readout method using extrinsic Raman labels adsorbed on immunogold colloids, Anal Chem **71** (1999) 4903– 4908.
[151] Isola, N. R., Stokes, D. L., Vo-Dinh, T., Surface-enhanced Raman gene probe for HIV detection, Anal Chem **70** (1998) 1352– 1356.
[152] Dou, X., Takama, T., Yamaguchi, Y., Hirai, K., Yamamoto, H., Doi, S., Ozaki, Y., Quantitative analysis of double-stranded DNA amplified by a polymerase chain reaction employing surface-enhanced Raman spectroscopy, Appl Opt **37** (1998) 759–763.
[153] Richterich, P., Church, G. M., DNA sequencing with direct transfer electrophoresis and nonradioactive detection, Methods in Enzymol **218** (1993) 187–222.
[154] Nirmal, M., Dabbousi, B, O., Bawendi, M. G., Macklin, J. J., Trautman, J. K., Harris, T. D., Brus, L. E., Fluorescence intermittency in single cadmium selenide nanocrystals, Nature **383** (1996) 802–804.
[155] Han, M. Y., Gao, X. H., Su, J. Z., Nie, S., Quantum-dot-tagged microbeads for multiplexed optical coding of biomolecules, Nat Biotechnol **19** (2001) 631–635.
[156] Schulze, H. G., Blades, M. W., Bree, A. V., Gorzalka, B. G., Greek, L. S., Turner, R. B., Characteristics of backpropagation neural networks employed in the identification of neurotransmitter Raman spectra, Appl Spectrosc **48** (1994) 50–57.
[157] Lee, N. S., Hsieh, Y. Z., Paisley, R. F., Morris, M. D., Surface-enhanced Raman spectroscopy of the catecholamine neurotransmitters and related compounds, Anal Chem **60** (1988) 442–446.
[158] McGlashen, M, L., Guhathakurta, U., Davis, K. L., Morris, M. D., SERS microscopy: laser illumination effects, Appl Spectrosc **45** (1991) 543–554.

[159] Kneipp, K., Wang, Y., Dasari, R. R., Feld, M. S., Near-infrared surface-enhanced Raman-scattering (Nir-Sers) of neurotransmitters in colloidal silver solutions, Spectrochim Acta A **51** (1995) 481–487.
[160] O'Neal, D. P., Motamedi, M., Chen, J., Cote, G. L., Surface-enhanced Raman spectroscopy for the near real-time diagnosis of brain trauma in rats, Proc SPIE **3918** (2000) 191–196.
[161] Efrima, S., Bronk, B. V., Silver colloids impregnating or coating bacteria, J Phys Chem B **102** (1998) 5947–5950.
[162] Sockalingum, G. D., Lamfarraj, H., Beljebbar, A., Pina, P., Delavenne, M., Witthun, F., et al., Vibrational spectroscopy as a probe to rapidly detect, identify, and characterize micro-organisms, Proc SPIE **3608** (1999) 185–194.
[163] Talley, C. E., Jusinski, L., Hollars, C. W., Lane, S. M., Huser, T., Intracellular pH sensors based on surface-enhanced raman scattering, Anal Chem **76** (2004) 7064– 7068.
[164] Bishnoi, S, W., Rozell, C. J., Levin, C. S., Gheith, M. K., Johnson., B. R., Johnson, D. H., et al., Nano Lett **6** (2006) 1687–1692.
[165] Schwartzberg, A. M., Oshiro, T. Y., Zhang, J. Z., Huser, T., Talley, C. E., Improving nanoprobes using surface-enhanced Raman scattering from 30-nm hollow gold particles, Anal Chem **78** (2006) 4732–4736.
[166] McHugh, C. J., Keir, R., Graham, D., Smith, W. E., Selective functionalisation of TNT for sensitive detection by SERRS, Chem Commun **6** (2002) 580–581.

Leon Marteaux

10 Si-based inorganic microencapsulation

10.1 Introduction

Microencapsulation is a process of enclosing micrometer-sized particles of solids, liquids or gases in an inert shell, which serves to isolate and protect the particles from the external world [1]. The first microencapsulation process was invented in 1953 by B.K. Green and L. Schleicher [2] working in the laboratories of the National Cash Register Company, as a way to encapsulate leuco dyes for carbonless copy paper (CCP). Since then, many encapsulation methods have been developed:

- Physical: for example pan coating, air suspension coating, centrifugal extrusion, vibration nozzle and spray-drying.
- Physicochemical: e.g. ionotropic effects, gelation or coacervation.
- Chemical: for example *in situ* polymerization, matrix polymerization, interfacial polycondensation or interfacial crosslinking.

Today, predominantly organic materials, such as, gelatin, formaldehyde-urea, polyurea, polyacrylates, polystyrene, polysaccharides etc., are used to encapsulate actives of interest within a core-shell type microcapsule or a matrix microsphere. A short history of organic based microencapsulation technology has been compiled by C. Thies [3] and key milestones are shown in Table 10.1.

Table 10.1: Key milestones in physicochemical microencapsulation technology.

Year*	Inventor/Company	Wall Chemistry	Ref
1953	Green/NCR	Simple coacervation of gelatin with sodium sulfate	[2]
1953	Green and Schleicher/NCR	Complex coacervation of gelatin with gum arabic	[4]
1962	Miller and Anderson/NCR	Ethylcellulose/polyethylene wax with hot cyclohexane	[5]
1962	Ruus/Moore	Polyester, polyamide, polyurea,	[6]
1962	Vrancken/Gevaert	W/O/W microcapsules	[7]
1963	Mackinney/IBM	Interfacial polycondensation	[8]
1965	Chang/McGill University	Nitrocellulose	[9]
1965	Morgan/DuPont	Polyamide	[10]
1966	Matson/3M	Urea with formaldehyde (aminoplast)	[11]
1968	Vandegaer/Pennwalt	Polyurethane, polyurea, polyester, polycarbonate	[12]
1978	Scher/Stauffer	Polyurea produced in the oil dispersed phase	[13]
1979	Beestman/Monsanto	Polyurea and lignosulfonate emulsifiers	[14]
1981	Lim/Virginia University	Cell encapsulation with alginate and poly-L-Lysine	[15]

*Prior article date.

https://doi.org/10.1515/9783110642070-010

Until the end of the 20th century, the biggest microcapsules consumers were the CCP and agrochemical markets. The commercial decline of the former forced the development of new opportunities. Today microencapsulation is regaining interest by providing innovative solutions for fragrance-controled release in household care, self-healing in coatings, drug delivery in health care, adhesives in automotive etc.

Despite the various organic chemistries available to encapsulate, technical gaps such as shell permeability, mechanical strength, chemical stability, limited functionalization, poor toxicological profile and high cost in use are still frequent and unacceptable for some markets.

Interestingly, metal alkoxydes in general, and silicon alkoxides, in particular were never considered for microencapsulation purposes before the late 1980s. Silicon alkoxides have considerable advantage vs. other metal alkoxides in terms of cost and control of the rate of hydrolysis and condensation reactions. The genesis of Si-based inorganic microencapsulation finds its roots in the usage of surfactants that template the formation of metal oxides layers around their hydrophilic heads. Beck et al. [16]

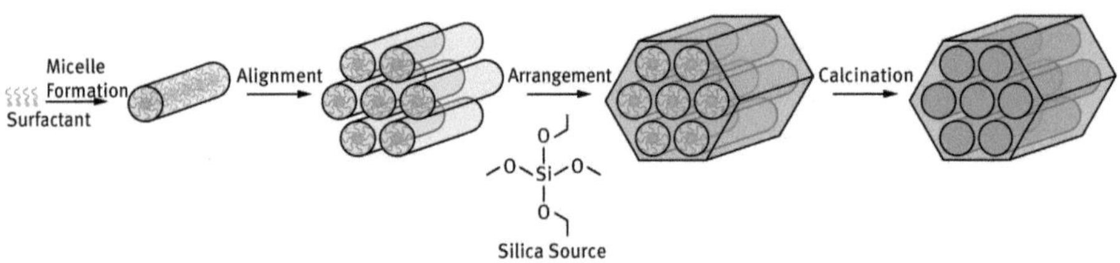

Fig. 10.1: Synthesis of mobile composition of matter 41 (MCM-41). Source: Hermann Luyken.

were the first group building ordered mesoporous material (MCM-41) from surfactant templating in alkaline conditions (Figure 10.1).

The missing link between surfactant templating and microencapsulation was found by G.D. Stucky et al. in 1996 [17]. They combined long-range O/W emulsion and O/W interface physics with the shorter range cooperative assembly of silica and surfactants at the O/W interface to create ordered composite mesostructured phases that are also macroscopically structured. They also used acid-prepared mesostructures synthesized from very acidic solutions below the pH of the isoelectric point of silica. The oil phase comprised of a volatile organic compound (mesitylene) and tetraethyl orthosilicate (TEOS). They suggested a model for the formation of the shell where TEOS contained in the oil droplet is hydrolyzed under acidic condition at the interface and forms the mesostructure under the influence of the surfactant. This was the very first *in situ* way, i.e. mixing the active with shell precursors before emulsification to make silica-based core-shell microcapsules and, after mesithylene removal, silica hollow spheres. However, for that group, microencapsulation was not the ultimate goal but only a step before calcination to ultimately obtain highly structured mesoporous materials.

10.2 Chemistry

10.2.1 The water glass process

Sodium silicate, also known as water glass or liquid glass, are compounds with the formula $Na_2(SiO_2)_nO$. They are available in liquid or solid forms and obtained from sodium carbonate and silica (eq. 10.1).

$$Na_2CO_3 + SiO_2 \rightarrow Na_2SiO_3 + CO_2 \tag{10.1}$$

Its main applications are in detergents, paper, water treatment, and construction materials. Many authors are using sodium metasilicate, Na_2SiO_3, as a precursor to encapsulate actives into silica in the presence of a strong acid (eq. 10.2).

$$\text{Active} + Na_2SiO_3 + HCl - (x\text{-}1)\,H_2O \rightarrow \text{Active} + SiO_2{\cdot}xH_2O + 2\,NaCl \tag{10.2}$$

This route is widely used and is very cost effective, but has major constraints in terms of pH and pI and can be a fatal flaw for biological stability.

10.3 The sol–gel process

More expensive, but more versatile and controlable than the water glass route, the so-called "sol–gel process" has been the topic of countless publications and text books, illustrating the specificity and the complexity of the process [18, 19]. This can be summarized as the hydrolysis and condensation of alkoxysilanes (eq. 8.3).

$$\text{Active} + Si(OR)_4 + 2H_2O \rightarrow \text{Active} + SiO_2 + 2\,ROH \tag{10.3}$$

Until the end of the 1990s, most of the fundamental understanding has been gained to support the manufacturing of silica gels, aerogels, glasses, organo-modified silicate, ceramics, enzyme immobilization and entrapment, membranes, coatings, etc. in water-depleted conditions. As a result, the sol–gel route found more and more applications as protective and smart coatings, for separative chromatography, catalysis, diagnostics, biotechnology, building waterproofing, optical lenses, restoration and controled release [20]. However, the fundamental understanding of the sol–gel route from metal alkoxides in the presence of a large excess of water was not been a center of interest and there is still a large gap existing today. Indeed, only a few publications can be found about the hydrolysis and condensation of alkoxysilane in an O/W emulsions such that a shell is specifically built at the O/W interface.

In order to obtain the tightest shell material possible with an acceptable toxicological profile and encapsulation kinetic, the usual choice is to start from tetraethylorthosilica (TEOS) instead of tetramethoxysilane (TMOS) as precursor. While TEOS is water-insoluble, its hydrolysis product, the orthosilicic acid

($Si(OH)_4$), is highly soluble in water. The total conversion of TEOS into silica (SiO_2) is sequentially obtained by hydrolysis (a) and condensation (b) in equation 10.4.

$$
\begin{array}{lcl}
\text{(a) TEOS} + 4\,H_2O & \rightarrow & Si(OH)_4 + 4\,C_2H_5OH \\
\text{(b) } Si(OH)_4 & \rightarrow & SiO_2 + 2\,H_2O \\
\hline
\text{TEOS} + 2\,H_2O & \rightarrow & SiO_2 + 4\,C_2H_5OH
\end{array}
\quad (10.4)
$$

The use of this chemistry is delicate because the structure and porosity of the silica produced depends on many physical parameters such as temperature, pH, ionic strength, etc. [18]. The hydrolysis and condensation reactions described above are further complicated by the presence of a surfactant to template the silica shell, as well as the presence of a dispersed oil phase in a large excess of water.

Due to its sequential nature, the hydrolysis and condensation of TEOS can follow multiple pathways [18]. From TEOS to silica, no less than 13 different intermediates species belonging to the categories of ethoxysilanes, ethoxysiloxanes or ethoxyhydroxysiloxanes can be generated. Their structures and molecular weights determine the properties of the intermediates (SiO_2 sols) or final products (SiO_2 gels). The pathway starts from TEOS, a water-insoluble liquid that can hydrolyze into orthosilicic acid ($Si(OH)_4$) a highly water-soluble species. Condensation of the orthosilicic acid leads to higher molecular weight species having reduced water solubility, eventually leading to the formation of precipitates. The water-insoluble condensate is a negatively charged polyelectrolyte at pH values above 4.5, the pKa of silanol [21]. The complete reaction, likewise an emulsion polymerization process, involves at least one phase transfer of the precursor. Because the hydrolysis of TEOS generates ethanol, a good solvent for both TEOS and water, TEOS becomes increasingly soluble into the water/ethanol continuous phase during the hydrolysis process. The three-dimensional (3D) structure of the resulting silica has been found to be dependent on a region in the ternary phase diagram in which the hydrolysis/condensation reaction was conducted.

The pH also influences the relative rates of the hydrolysis, condensation, and dissolution reactions and has been summarized in the familiar Iler graph [18]. It follows from this graph that the condensation rate in solution is at a maximum for pH ranging from 6 to 7 and at minimum for the pH ranging from 1.5 to 2. In alkaline media the condensation rate of silicic acid falls sharply and at pH above 11 a reverse depolymerization or hydrolysis of Si-O-Si bonds takes place. It can be anticipated that this differentiation of the reaction kinetic can lead to differentiated silica structures [22].

A key aspect of this chemistry is the combined impact of pH and ionic strength on the spatial organization of the silica produced [18]. Conducting the hydrolysis and condensation at basic pH and low pI leads to the synthesis of the so-called "Stober" silica [23]. Particles grow in size with a decrease in number leading to a colloidal silica sol of individual spherical nanoparticles. In acid solution or in the presence of flocculating salts, particles aggregate into 3D fractal networks and form gels [18].

Finally silica starts to solubilize at a basic pH of 10 at room temperature [18]. In contrast, silica is stable at very acidic pH which is an asset versus organic shell materials.

10.4 The main structures and their production process

The purpose of this paragraph is to describe the different processes that researchers have found to encapsulate actives with inorganic materials. The processes of producing hollow glass spheres at temperatures above 500 °C is not part of this review as no active materials, except some metal oxides, can be usefully encapsulated [24] that way. On the contrary, both "water silica" and "sol–gel" routes are useful for the encapsulation of hydrophilic and lipophilic active materials.

The tracking of patents and publication literature in the field of Si-based inorganic microencapsulation is complex due to the number of terms used by authors to describe inorganic microencapsulation. The most-used attribute to describe the type of inorganic microcapsules is "sol–gel". However, other key words like ceramispheres can be found too. One way to approach the activity in the field is to conduct a literature search on silica microcapsules. As shown in Figure 10.2, the intellectual property and literature landscape is showing increasing activity which began in industry and is now endorsed by academia.

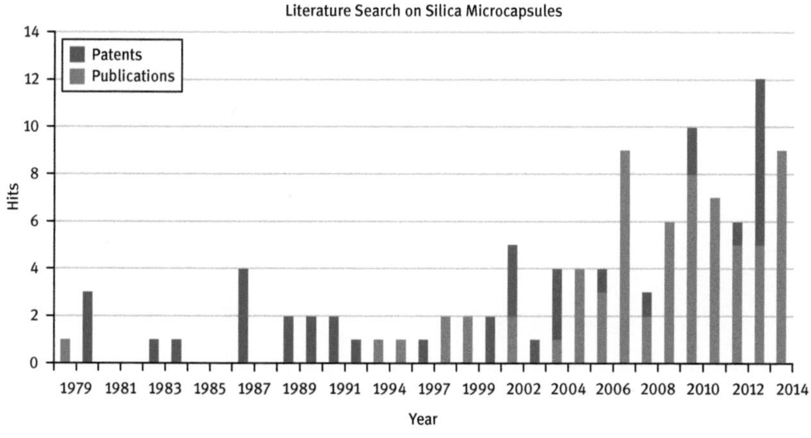

Fig. 10.2: Literature search on "silica microcapsules". Source: Scifinder.

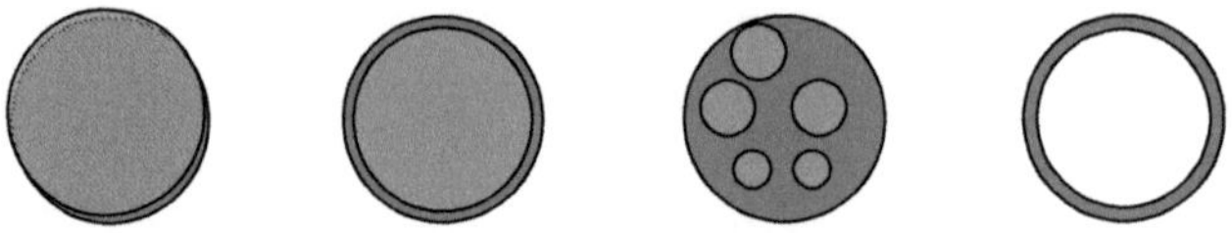

Fig. 10.3: (a) Nano/microsphere, (b) core-shell nano/microcapsules, (c) polynuclear nano/microcapsule, (d) hollow spheres.

Depending on their production process, microcapsules adopt different types of morphologies. Generally we can distinguish microspheres, core-shell microcapsules, polynuclear microcapsules and hollow spheres (Figure 10.3).

10.5 Active-containing silica microsphere: encapsulation of hydrophilic and lipophilic actives

Microspheres can be broadly defined as submicron spherical polymer particles wherein an active is homogeneously dispersed. Methods of preparation and use of silica microspheres for encapsulation purposes has been mostly reported in academic publications (Figure 10.4).

Generally microspheres are the easiest type of microcapsules to produce. The usual method of microspheres preparation consists in first solubilizing or dispersing the active to be encapsulated into the monomer or polymer matrix. Then this mixture is dispersed into an aqueous or an organic solvent phase with or without the help of an emulsifier [25]. However, for making silica microspheres, the phase behavior of

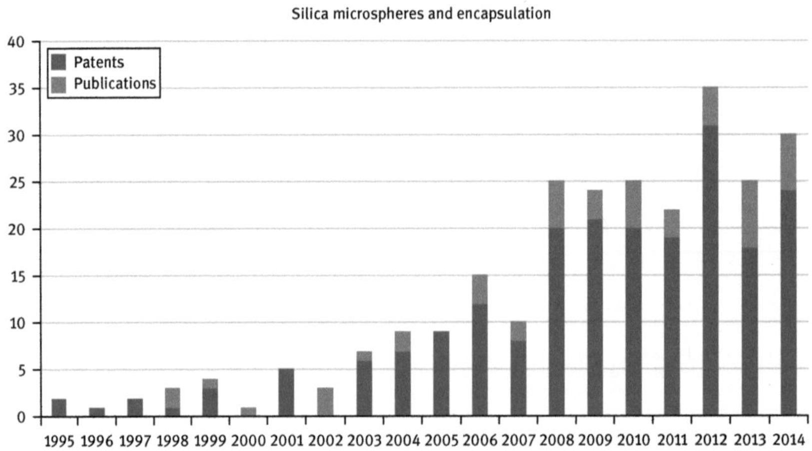

Fig. 10.4: Occurrence of publications in the field of "silica microspheres and encapsulation". Source: SciFinder.

precursors leads to different methods. Silica microspheres can be obtained from grinding of monoliths, inverse micelles or water in oil emulsions (W/O).

10.5.1 Silica microspheres from monoliths

These silica monoliths can be obtained by the "water glass" route or by the "sol–gel" process. The water glass route starts with a liquid sol that generally contains sodium silicate and the hydrophilic active material to be entrapped and leads to gels. The extreme pH starting condition and the final high ionic strength (pI) penalizes the use of that route.

The "sol–gel" route is more robust and less demanding for labile molecules like enzymes or living bodies if the released alcohol can be rapidly eliminated from the reaction environment. First the active is mixed with Si alkoxide precursors and after their hydrolysis and condensation process, a monolith silica or organically modified silica (Ormosil) matrix is obtained [26]. Next, the monolith is ground into microspheres. The main issue with that approach is that the entrapped active can be exposed to the environment and lose its protection. Another constraint of this encapsulation approach is that the molecular conformation of the active can change due to the shrinkage of the xerogel upon the drying process of the wet gel. However, shrinking can be reduced by reducing the amount of silanol-silanol interactions by hydrophobization to obtain an ambigel. Another strategy is the use of supercritical drying by CO_2 at 31 °C and 1072 PSI to obtain an aerogel (Figure 10.5).

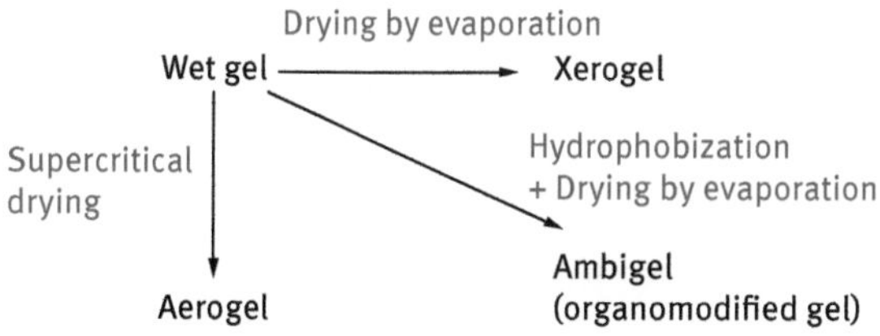

Fig. 10.5: Strategies to reduce shrinkage of wet gels.

The encapsulation of enzymes by this route found many industrial applications in medical diagnostics [27] as biosensors for glucose, cholesterol, urea, lactate, assays, etc. The end-points can be electrochemical or optical. In the former case the gel contains additional conducting particles like graphite, metal powders, mediators or co-reagents and a current is measured. In the latter case the enzymes are labeled with chromophoric or fluorescent groups or a molecule reacting to changes in pH or O_2 levels and a light emission or absorption is measured. The technology is also used for the synthesis of chiral compounds, chromatographic columns, and biocompatible implants and even in ammonia-free hair colorants [24]. However, gels have constraints in terms of loading of active ingredients and the small surface of exchange significantly reduces their controled delivery kinetic. To respond to these limitations, the use of a colloidal delivery system like microspheres attracted attention.

10.5.2 Silica microspheres from inverse micelles templating

While Nakahara synthesized spherical porous silica particles from nonionic surfactant micelles back in 1978, it was only in 1995 that he applied the preparation method to make active-containing inorganic microspheres [28]. The hydrophilic core materials were released at a rate controled by the size of micropores in the outer wall of microcapsules.

10.5.3 Silica microspheres from W/O emulsions templating

Another route to entrap water-soluble actives into a silica microsphere is to start from a W/O emulsion, wherein the water phase contains the active to be entrapped and the hydrolyzed Si alkoxide precursors [29]. Pope et al. encapsulated living tissue cells into a silica microsphere in the absence of surfactant. Using surfactant-made W/O emulsions as a template to entrap hydrophilic or hydrophobic actives opens additional features such as smaller and narrower distribution microcapsules sizes as well as higher content in the slurry.

The W/O route to entrap active materials has been extensively studied and patented in 2000 by Barbé et al. [30]. In that process the sol–gel precursors are dispersed into the drug containing water (Figure 10.6).

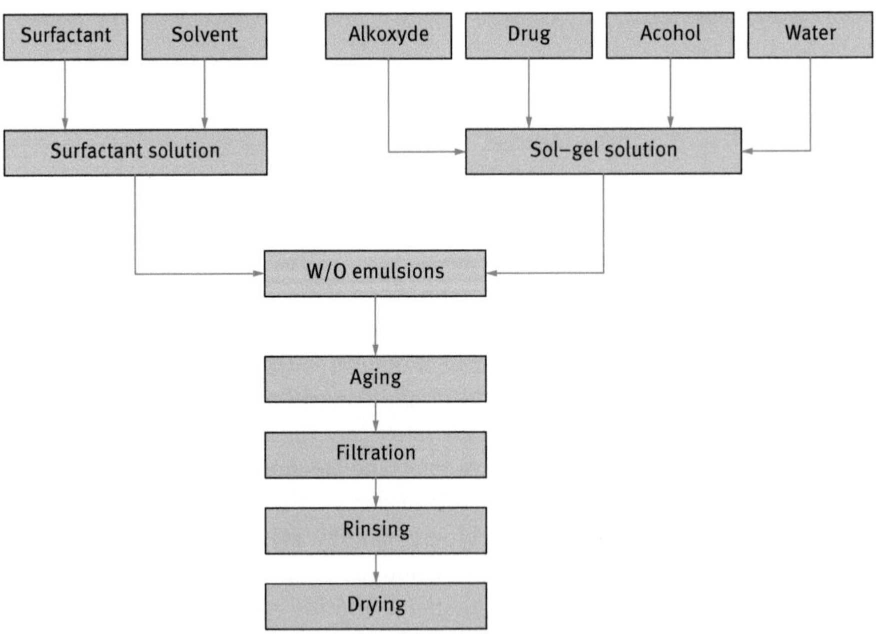

Fig. 10.6: Reproduced from EP 1257259B1 (assignee: ANSTO/Ceramispheres).

Nonionic surfactants with a hydrophilic-lipophilic balance (HLB) fewer than 10 are used to orientate the interface in the W/O direction. When the hydrolysis and condensation of alkoxides occurs in the water phase, a gel, i.e. a tridimensional network of silica or organo-modified silica (Figure 10.7), will homogeneously entrap the drug in the network.

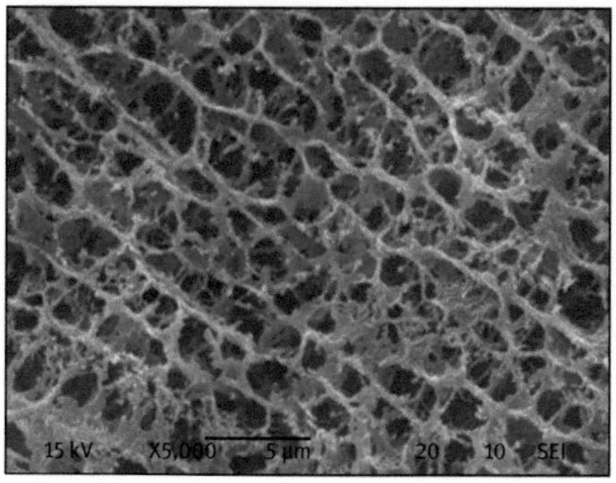

Fig. 10.7: Tridimensional colloidal silica gel network obtained at acidic pH.

The size of the microcapsules can be controled by the amount of water and the solvent/surfactant ratio, such that a size range 10–100 µm is claimed. In a later publication Barbé et al. discovered that at acidic pH, where the hydrolysis rate is fast and condensation is the limiting step, this process leads to homogeneous silica particles with mesopore sizes 2–6 nm [31]. In basic conditions the limiting hydrolysis step followed by fast condensation produces an in homogeneous system, characterized by large silica micropore size distributions from 3 to 11 nm.

If the hydrolysis and condensation of alkoxides is located at the W/O interface then the process leads to hollow silica or organo-silica spheres. In 2008, Wang and co-workers published a paper describing the direct microencapsulation of two water-soluble model drugs, gentamicin sulfate (GS) and salbutamol sulfate (SS), into silica microcapsules using a sol–gel process and TEOS as precursor in W/O emulsion [32]. The water phase was acidified with an aqueous solution of hydrochloric acid (HCl) and contained Tween 80, GS and SS. The oil phase was a cyclohexane solution containing Span 80. After 24 h of encapsulation the microcapsules have been filtrated and washed with cyclohexane. The silica microcapsules were uniform spherical particles with a size range of 5–10 µm, and had a specific surface area of about 306 m^2/g. *In vitro* release behavior of drugs in simulated body fluid revealed that the system exhibited excellent sustained release properties.

10.6 Core-shell micro and nanocapsules from O/W emulsions templating: encapsulation of lipophilic actives

After the first publication Stucky et al. [17] describing the use of O/W emulsions to template the building of water-insoluble liquid-containing core-shell microcapsules by mainly industrial developments took place to encapsulate lipophilic actives up to 2002 before being endorsed by academics (Figure 10.8).

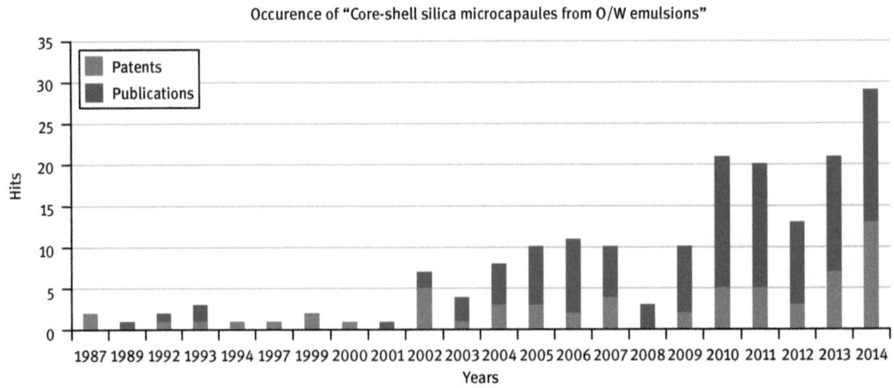

Fig. 10.8: Occurrence of publications in the field of "silica microspheres and encapsulation". Source: SciFinder.

In February 1998 Yoshioka et al. encapsulated organic sunscreens by using surface-active hydrolyzed protein functionalized silanes [33]. No surfactant templating per se was used, instead the surface-active silanes play both the role of emulsifier and shell precursor and formed a capsule after the hydrolysis and condensation at the active/water interface.

One month later, Dauth et al. filed a patent [34] describing the encapsulation of a nonionic surfactant stabilized emulsion. However, they obtained poor microencapsulation yields from 38 to 87%.

In August 1998, Sol-Gel Technologies Ltd., a spin-off of the Hebrew University of Jerusalem, filed a patent application for a process of making sol–gel microcapsules [35]. The process consists of blending the lipophilic active to be encapsulated with sol–gel precursors to make core-shell microcapsule (Figure 10.9).

This technique is useful for the entrapment of a water-insoluble liquid material called the dopant. The dopant is first mixed with the sol-gel precursor. The sol–gel solution is first emulsified with the help of cetyltrimethyl ammonium chloride. The volume fraction of the emulsion droplets is in the range of 10%. The hydrolysis and condensation of the sol–gel precursor at the O/W interface is building an amorphous silica shell. The microencapsulation process is first conducted at acidic pH to accelerate the hydrolysis and then the pH is increased to complete condensation. A film-forming polymer like polyvinylpirolidone is added to the suspension. In a third step, the suspension of core-shell microcapsules is spray dried or freeze dried. The microcapsule powder is washed and dispersed in water to obtain a final volume fraction of about 30%. This process is a true sol–gel process in the sense that it goes through a gelation step and the removal of the mother liquor. In this case, one deals with an *in situ* route, i.e. the shell

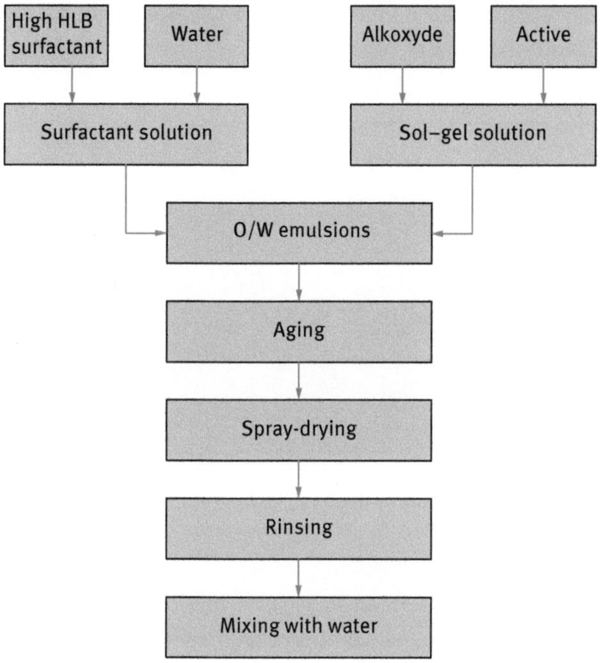

Fig. 10.9: *In situ* process of making core-shell microcapsules. (Reproduced from US 6303149B1.)

precursors are added into the dispersed phase before emulsification. The hydrolysis and condensation of the alkoxysilane occurs at the O/W interface. Consequently, the shell layers are built from the outside to the inside of the oil droplet.

In 2002, the *ex situ* route, i.e. the shell precursors are added in the continuous phase, was filed by Dow Corning Corporation in EP1471995B1 [36]. This is generated from a positive zeta potential O/W emulsion in a one-step process without gelation and without removal of the mother liquor (Figure 10.10).

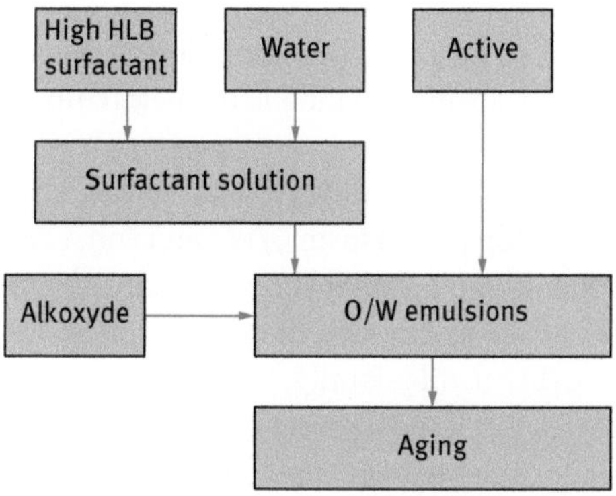

Fig. 10.10: Ex-*situ* process of making core-shell microcapsules. (Reproduced from EP 1471995B1.)

Optionally, the core-shell microcapsules can be harvested in a powdery form by, for example, spray-drying or freeze-drying (Figure 10.11).

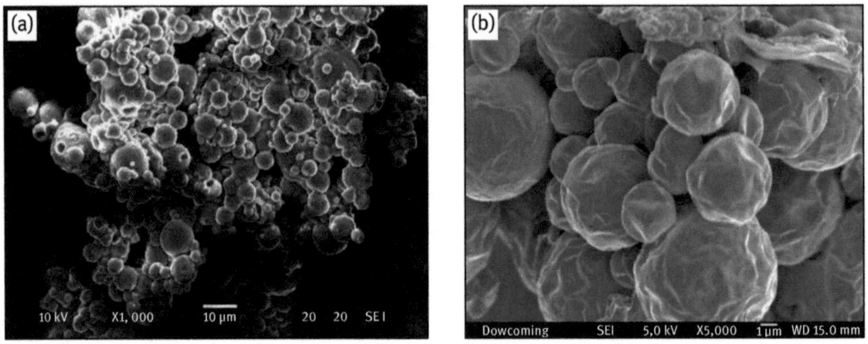

Fig. 10.11: Scanning electron microscope (SEM) of (a) spray-dried (a), (b) freeze-dried microcapsule suspensions.

In 2003, Unitech Co. Ltd., a spin-off of the Korean Reasearch Institute of Chemical Technology, disclosed a process for preparing silica microcapsules [37]. The process is comprised of the following steps: (i) dissolve TEOS into an aqueous solution containing a hydrolysis catalyst; (ii) dissolve the active to be encapsulated; and (iii) add aminopropyltrialkoxysilane (APS) as a gelling agent. Then, emulsify in a separate container with the active in a nonionic surfactant in a solution of opposite polarity. This process is a two kettle *ex situ* process with an O/W cationic interface provided by the APS. The exemplified actives are UV sunscreens.

In March 2005, Aquea Scientific corporation disclosed body wash compositions containing cationic *in situ* sol–gel microcapsules [38].

In 2005, the Australian Nuclear Science & Technology Organization disclosed a process of making particles comprising a releasable dopant [39]. They describe an *ex situ* method using trialkoxysilanes followed by the addition of aminopropyl trimethoxysilane.

In 2007, BASF disclosed UV sunscreens containing microcapsule compositions made by the *in situ* method [40].

In 2008, the company Ets Robert Blondel filed an *ex situ* process using a polymeric cationic emulsifier, claiming a faster microencapsulation kinetic using mixture of formic acid and acetic acid [41].

Still in 2008, Dow Corning disclosed a process for preparing cationic silicate shell microcapsules by adding a water reactive silicon compound to an O/W emulsion, comprising of a tetraalkoxysilane and an alkoxysilane with an amino or quaternary ammonium substituted alkyl group [42]. The purpose is to improve the deposition of the microcapsules onto negatively charged surfaces like textiles fabrics, hair fibres and skin.

Five days later, Microcapsules Technologies disclosed a process of making microcapsules composed essentially from silsesquioxane homo or copolymers [43]. In this application, methyltriethoxysilane (MTES) and methyltrimethoxysilane (MTMS) are used as *in situ* sol–gel precursors optionally in combination with ethylpolysilicate pre-polymers. The O/W emulsion is made from nonionic or anionic protective colloids. The option to use quaternary ammonium is disclosed in the application.

In 2008, IFF disclosed fragrance containing microcapsules obtained by the *in situ* and *ex situ* method [44].

In 2009, Biosynthis basically utilized the same process of making microcapsules as Microcapsules Technologies, but added polyquaterium 80 at the end, a silicone-based quaternary ammonium acetate ABA copolymer to reduce skin permeation of organic sunscreens [45].

In July 2009, Altachem disclosed an *ex situ* method for the preparation and the use of leach-proof microcapsules [46]. The inventor found that the partial alkylation of the silica shell from 2% to 25% improves the imperviousness of the benzoyl peroxide (BPO) catalyst containing microcapsule useful for the preparation of one component PU foam.

In July 2009, the CNRS filed a method for preparing core-shell materials and the use thereof for the thermos-stimulated generation of substances of interest [47]. The core material predominantly contains crystallizable oil, having a melting temperature below 100 °C, emulsified with the help of amphiphyle nano-silica. The Pickering emulsion obtained is further encapsulated with TEOS at strongly acidic pH of 0.2 in presence of cetyltrimethylammonium chloride.

In 2009, Sol-Gel Technologies Ltd. filed a patent disclosing the use of metal alkoxide nanoparticles, typically Ludox®™ 50, colloidal silica along with a sol–gel precursor to make thicker and therefore more impervious shell walls [48]. Ludox®™ 50 colloidal silica particles have a nominal particle size of 22 nm and are negatively charged.

In 2010, BASF disclosed a patent describing the microencapsulation of fragrances, perfumes or flavors first dissolved in paraffin or polyvinylether waxes [49]. The blend is later mixed with sol–gel precursors and emulsified. This *in situ* process is making core-shell microcapsules that have a payload smaller than 80%, wherein the core material comprises at least one fragrance, perfume or flavor and the shell at least one inorganic/hybrid material.

Still in 2010, the University of Tours François Rabelais patented an improvement of the shell imperviousness compared to the Robert Blondel's process [50]. This is obtained by (i) emulsifying the active-containing oil phase in an acidic aqueous phase at 50 °C, (ii) increasing pH, (iii) decreasing pH and (iv) finally increasing the pH to obtain a better condensation. Optionally, a hydrolyzed silane can be added before the last pH increase to form double layer shells.

In 2011, Givaudan filed an *ex situ* route to encapsulate perfume [51]. In order to confer positive zeta potential to the O/W interface, aminopropyltriethoxysilane is systematically added.

In 2012, Les Innovations Materium filed an *in situ* process able to build thick silica or organo-modified shells from 50 nm to about 500 µm [52]. The templating can be achieved by a nonionic surfactant like polyoxyethylene (20) sorbitan monooleate. The shell can be further functionalized by a trialkoxysilane.

10.7 Core-shell micro and nanocapsules from O/W/O emulsions templating: encapsulation of lipophilic actives

In 2006, Barbé and co-workers disclosed a process of preparing particles with hydrophobic material therein [53], using an O/W/O multiple emulsion. The hydrophobic material to be encapsulated is micellized in high HLB surfactants like Tween 21®. The swollen micelles are mixed with the hydrophilic phase containing the sol–gel precursor. Low HLB surfactants like Span 80 are used to emulsify the O/W emulsion into cyclohexane. After the ageing, i.e. the complete hydrolysis and condensation of the sol–gel precursors, the O/Silica/Cyclohexane suspension is filtrated and the recovered microcapsules washed with NaCl and water solutions.

10.8 Core-shell micro and nanocapsules from W/O/W emulsions templating: encapsulation of hydrophilic actives

Because of the limited surfaces developed by sol–gel monoliths, they found limited industrial applications wherein transformation rates are critical. One way to meet this requirement is to significantly increase the interfacial exchange surface between the biocatalyst and its substrate medium. An option is to microencapsulate the hydrophilic active into a nano or microcapsule in suspension in the substrate media. One solution could be a W/O/W multiple emulsion as the template. However, W/O/W multiple emulsions are very unstable. Indeed, on top of the intrinsic entropic instability of emulsions they have to face an osmotic pressure gradient between the internal and the external water phases. One way to mitigate the later is to increase the elastic modulus (G′) of the oil phase.

In 2006, Nakahara et al. [54] disclosed a method of making hollow silica particles via the water glass route, comprising of a silicon shell with macropores starting from a W/O/W emulsion. The internal water phase is composed of water-soluble silicates and the actives to be encapsulated and is emulsified into a continuous oil phase. The obtained W/O emulsion is further emulsified into a water external phase. Huge biomolecules, cells, or viruses can be included and claimed to be preserved for a long time in the hollow particle.

Another approach disclosed by Dow Corning Corporation [55] is to use sol-gel precursors like alkoxysilanes as the initial oil phase. They start from a W/Alkoxysilane/W multiple emulsion and end with a W/silica or organo-modified silica/W polynuclear microcapsule suspension (Figure 10.12).

The inventors describe the use of this process to encapsulate biocatalysts. The internal water phase contains the biocatalyst, preferably water-soluble enzymes and

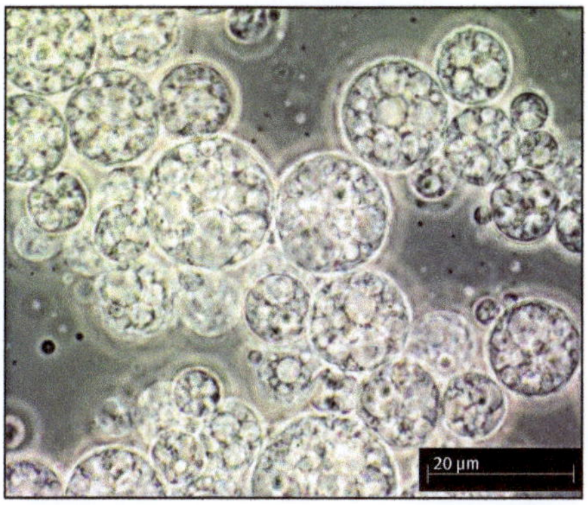

Fig. 10.12: W/Silica/W polynuclear microcapsule suspension.

its co-factor, and the external phase, the substrate. The goal is that the polynuclear microcapsule is acting as a microbioreactor, wherein the substrate can diffuse in the internal water phase and be transformed by the biocatalyst in a product that can diffuse out to the external water phase. The internal water phase contains *Aspergillus niger* catalase, an oxido-reductase enzyme, catalyzing the transformation of hydrogen peroxide into water and oxygen (eq. 10.5):

$$\begin{array}{lcl} H_2O_2 + \text{Fe(III)-C} & \rightarrow & H_2O + \text{O=Fe(IV)-C} \\ H_2O_2 + H_2O + \text{O=Fe(IV)-C} & \rightarrow & O_2 + \text{Fe(III)-C} \\ \hline 2H_2O_2 & \rightarrow & 2O_2 + O_2 \end{array} \tag{10.5}$$

Catalase contains four sub units of polypeptide chains and four porphyrin hemes for a total Mwt. of about 345,000 g/mole (Figure 10.13). Its Stokes radius is 5.83 +/−0.49 nm. Its reaction rate is only limited by substrate diffusion allowing ~200.000 reactions/s.

The authors observed non-measurable loss of catalase from the internal water to the external water phase upon ageing. All the microencapsulated catalase was fully

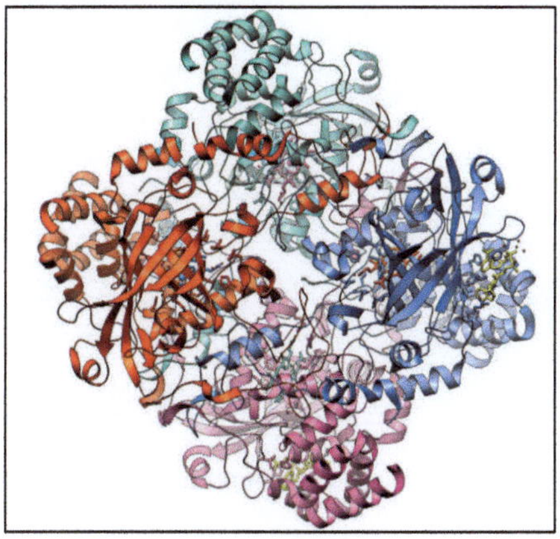

Fig. 10.13: Catalase from human erythrocyte. Source: Vossman.

encapsulated at pH 4, 7 and 8.9. The enzymatic activity of the suspensions obtained at pH 4, 7 and 8, 9 have been monitored upon shelf life. The inventors found that the microencapsulation of catalase from *A. niger* into silica polynuclear microcapsules extends its half-life time from 2 weeks RT to 1 year RT.

10.9 Triggers

The technical potential and utility of a microencapsulation technology resides in the triggers that can be used to release the encapsulated active. Silica and organo-modified silica have, in that respect, many advantages vs. organic shell materials. Some triggers developed in this chapter are useful for the delivery of actives from all type of Si-based nano and microcapsules. Others are specific to core-shell structures.

10.10 Trigger mechanisms for breaking capsules

10.10.1 Shear

Shear sensitivity of microcapsules is mainly correlated to their sizes and their mechanical strength. The latter depends, among other parameters, on the payload, the viscosity of core material and the mechanical strength of the shell material. It is generally acknowledged that microcapsules with sizes below 10 μm are not shear-sensitive. Using shear as a trigger is therefore easily accomplished, providing that large microcapsule sizes are acceptable in the application. One illustration of this phenomenon can be observed under an optical microscope with a vitamin A palmitate containing microcapsules suspension. The compression of the cover slip breaks the microcapsules standing between the microscope slide and the cover slip (Figure 10.14).

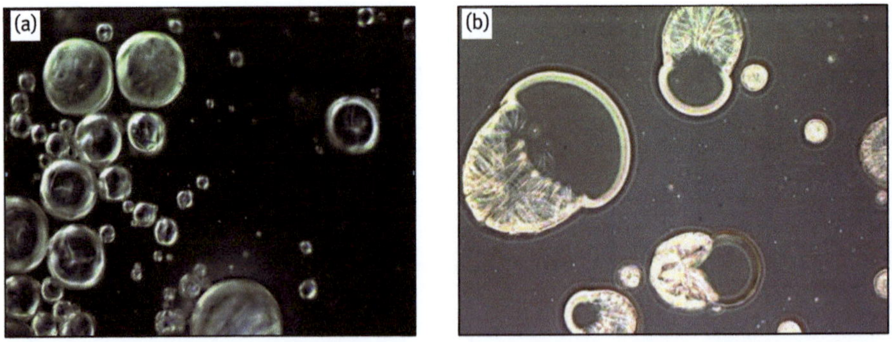

Fig. 10.14: Vitamin A Palmitate containing microcapsules (a) before and (b) after glass slides compression (average microcapsule size = 60 μm).

10.10.2 pH > 10 (silica dissolution)

The silica dissolution at basic pH [18] can be used as a trigger mechanism to release an active. Such a trigger can be used in cementitious matrices in the construction industry.

10.10.3 Osmotic pressure

In colloidal systems at equilibrium, such as microcapsule suspensions, chemical potentials always tend to equalize. Because of the chemical composition difference between each side of the microcapsule shells, the overall chemical potential must be compensated by the osmotic pressure. The later can be stronger than the mechanical resistance of the shell. Depending on the gyration radius of the active molecule, the silica shell porosity can be designed as an impervious, semi-permeable or permeable membrane.

When a low Mwt solvent is added to the suspension of microcapsule core material and continuous phase, the solvent is able to diffuse quickly through the shell and burst the microcapsule rapidly. As observed (Figure 10.19) this can be done with the use of ethanol in an organic sunscreen containing microcapsule suspension. Indeed, ethanol does fulfil the requirements for microcapsule breakage by osmotic pressure as it is a smaller molecule than the silica shell pores estimated in the range ~2–3 nm and soluble in both water and the organic sunscreen. For instance, post-addition of only 10% ethanol to porous ethylhexylmethoxycinnamate (EHMC) containing silica shell microcapsule leads to a significant leakage of the encapsulated EHMC as measured by dialysis (Figure 10.15).

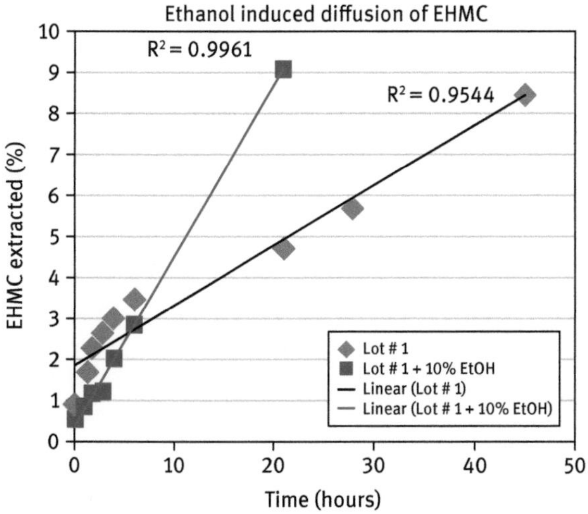

Fig. 10.15: Impact of a post addition of EtOH to core-shell microcapsule suspension.

10.11 Specific to core-shell nano and microcapsules

10.11.1 Drying of the microcapsule suspension

The drying of the microcapsule suspension by evaporation of the continuous hydro-alcoholic phase is a very useful trigger. The drying process leads to the concentration of the core-shell microcapsules until their close packing and aggregation (Figure 10.16). At that stage a Laplace pressure due to the capillary forces occurring between them can be stronger than the mechanical strength of the silica or organo-modified silica shell. In that case the drying ends with the breakage of the shell and the release of the encapsulated active.

The process can be easily observed under an optical microscope (Figure 10.17).

This concept has been further developed by the Dow Corning Corporation [56], where curable compositions have been separately encapsulated in silicate shell microcapsules and mixed together as a one part water-based slurry. The latter shows extended bath-life times and when it dries, the two compositions react to form a cured siloxane composition.

10.11.2 Heat

Another important trigger that can be used in different settings is heat. As metal oxides in general, and silica in particular, have high T_g in the range of 520–600 °C [32] they are not able to melt at low temperatures like waxes or low T_g organic polymers.

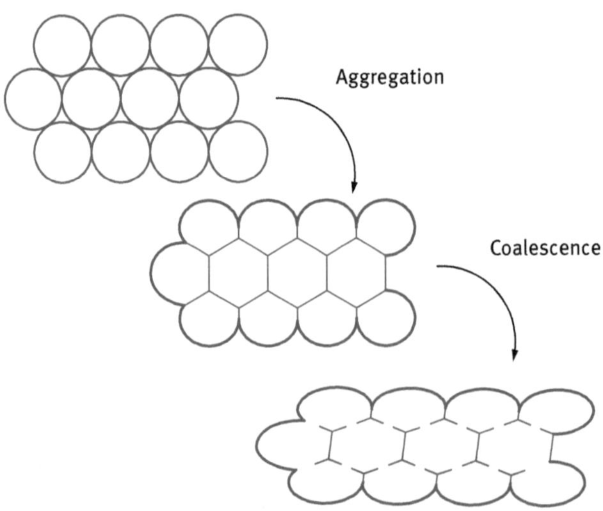

Fig. 10.16: The coalescence process.

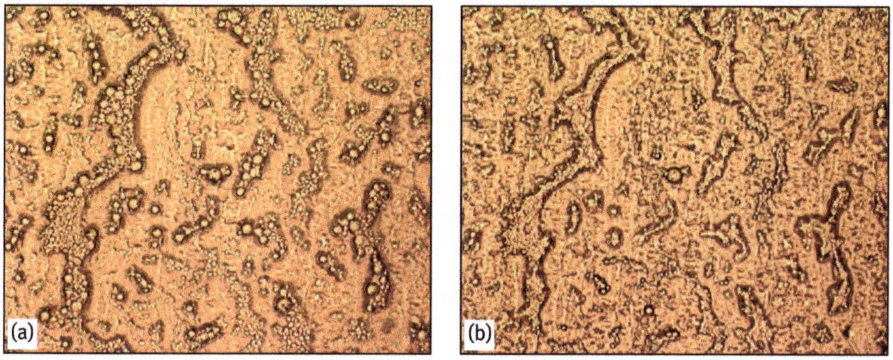

Fig. 10.17: Coalescence of core-shell microcapsules from O/W emulsion (a) before and (b) after drying of a 120 μm film on glass.

Another patented [57] strategy for rendering the microcapsules heat sensitive is to co-encapsulate the active of interest with a "blowing aid" that has a low boiling point (Figure 10.18a). As expected, the burst temperature of the microcapsule, measured by headspace GC/MS, depends on the vapor pressure of the blowing aid and is thus strongly correlated to its boiling point, i.e. the temperature the co-encapsulated blowing aid starts to boil at atmospheric pressure (Figure 10.18b).

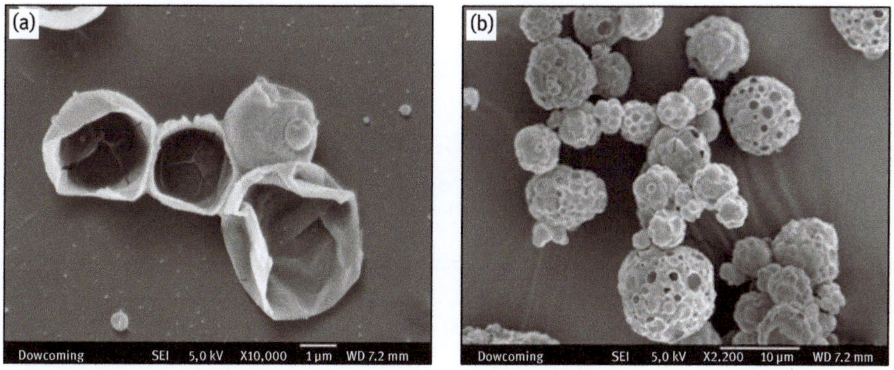

Fig. 10.18: Scanning electron microscope (SEM) imaging of burst core-shell microcapsules containing (a) hexamethyldisiloxane and (b) vinylterminated polydimethylsiloxane in scanning electron microscope (SEM).

10.11.3 Good solvents

If the silica shell porosity is designed such that the core material can diffuse throughout it, then a passive diffusion can be triggered by the presence of a good solvent for the core material. One example to illustrate that trigger is the extraction of

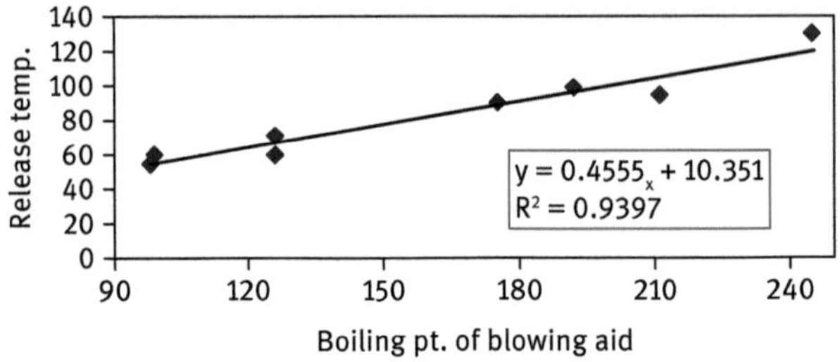

Fig. 10.19: Correlation between burst temperature and the boiling point of blowing aid.

microencapsulated ethylhexylmethoxycinnamate, a UV-B sunscreen, from a core-shell microcapsule aqueous slurry by light mineral oil (Isopar L) by dialysis. As anticipated from a core-shell microcapsule, EHMC shows a zero order controled delivery kinetic (Figure 10.20).

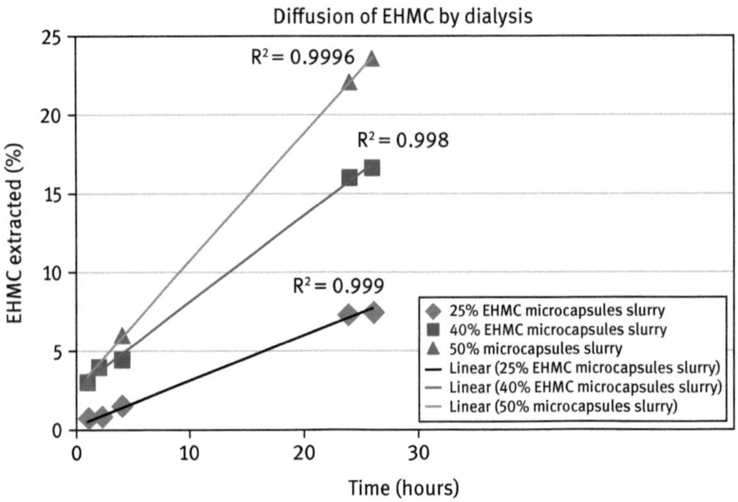

Fig. 10.20: Zero order delivery of EHMC from core-shell aqueous slurry induced by mineral oil in a dialysis device.

10.11.4 Sonication

Sonication, in addition to a low Mwt. good solvent for the core material, is a common method used to break microcapsules and quantitatively recover their content. Sonicating for about 15 min with an ultrasonic probe like Bioblock Scientific, 88169, Type T460H at a HF-Frequency of 35 kHz, leads to the complete breakage of the shell.

10.11.5 Vacuum

Even though using vacuum is not considered as a common trigger, microcapsules have been shown to burst following application of vacuum.

10.12 Industrial examples

10.12.1 Core-shell microcapsules from O/W emulsions

10.12.1.1 Sun protection

The first commercial application for silica-based microcapsules was to encapsulate organic sunscreens (OS). OS have an unpleasant greasy feel and there is a need to prevent direct contact between the skin and potentially irritant OS. As they are well known to penetrate the skin, entrapping them into a larger object significantly reduces the risk of percutaneous permeation [33]. More specifically, the mixture of ethylhexylmethoxycinnamate (EHMC) – the most widely used UV-B sunscreen – and benzyldimethoxybenzoylmethane (BMBM) – the most widely used UV-A sunscreen – is photo instable. All of these technical challenges can be overtaken by a silica-based core-shell microcapsule. Indeed OS are polar oils that are good solvents for organic based polymers and wall material in general. Their dosage level in a sun protection product is in the range 5–15%. As a consequence, only high payload microcapsules like core-shell type can be envisaged. The shell must be thin, to allow a UV beam to cross it and be absorbed by the OS, and impervious at the same time.

In 2002, Seiwa Kasei launched Silasoma, an encapsulated EHMC and BMBM. Microcapsule size is typically 2 µm and they are useful for protecting both skin and hair.

In mid-2002, Merck launched Eusolex® UV-Pearls™ under license from Sol-Gel Technologies Ltd. The Eusolex® UV-Pearls™ product range is composed of Eusolex® UV-Pearls™ OMC and Eusolex® UV-Pearls™ 2292 containing EHMC and preserved with parabens and chlorphenesin, respectively. Eusolex® UV-Pearls™ BO-2 and Eusolex® UV-Pearls™ BO contain a blend of Octocrylene and BMDBM preserved with parabens and chlorphenesin, respectively. Eusolex® UV-Pearls™ suspensions typically contain 40% of OS and microcapsules sizes are typically one micron.

Additional patent applications from key OS manufacturers like DSM and BASF emphasizes the good fit of Si-based microencapsulation for safer sun protection.

10.12.2 Construction chemicals

The encapsulation in acrylic polymers of phase change materials (PCM) that mitigate brutal temperature change thanks to their enthalpy of fusion, reached the market.

Micronal PCM™ from BASF is one example. Zhang et al. used TEOS to encapsulate n-octadecane [58]. They proved by FTIR that a silica shell was successfully built onto the core of PCM material. They synthetised at pH 2.45 PCM microcapsules in the 7–16 μm range. X-ray scattering indicates that the n-octadecane retains a good crystallinity. Thermogravimetric analysis shows that the silica microcapsules have good thermal stability. By controling the loading of the core material and the acidic pH of the reaction solution during the sol-gel process, the silica-microencapsulated PCM can achieve good phase-change performance, high encapsulation efficiency, and good antiosmosis properties. During the cooling process, the authors observed that the silica shell contribute to an improvement of the melting temperature and a widening of phase transition temperature range.

Shi and co-workers assessed in the lab new self-healing materials, called passive smart microcapsules, which hold promise for "crack-free" concrete or other cementitious composites [59]. Cement healing agents and catalysts in methylmethacrylate monomer and triethylborane, respectively, have been encapsulated separately from TEOS in an O/W emulsion. Sulfonated polystyrene particles are used as the template for an interfacial self-assembly process of the TEOS/Core mixture. The microcapsules are then dispersed in fresh cement mortar along with carbon microfibers. Microcapsule breakage is triggered by crack propagation, releasing the healing agent and the catalyst into the microcracks.

Lecomte et al. disclosed two patents covering processes for increasing hydrophobicity of porous materials, clay, bricks, gypsum or lime substrates [60, 61]. The inventions disclosed the treatment of construction substrates with *ex situ* prepared core-shell microcapsules to provide superior water-repellent performances compared with existing organopolysiloxane or alkoxysilane emulsions. Core materials are organopolysiloxane for porous material treatment and organosilane (e.g. aminoalkylalkoxysilane, octyltriethoxysilane, or tetraalkoxysilane) and a branched siloxane resin for cementitious substrates, clay-based bricks, gypsum-based substrates, lime-based substrates or wood-based substrates.

De Schijver et al. disclosed a method of making hollow mesopous silica spheres containing benzoyl peroxide, a catalyst for two component PU foam. The purpose is to formulate a one component PU foam formulation wherein the encapsulated BPO is released by burstingthe leach-proof organically modified silica microcapsule upon dispensing out of a pressurized can.

10.12.3 Textiles

Dow Corning Corporation launched the DS 9000 Multifunctional Additive, a polydimethylsiloxane-containing microcapsule suspension for textile treatment obtained by the *ex situ* process. The suspension is claimed to provide the treated fabric with superior hydrophobicity, quick drying, softness and luxurious hand product by cold

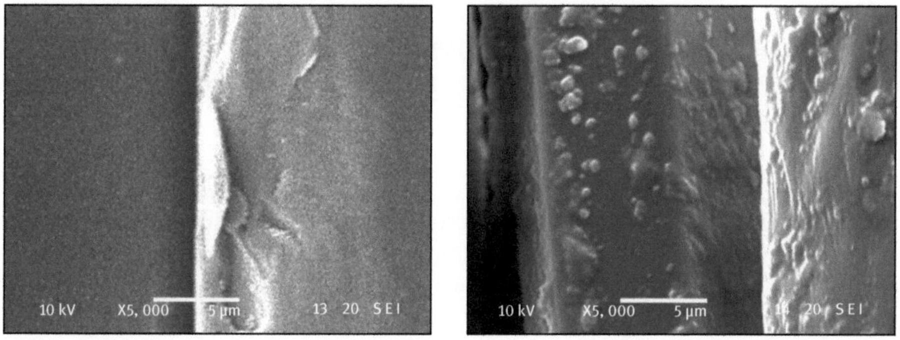

Fig. 10.21: Multifunctional Additive Trevira® CS polyester fibers. (a) Untreated and (b) treated with 40 g/L of DS 9000.

activation treatment. The treatment is of particular interest to reduce the harshness of flame retardant treated fabrics without impacting their fire resistance (Figure 10. 21).

10.12.4 Pharmaceuticals

Sol-gel Technologies Ltd. is very active in the field of the skin acne and rosacea therapy. Therefore, they use their patented *in situ* process to microencapsulate benzoyl peroxide (BPO) crystals. BPO is a skin irritant and the porosity of the silica shell of the microcapsule is designed to limit its direct contact while controling its delivery. The latter is claimed to be due to a slow dissolution of the BPO crystals by the skin lipids to the sebaceous follicles. A first product, Cool Pearls BPO anti-acne kit, was commercialized in 2009 by a dermo-pharmaceutical company in the U.S. The BPO-based product aimed to treat rosacea obtained positive results in a phase II clinical trial.

10.13 Core-shell microcapsules from W/O emulsions

10.13.1 Coatings

Based on US20120085261 [62], Ceramispheres launched Inhibispheres™ submicron particles providing specific functionalities to coating formulations. Active materials, such as corrosion inhibitors, biocides, fungicides, etc., can be incorporated inside sol-gel micro particles that can be mixed into a paint or coating formulation. The particles are shear-resistant and survive most paint processing techniques, such as milling and extrusion and do not adversely affect the key properties of the coating. Both water-soluble and poorly water-soluble active materials can be encapsulated, and Inhibispheres™ provides sustained release of the active from the coating over extended periods of time. Inhibispheres™ are compatible with both solvent- and water-based paint as well as powder coating.

10.13.2 Fermentation

Biotech applications in general, and in fermentation processes in particular, should attract more attention in the future. In addition to enzyme stabilization [51], the publications demonstrating the interest to microencapsulate living cells and organites by sol-gel precursors are growing fast. For instance Mellati and co-workers [63] used a W/O emulsion wherein living *Saccharomyces cerevisiae* (S.c.) was first mixed with TMOS as a precursor and emulsified into a vegetable oil. Bioactivity of immobilized yeast in microcapsules was investigated by measuring the amount of CO_2 released. The fermentation kinetic expressed by the production of CO_2 by encapsulated yeast increased about 150% more than free yeast. Repetition of the tests has proved that microcapsules saved their bioactivity for a month. Conversely, particle size was decreased from 175 to 110 μm by increasing the mixer rate from 600 to 1200 rpm during gelation and smaller particles showed more bioactivity up to 30%.

10.14 Key features of Si-based microcapsules

The key features of the main Si-based microcapsules and the most representative organic based microencapsulation technologies are compared in Table 10.2.

10.15 Conclusions and perspectives

Inorganic microencapsulation with silicon-based materials obtained by the water glass route or the sol-gel route has unique features. Different types of microspheres, hollow particles, core-shell or polynuclear microcapsules with size ranges from 10 nm to 80 μm can be obtained. Therefore, shear-resistant or shear-sensitive microcapsules can be designed. In the case of porous silica shells, the wide range of reservoir sizes enable the fine controled delivery of active material over an extended period of time. As it can be run at room temperature and neutral pH, the sol-gel route is a mild encapsulation process particularly adapted for volatile compounds and labile molecules like proteins, enzymes and living cells. To complete this environmentally friendly profile, it is worth mentioning that no aldehydes are generated in that chemistry. The hydrolysis and condensation of alkoxysilane has no effect on the molecular structure of organic actives to be protected, which is an asset when active pharmaceutical ingredients are involved. The use of surfactants to template the microcapsules

Table 10.2: Key features of main Si-based microencapsulation technologies vs. main organic ones.

	Complex coacervation	Interfacial polymerization	Urea-formaldehyde	Core-shell microcapsules	Polynuclear from W/O/W
Core					
$H_2$0 soluble solids	No	No	No	No	Yes
Insoluble solids	No	No	SOME	Yes	No
Sherical MICS	THIN CORNERS	Yes	Yes	Yes	Yes
Polar liquids	Yes	No	No	Yes	Yes
Non-polar liquids	No	Yes	Yes	Yes	No
Labiles, e.g. ENZ, fragrances.		Yes	No	Yes	Yes
Gas	No	No	No	No	No
Mic size (µm)	2–200	2–200	2–200	0·2–200	2–200
Wall	Gelatin and arabic gum	PA, pUREA PU, PE, PC	Urea and formaldehyde	Silica or organo-modified	Silica or organo-modified
Melt	No	No	No	No	No
Water-soluble	No	No	No	No	No
Capsules					
Payload over 90% Size distribution	Yes	Yes	Yes	Yes	
Process	Polydisperse	Polydisperse	Polydisperse	Monodisperse	Monodisperse
Batch/ continuous	Batch	Batch	Batch	Both	Both
Largest batch	1000 kg	20 tons		>1 tonne	>1 tonne
Batch time	16 h	10 min – 2 h	1–4 h	<16 h	<16 h
Issues	Aggregation Core wetability. RM cost and supply Process control Gas permeability	Few wall mat. Core wettability Gas permeability	Few wall mat. Aggregation Gas permeability	Gelation gas permeability	Gelation gas permeability

Tab. 10.2 *(continued)*

	Complex coacervation	Interfacial polymerization	Urea-formaldehyde	Core-shell microcapsules	Polynuclear from W/O/W
Cannot do	H_2O soluble core Polar liquid Gas	Solid cores Thick walls Melt walls Polar oils Gas pH-sensitive cores	H_2O soluble cores Polar liquids Gas	H_2O soluble cores Gas	Gas
Advantages	In sol. Hydroph-imper. wall Proven process Good barriers	Well developed process Large batches	Well developed process Easy process	Liquid cores Easy process	LIQUID cores
Excellent for	Hydrophobic liquid core Core < 20 MIC	Drop below 20 MIC. Hydrophobic liq. Compl. Insol. Wall Flexible walls	Drop below 20 MIC Hydrophobic liq. Compl. Insol. Wall Thin wall	Labile mat Big part and large walls	Water-soluble actives
Toxicological profile	Glutaraldehydes	Organic monomers	Therm. Stable wall [FORMALDEHYDE] >80 PPM	Ethanol Silica IS Gras by FDA	Ethanol Silica IS Gras by FDA
Energy consumption	Heating (50 °C) and cooling (5 °C)	Cooling or heating	Heating to 55 °C for 2 h;	None	None
Triggers	Shear/pressure	Shear/pressure release Good solvent for wall material	Shear/pressure	Shear/pressure T° drying, good solvents, Vacuum and U.S.	Shear Water T°
Usage	CCP	Pesticides, CCP, insecticides	Pesticides Fragrances Adhesives, CCP	Textiles UV protection	Enzymes

is a way to ensure high encapsulation yield. Basically, all the emulsified actives are encapsulated and the process is very efficient to separate incompatible compounds. The multiple processes that we have reviewed demonstrate the flexibility in the type of actives that can be entrapped. From water-soluble actives like enzymes to apolar oils such as polydimethylsiloxane via polar oils like OS and volatile compounds such as fragrance, the range of active that can be encapsulated seems to be endless. The payload for the core-shell type of microcapsule is usually above 95%. Depending on their molecular weight and the porosity of the shell the encapsulation can be permanent or the active release according to a zero order delivery. In that case the amount of active diffusing out of the shell is constant upon time until the reservoir is nearly empty. Triggers like water evaporation, heat, osmosis, pH, sonication and vacuum can be used to burst microcapsules or to onset active release. The active content in the suspension can reach high solid content in the range of 50%. The limitation is the potential shear sensitivity of the capsules against high viscosity slurries. Finally, the suspensions can be spray or freeze dried to obtain microcapsule in powdery form. All these features put together draw a microencapsulation technology profile which should be taken into account by researchers at the forefront of technological challenges to be overcome.

As the prior article search indicates, and despite the fact that interest is steadily growing, inorganic microencapsulation is still a juvenile topic that is lacking fundamental understanding and deep characterization of the encapsulant material. Indeed, the literature search indicates that encapsulation performances, when assessed, are generally obtained indirectly by a dosage of the released active and/or an overall performance of the material containing the capsules vs. the reference. Too often authors jump from the method of making the capsules to the performance of the final material without a deep characterization of the capsules. There is a need for developing analysis techniques to enabling the study of the molecular architecture of the barrier material *in situ* and online.

10.16 References

[1] Ghosh, S., Functional coatings and microencapsulation: a general perspective. In: Functional Coatings, Weinheim, Germany, Wiley-VCH, 2006.
[2] Green, B. K., Schleicher, L., Manifold record material, US Patent 2730456; 10 January 1956.
[3] Thies, C., A short history of microencapsulation technology, in Arshady, R., (editor). Microspheres, Microcapsules & Liposomes. Vol 1, Preparation & Chemical Applications. Citus Book, London, UK, 1999, 47–54.
[4] Green, B. K., Oil-containing microscopic capsules and method of making them, US Patent 2800457, 23 July 1957.
[5] Miller, R. E., Anderson, J. L., Microencapsulation process and its product, US Patent 3155590, 3 November 1964.
[6] Ruus, H., Method of encapsulation, US Patent 3429827, 25 February 1969.
[7] Vrancken, M. N., Claeys, D. A., Process of encapsulating water phase and compounds in aqueous phase by evaporation, US Patent 3523906, 11 August 1970.

[8] Mackinney, H. W., Microencapsulation by interfacial polycondensation, Polym Eng Sci SPE transaction **3** (1963) 71.
[9] Chang, T. M. S., Semipermeable microcapsules, Sciences **146** (1964) 524–525.
[10] Morgan, P. W., Condensation polymers by interfacial and solution methods, USA, Interscience, New York, 1965.
[11] Matson, G. W., Microcapsule containing paper, US Patent 3516846, 25 July 1966.
[12] Vandegaer, J. E., Encapsulation by interfacial polycondensation, US Patent 3577515, 4 May 1971.
[13] Scher, H. B., Encapsulation process and capsules produced thereby, US Patent 4285720, 25 August 1981.
[14] Beestman, G. B., Deming, J. M., Encapsulation by interfacial polycondensation and aqueous herbicidal composition containing microcapsules produced thereby, US Patent 4280833, 28 July 1981.
[15] Lim, F., Moss, R. D., Microencapsulation of living cells and tissues, J Pharm Sci **70** (1981) 351–354.
[16] Beck, J., Chang, C. D., Lawton, S. L., Leonowicz, M. E., Lissy, D. N., Rubin, M. K., Synthetic porous crystalline material, its preparation, and use, WO 9111390 A2, 8 August 1991.
[17] Schacht, S., Huo, Q., Voigt-Martin, I. G., Stucky, G. D., Schüth, F. D., Oil–water interface templating of mesoporous macroscale structures, Science **273** (1996) 768–771.
[18] Iler, R. K., The chemistry of silica, USA, Wiley, New York, 1979.
[19] Brinker, C. J., Scherrer, G. W., Sol–Gel Science. The Physics and Chemistry of Sol–Gel Processing, Boston, MA, USA, Academic Press, 1990. p. 153.
[20] Pagliaro, M. Fidalgo, A., Pandarus V., Béland, F., Ilharco, L. M., The sol–gel route to advanced silica-based materials and recent applications, Chem Rev **113** (2013) 65–92.
[21] Leung, K., Criscenti, L. J., Elucidating the bimodal acid-base behavior of the water–silica interface from first principles, J Amer Chem Soc **131** (2010) 18358.
[22] Handy, B., Baiker, A., Influence of preparation parameters on pore structure of silica gels prepared from tetraethoxy orthosilicate, Stud Surf Sci Catal **63** (1991) 239–246.
[23] Stober, W., Fink, A., Bohn, E., Controlled growth of monodisperse silica spheres in the micron size range, J Interface Sci **26** (1968) 62–69.
[24] Veatch, F., Method of producing hollow glass spheres, US Patent 2978339, 22 October 1957.
[25] Matsuno, M., Watanabe, A., Silica microcapsules containing pigments, JP Kokai, July 1 1986.
[26] Avnir, D., Doped sol–gel glasses for obtaining chemical interactions, US Patent 5292801, 8 March 1994.
[27] Pierre, A. C., The sol–gel encapsulation of enzymes, Biocatal Biotransfor **22** (2004) 145–170.
[28] Nakahara, Y., Inorganic microcapsules for controlled release systems, Adv Sci Tech, (Faenza, Italy) **10** 239–250.
[29] Pope, E. J. A., Encapsulation of living tissue cells in a microsphere, WO Patent 9909142, August 1997.
[30] Barbe, C., Controlled release ceramic particles, compositions thereof, process of preparation and methods of use, AU Patent 20005733, 21 February 2000.
[31] Barbé, C., Kong, L., Finnie, K., Calleja, S., Lin, H., Draberek, E., et al., Sol–gel matrices for controlled release: from macro to nano using emulsion polymerisation, J Sol-Gel Sci Tech **46** (2008) 393.
[32] Wang, J-X., Wang Z-H., Chen J-F., Yun, J., Direct encapsulation of water-soluble drug into silica microcapsules for sustained release applications, Mater Res Bull **43** (2008) 3374–3381.
[33] Yoshioka, M., Microcapsules having a specific wall and method for producing the same, Patent JP 4106398, 6 February 1998.
[34] Dauth, J., Daubzer, B., Method of preparation of microencapsulated products with walls in organopolysiloxane, EP0941761B1, March 12 1998.

[35] Magdassi, S., Method for the preparation of oxide microcapsules loaded with functional molecules and the products obtained thereof, US Patent 6303149B1, August 11 1999.
[36] Marteaux, L., Encapsulation process and encapsulated compositions, EP1471995, 7 February 2002.
[37] Seok, S., Process for preparing silica microcapsules, US Patent 6855335, 7 October 2003.
[38] Traynor, D., Sunscreen compositions and methods of use, US Patent 7001592B2, 8 March 2005.
[39] Finnie, K. A., Barbe, C., Kong, L., Particles comprising a releasable dopant therein, WO Patent 2006/133519, 5 June 2005.
[40] Pfuecker, F., Mueller, B., Witte, G., Andre, V., UV-Filter capsules, EP Patent 2170253, April 7 2007.
[41] Bone, S., Method of encapsulating a lipophilic or hydrophilic product in a polysiloxane membrane, EP2080552A1, 21 January 2008.
[42] Marteaux, L., Deklippel, L., Dimitrova, T., Elms, R., Galeone, F., Lenoble, B., Process for preparing silicate shells, WO Patent 2010045446, 15 October 2008.
[43] Habar, G., Microcapsules having shells essentially composed of silsequioxane of homopolymers or copolymers, F.R. Patent 2937248, 20 October 2008.
[44] Popplewell, L., Microcapsules containing active ingredients, US Patent 328340, 4 December 2008.
[45] Habar, G., Bernoud, T., Silicone microcapsules having cationic function, WO Patent 2012/004461, 26 May 2009.
[46] De Schrijver, A., Leach-proof microcapsules, the method for preparation and use of leach-proof microcapsules, WO 2011003805 A2, 9 July 2009.
[47] Schmitt, V., Destribats, M., Backov, R., Core-shell material, method for preparing same, and use thereof for the thermostimulated generation of substances of interest, WO Patent 2011012813, 31 July 2009.
[48] Toledano, O., Sertchook, H., Loboda, N., Abu-Reziq, R., Microcapsules comprising active ingredients and a metal oxide shell, a method for their preparation and uses thereof, US Patent 2011/0177951, 2 August 2009.
[49] Dreher, J., In-situ sol-gel encapsulation of fragrances, perfumes or flavours, W.O. 2011/124706, 8 April 2011.
[50] Viaud-Massuard, M-C., Procédé de fabrication de microcapsules polysiloxane fonctionnalisée et peu poreuse, FR Patent 2965190, 29 September 2010.
[51] Bone, S., Microcapsules, a process of making such microcapsules and compositions utilizing such microcapsules, WO Patent 2013/083760, 7 December 2011.
[52] Gosselin, M., Silica microcapsules, process of making the same and uses thereof, US Patent 2014/0341958, 20 November 2014.
[53] Barbé, C., Kong, L., Particles having hydrophobic material therein, US Patent 20090246279, 19 June 2006.
[54] Fujiwara, M., Nakahara, K., Hollow particle comprises silicon shell having macropores. JP Kokai 20065150, 28 February 2006.
[55] Marteaux, L., Zimmerman, B. L., Polynuclear microcapsules, US Patent 8435560, 21 December 2007.
[56] Marteaux, L., Kretschmer, A., Marteaux, L., Simonnet, J-T., Zimmerman, B. L., Microcapsules containing curable siloxane, US Patent 8487020, 28 June 2010.
[57] Zimmerman, B. L., Marteaux, L. A., Suspensions of silicate shell microcapsules for temperature controlled release, EP Patent 2367619, 17 December 2008.
[58] Zhang, H., Wang, X., Wu, D., Silica encapsulation of n-octadecane via sol–gel process: A novel microencapsulated phase-change material with enhanced thermal conductivity and performance, J Colloid Interf Sci **343** (2010) 246–255.

[59] Shi, X., Yang, Z., Hollar, J., He, X., Shi, X., Laboratory assessment of a self-healing cementitious composite, Transport Res Rec, Nanotechnology in Cement and Concrete **2142** (2010) 9–17.
[60] Lecomte, J-. P., Water repellent organopolysiloxane and treatment of porous materials, WO Patent 2013164381, 7 November 2013.
[61] Lecomte, J-P., Water repellent organosilicon materials for clay bricks, gypsum- or lime substrates, WO Patent 2013166280, 7 November 2013.
[62] Barbe, C., Caldeira, N. M., Campazzi, E., Kong, L., Finnie, K. S., et al., Ceramic particles and coating composition including said particles, US Patent 20120085261, 12 December 2011.
[63] Mellati, A., Attar, H., Farahani, M. F., Microencapsulation of Saccharomyces cerevisiae using a novel sol-gel method and investigate on its bioactivity, Asian J Biotechnol 2010;2:127 132.
[64] Griffin, W. C. C., Classification of Surface-Active Agents by 'HLB', J Soc Cosmet Chem **1** (1949) 311–326.

Bartosz Tylkowski, Magdalena Olkiewicz, Xavier Montane, Adrianna Nogalska, Monika Haponska, Josep M. Montornes, Jolanta Kowalska and Eligio Malusá

11 Encapsulation technologies in agriculture

11.1 Introduction

Every day, more than 22 million farmers and agricultural workers in 28 European Union (EU) member countries must face an enormous responsibility and challenge to produce enough quantities of high-quality and safe food for more than 500 million EU citizens and for the consumers in all the countries importing from the EU, particularly the US. According to recently published reports, the world's population will grow to an estimated 8 billion people by 2025 and 9 billion by 2050, and it is widely recognized that global agricultural productivity must increase to feed such a rapidly growing world population [1, 2]. Therefore, sustainable crop production needs to be secured and increased. On the other hand, agricultural practices are often in the public eye due to their impact on climate change, as well as because of energy and resources' constraints (e.g. water and soil), which are placing unprecedented pressure on the environment and the availability of natural resources [3].

Fertilizers, pesticides, and other agrochemicals play an important role in enhancing crop productivity. Fertilizers are products rich in plant nutrients which are applied to soil or plant tissues in order to sustain and stimulate plant growth.

Fertilizers are available as organic and inorganic compounds [4]. Nitrogen is taken up by most plants, only in two forms: nitrate (NO_3^-) and ammonium (NH_4^+) ions. According to officially published reports and statistics [5–8], ammonium nitrate and calcium ammonium nitrate (a mixture of approximately 80% ammonium nitrate and 20% calcium carbonate) are the most commonly used nitrogen-containing fertilizers in Europe. Nevertheless, worldwide, about one-half of the nitrogen applied as fertilizer is as carbamide (urea). It is the most concentrated solid nitrogen fertilizer (46% N) and it is converted quickly into ammonium ions by an enzyme, urease, which occurs in many soils. Phosphorus is mostly taken up by plant roots as the dihydrogenphosphate ion, $H_2PO_4^-$. The mostly applied phosphorus fertilizers are triple superphosphate, which is water soluble calcium dihydrogenphosphate, $Ca(H_2PO_4)_2$, and ammonium phosphate. According to a patent research study, phosphate rock (apatite) – reach in various forms of water insoluble tricalcium phosphate $Ca_3(PO_4)_2$ – is the starting raw material typically used for a production of mentioned water-soluble phosphates. Potassium is taken up as a positively charged potassium ion K^+, and it is mostly applied as potassium chloride. According to the report published by Yang et al. [9], mineral fertilization plays a crucial role in guaranteeing food security around the world, which contributes almost 50% of the increase in crop yield [10].

https://doi.org/10.1515/9783110642070-011

Pesticides are used in crops to eliminate, prevent, or control pests such as insects and mites, fungi, weeds, and other damaging organisms (e.g. slugs and snails) [11]. Approximately 4.6 million tons of pesticides are applied to prevent and control pests every year all over the world. According to the Food and Agriculture Organization of the United Nations statistics, pest and pathogen control with pesticides has restored approximately 30% of whole output of agricultural products all over the world [12]. However, the wide and indiscriminate use of pesticides has induced both environmental and human safety concerns. These derive from the fact that about 90% of the applied product might run off into the environment, or can be found as residues in agricultural products, and redistribute in the ecological cycle after application [12, 13]. Inefficient application of pesticides causes a series of ecological environmental problems, such as induction of pathogen and pest resistance, non-point pollution, water eutrophication, bioaccumulation in the food chain, and loss of biodiversity. The loss and decomposition rate of pesticide on crops is typically up to 70%, caused by runoff, spray drift, and rolling down during field application. The current utilization of biological target uptake is only less than 0.1% after dust drift and rainwater leaching [14].

It has been demonstrated that encapsulating pesticides, fertilizers, and other agrochemicals (e.g. plant strengtheners) allows growers to precisely control the conditions under which the active ingredient is released [15, 16], thus increasing the efficiency of the product and reducing the amount needed to be applied. Furthermore, encapsulation technology has a significant impact by reducing leaching [17], thus limiting the environmental impact on groundwater and surface water [18].

11.2 Fertilizer encapsulation

In order to improve fertilizer quality as well as to protect the environment and the ecosystem, during the last years, scientists have put a special attention on the development of new technologies for plant nutrient delivery in a slow or controlled manner in the water or soil. The use of controlled-release fertilizers (CRFs) increases fertilizer efficiency, reduces nutrient loss and soil toxicity, minimizes the potential for negative effects associated with overdosage, and reduces the frequency of the applications in accordance with normal crop requirements. Generally, there are 3 types of CRFs [19]: chemically controlled releasing products, polymer-coated fertilizers (PCFs), and microspheres containing fertilizers.

11.2.1 Chemically controlled releasing products

The first major category of such fertilizers is accomplished by means of chemically controlled releasing products, such as urea-formaldehyde. The release rate of nutrients by such kind of fertilizers is controlled by the degradation rate of the associated substance, which in turn is affected by various factors, such as molecular weight of the polymer, pH, temperature, ions and microorganisms in the soil, etc. Urea-formaldehyde (UF) is a long-chain polymer derived from a reaction between urea and formaldehyde. Urea is primarily synthesized by reacting CO_2 with anhydrous ammonia under high temperature and pressure. The molten urea product is processed into pills or granules for use as a fertilizer. Owing to higher nitrogen content, urea supplies more nitrogen per ton of applied solids as compared to other nitrogenous fertilizers. UF can be degraded by microorganisms; however, the macromolecules and crystalline regions in UF can hardly be decomposed by microbial action in a short time. Therefore, the release of the nutrient generally does not fully match with the uptake requirements of the crops. Thus, the share of UF in the global market for slow-release fertilizers has been falling significantly. Furthermore, the large-scale preparation of UF capsules is practically limited due to the high (63 wt.%) water content in the raw formaldehyde material [20]. Recently, Elhassani and coworkers [21] synthesized improved encapsulated fertilizers composed of urea, hydroxyapatite, and woodchip materials differently formulated. According to the authors, the immobilization of urea on hydroxyapatite leads to a functionalized fertilizer with slow-release properties owing to the strong interfacial interactions established between amine functions of the urea and carbonyl groups present on the surface of hydroxyapatite. The authors reported that the nonencapsulated urea/hydroxyapatite proved an advantageous slow-release pattern; however, it was not as effective as when the woodchips were used to prepare encapsulated pellets fertilizers. Through a water immersion experiment, the authors showed that the urea-modified hydroxyapatite –encapsulated with a sugarcane bagasse, cellulose, and lignin composite – displayed a slow and sustained release of nitrogen compared to the urea-modified hydroxyapatite during a period of more than 60 days.

11.2.2 Polymer-coated fertilizers

Another way of controlling the release of fertilizers is their encapsulation by coating method. Within this process, the fertilizers are core materials surrounded by inert polymeric shells. The release of the fertilizers is controlled by their diffusion through the shell. Due to the production cost and bio-economy requirements, the coating materials should be cheap and exhibit a good coating property. Furthermore, they should pass the degradation test in soil, during which they cannot produce any toxic substance that could affect the crop. The type of coating is responsible for the mechanism of release of elements from encapsulated fertilizer. According to the European Standard EN 13266 issued by the European Committee for Standardization, a fertilizer is commonly referred to as a slow-release fertilizer or CRF if the following requirements are met: no more than 15% of the nutrients are released within 24 h and no more than 75% of the nutrients are released within 28 days [22]. PCFs are considered as the most sophisticated and advance means of controlling nutrient release and fertilizer longevity. Jarosiewicz and Tomaszewska [23] have encapsulated fertilizers by applying a phase inversion technology for the coating using the following polymers: cellulose acetate (CA), polysulfone (PSF), and polyacrylonitrile (PAN). The authors decided to use these polymers because CA is a classic membrane material employed by the pioneers of modern membrane technology to create asymmetric membranes. It is a highly biodegradable polymer due to the nature of its cellulose backbone. It means that it is highly susceptible to microbial attack. The membranes and capsules fabricated from CA possess hydrophilic behavior, but they exhibit a low-temperature resistance and they are pH-sensitive. It has been reported that at pH 4–5, their lifetime is approximately 4 years, while at pH 6, they are stable only for a maximum of 2 years. Furthermore, reported data showed that at very basic (pH 9) or very acidic (pH 1) conditions, the lifetime of CA-based membranes and microcapsules decreases to a few days. However, the latter conditions are never met in agricultural soils. PAN possesses a higher temperature resistance than CA does, and the microcapsules and membrane prepared from PAN can be used at 45 °C and pH from 1 to 10 [24]. PSFs are ranked second after CA as a commercial membrane material, with applications in both large-scale production and laboratory scale. Moreover, microcapsules prepared from PSF are resistant to both temperature and pH. Figure 11.1 shows the cross-section of the microcapsules containing fertilizers and prepared with CA (Figure 11.1a) and PAN (Figure 11.1b)

According to the authors, the hydrophobic/hydrophilic character of the material applied for the capsule preparation has a substantial influence on the nutrient release. For example, comparison of coatings prepared from PAN and PSF put into evidence that the release rate of NH_4^+ is approximately three times faster in the case of hydrophilic coatings (PAN) in comparison to PSF, although the total porosity of the coatings was the same. Moreover, the authors observed that the release rate of components from the fertilizer coated with biodegradable coating (CA) was the highest. In the case of coatings prepared from PAN and PSF, the release rate of nutrients was much lower. These coatings are not biodegradable, but their presence in soil might improve the soil structure and quality. Abedi-Koupai and coworkers [25] coated Fe-based fertilizers with ethyl cellulose, glycerol monostearate, and ethylene vinyl acetates using an extrusion/spheronization technique. Among the selected polymers, only ethylene vinyl acetates seemed to be the best candidate for field application. The microcapsules based on it were able to slowly release Fe ions in water medium within 7 days. Thus, it seems that these microcapsules would have good controlled-release properties in the soil. Another well-known and tested substance for encapsulation is alginate [26]. Sodium alginate (SA) is a naturally occurring polysaccharide, which is commercially derived from several species of brown algae. It is a nontoxic and water-soluble polymer widely applied in pharmaceutical and food industries [27, 28]. It has been regarded as a brilliant polysaccharide for water

A)
B)
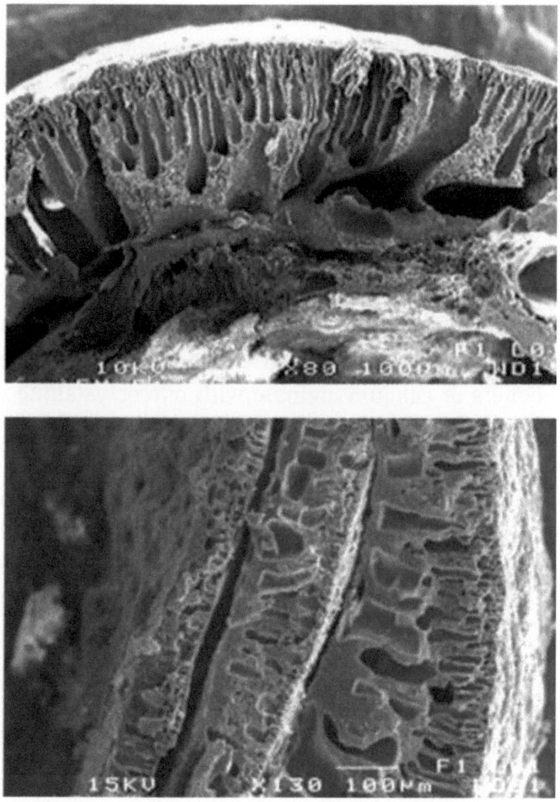

Fig. 11.1: Scanning electron microscope (SEM) cross-section of the microcapsules containing fertilizers and prepared with cellulose acetate (a) and polyacrylonitrile (b). Reprinted with permission from [23]. Copyright 2003 American Chemical Society.

and fertilizer management because of its unique hydrophilicity, biocompatibility, biodegradability, and nontoxicity [29]. Recently, Wu and coworkers [30] investigated encapsulation technology for microbial fertilizer to enhance the *Klebsiella oxytoca* Rs-5 bacterial survival rate under salinity stress. The authors used SA and calcium chloride (CC) to form microcapsule shells. The best microcapsule formulation for the release and activity of the bacteria resulted when the concentrations of SA and CC were 1.5–2.0% and 2.0%, respectively, and the volumetric ratio of SA and Rs-5 culture was 3:1. The bacteria released from microcapsules reached up to 10^{10} cfu/g when immersed in physiological saline solution for 3 weeks, providing the base for a higher cotton growth than the free cells under the same salty conditions in pot experiments. The alginate-based fertilizers have been also investigated to reduce eutrophication of aquatic environments [31]. It is well known that leaching from nitrogen-based fertilizers could cause eutrophication of aquatic environments, which ultimately leads to hypoxia. For example, the surface area of such oxygen-starved regions in the Gulf of

Mexico is rapidly increasing and greatly affects the marine ecosystem and development of the Gulf coast fishing industry. For this reason, Kay [31] developed an alternative to conventional fertilizer by incorporating a commercial fertilizer into a SA base to help retain the nitrogen at the site of application. Wang et al. [32] have been able to encapsulate 22.6% of nitrogen content of κ-Carrageenan in SA beads. The authors demonstrated that the nutrient N had a release value of 94.2% after being incubated in the soil for 25 days. The polymeric carrier could significantly improve the water-holding capacity and water-retention properties of soil. Moreover, it can prevent soil from becoming harder. The same authors have also studied encapsulation of water-soluble nitrogen fertilizer pellets with an inner coating of calcium alginate and chitosan-glutaraldehyde copolymers. According to the authors, a combination of chitosan-glutaraldehyde copolymers or calcium alginate with microcrystalline wax as a coating material could be a new approach of controllable and efficient release of the encapsulated nutrients. Indeed, they reported that a double coating could be considered as a convenient and eco-friendly method for the preparation of microcrystalline wax derivative of polysaccharide-coated fertilizer. Furthermore, chitosan is a bioactive polymer with a wide variety of applications due to its functional properties, such as antibacterial activity, nontoxicity, ease of modification, and biodegradability [33].

11.2.3 Microspheres containing fertilizers

Matrix-type formulation is the third major category of slow or controlled-release fertilizers due to simple fabrication processes. The active ingredients are dispersed in the matrix and diffuse through the matrix continual or intergranular openings, that is, through pores or channels in the carrier phase. Chitosan-based capsules containing fertilizers have been investigated by Faez and coworkers [34], who fabricated chitosan-based microspheres as a sustainable fertilizer delivery system. Figure 11.2 shows scanning electron microscope (SEM) micrographs of chitosan-based microsphere as a spherical material with some deformations at the breakpoint of the drops.

Recently, very interesting studies were published by França and coauthors [35], who have applied the spray drying method with two types of nozzles to understand the relationship between the structure (capsule or sphere) and the release process in soil and water. Achieved results put into evidence that the higher swelling degree and delayed release of nutrients for microcapsules were associated to the core-shell structure. The authors suggest that encapsulation technology providing the shell/core capsule structures should be explored by the agriculture industry instead of the microsphere ones. Indeed, it has been ascertained that a SA-based polymer encapsulating water and fertilizer is superior to a nonencapsulated commercial

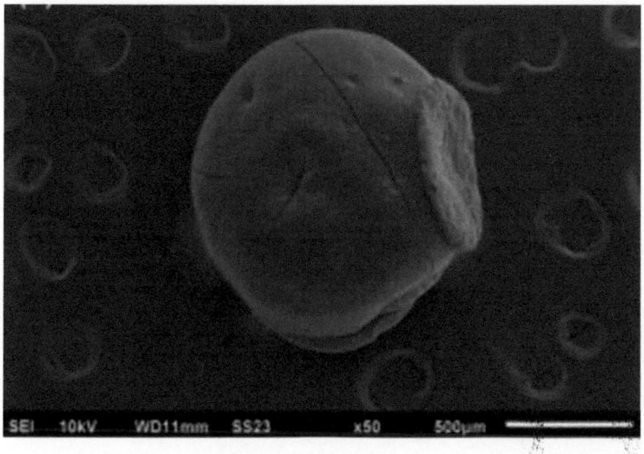

Fig. 11.2: SEM micrograph of chitosan-based microsphere prepared by Faez et al. [34].

formulation in extending agriculture applications. However, the release properties of pure Ca-SA microspheres regretfully often suffer from burst release and quick breakdown in the *in vitro* release process [29].

In order to overcome this challenge, Feng et al. [36] put forward a simple strategy for the fabrication of novel biodegradable yeast/SA/poly(vinyl alcohol) (yeast/SA/PVA) superabsorbent microspheres with a diffusion barrier induced through a thermochemical modification route. Figure 11.3 provides photographs (a) and SEM micrographs (b) of the yeast/SA/PVA microspheres with or without thermochemical modification. The variances of the size with and without modification suggests that the –COOH groups of citric acid used for the modification reacted with free –OH groups of the polymer matrix via esterification cross-linking, inducing the formation of a denser cross-linked network on the surface which "wrinkled" the microspheres. Various amounts of citric acid would bring about different esterification degrees of the diffusion barrier, i.e. with the increasing citric acid, much more –COOH groups participated in the thermochemical modification, resulting in different surface morphologies. Uniform and orderly folds are clearly visible on the surface of the microspheres without thermochemical modification, while irregular bulges and inhomogeneous grooves are observed on the surface of the modified microspheres.

Based on generated results, the authors indicate that the obtained microspheres with diffusion barrier could be considered as an outstanding water and fertilizer manager to efficaciously improve the utilization efficiency of water and fertilizer, having a great potential for agriculture applications. In addition, as a facile method to build a diffusion barrier, thermochemical modification with citric acid may promote the immediate applications of biodegradable superabsorbent microspheres in areas related to substance diffusion and transport in sustainable development of agriculture.

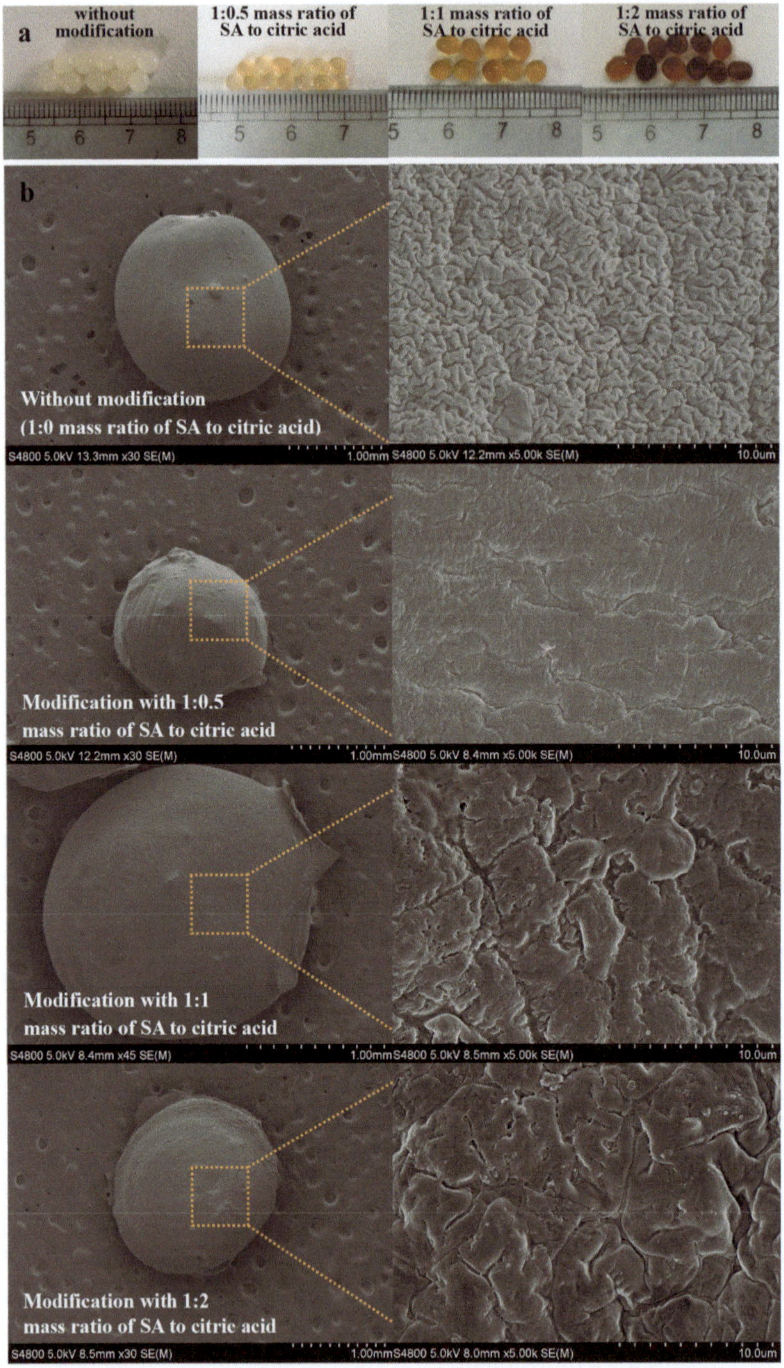

Fig. 11.3: (a) Photographs and (b) SEM images of the yeast/SA/PVA microspheres with or without thermochemical modification. Reprinted with permission from [36]. Copyright 2017 American Chemical Society.

11.3 Pesticide encapsulation

Recently, porous micro- and nano-materials have been developed as a tool for slow-release pesticides. They have attracted extensive attention because they offer higher pesticide utilization efficiency and fewer side effects to the environment versus traditional pesticides.

Chlorantraniliprole (CAP) has been widely investigated to prevent and control pests as a broad-spectrum pesticide with long residual activity, low toxicity, and no cross-resistance with other pesticides. Currently, only traditional formulations such as emulsifiable concentrate (EC), suspension concentrate (SC), and wettable powder (WP) of CAP are accessible. The major challenge is that poor solubility of CAP restricts its loading content and subsequent release from the microcapsule. Thus, Liu et al. [37] decided to focus their investigation on the development of novel, environmentally friendly, and efficient microcapsule formulations for CAP with a high loading content and controlled-release property. The authors prepared the CAP microcapsule by encapsulating the highly insoluble CAP inside the microcapsules through a premix membrane emulsification method (PME) combined with the solid-in-oil-in-water (S/O/W) double-emulsion method, as illustrated in Figure 11.4.

The authors fabricated polylactide microspheres with controlled surface morphology and size by changing the osmotic agents (bovine serum albumin and PVA) and process parameters of PME, respectively. The authors demonstrated that different microcapsules with tunable surface porosity and sizes could be prepared by simply manipulating the osmotic agent and the process parameters. Compared with both CAP technical and commercial formulations, the various CAP-loaded microcapsules exhibited a sustained release for a longer period. By varying the size and porosity of the microcapsules, the authors could achieve different responses in a controlled

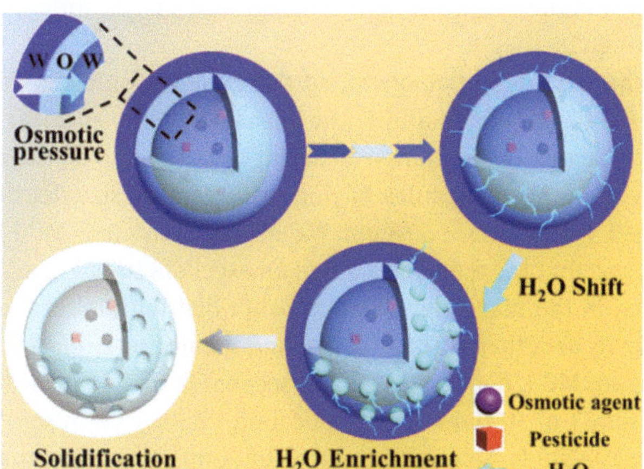

Fig. 11.4: Schematic description for the preparation of porous microcapsules via the osmosis induction method. Reprinted with permission from [37]. Copyright 2018 American Chemical Society.

release. They reported that the microcapsule with greater porosity and small size possesses a quick release of CAP. Moreover, the pest control efficacy of the porous microcapsule, having good ultraviolet and thermal stability, is superior to that of the current commercial EC, SC, and WP formulations.

Very interesting results were published by Kawashima and coworkers [38], who have performed encapsulation of fenitrothion, a broad-spectrum insecticide with low environment and mammalian toxicity, applying an interfacial polymerization method with polyisocyanate and ethylene glycol. The authors found that microencapsulated fenitrothion could be considerably effective to treat insect-proof plywood because of reduction of the dosage and keeping a residual efficacy even under severe storage conditions, independent of the kind of adhesives. The interfacial polymerization of isocyanate monomers for pesticide encapsulation was deeply studied by Scher [39], who demonstrated that the release rate of this microcapsule system strongly depends on microcapsule particle size (i.e. total surface area per pound of pesticide), wall thickness (weight percent of isocyanate monomers in organic phase), and wall permeability. The wall permeability can be varied by varying the crosslink density of the polyurea, by varying the wall microporosity (solvent effect), by limiting crosslinking (ammonia addition), and by varying the isocyanate monomer chemical composition. Furthermore, the author indicated that the slurries of microcapsules with different wall thicknesses and wall permeabilities can be fabricated using the interfacial polymerization with sequential dispersion of different organic phases into the same aqueous phase. Seyed Ali Hashemi and Mojgan Zandi [40] encapsulated liquid pesticide (Dursbantrade name for 0,0-3,5,6-trichloro-2-pyfidyl phosphorothioate, C91-I11C13NO5PS) in polyurea microcapsules using interfacial polymerization of polyisocyanate (such as toluene diisocyanate) and polyamine (such as diethylene triamine) as the monomers. The effect of various emulsifiers such as Tween 20, 40, and 80 was studied, and the best one was found to be Tween 80. More stable and spherical microcapsules were obtained by using coemulsifier and matrix forming agent and temperature adjusting.

Based on the work of Scher [39], who demonstrated that microcapsule size is an important and adjustable parameter which affects the release rate, Luo et al. [41] studied the influence of this parameter on the release behavior of a pesticide. In their work, phoxim-loaded polyurethane microcapsules having three various size distributions (average diameters of 1.39 µm, MC-S; 5.78 µm, MC-M; and 23.60 µm, MC-L) were tested. The results indicated that MC-S and MC-M had an excellent initial insecticidal activity, which occurred in the first 3 days, due to better distribution on the organism surface (more likely to be adhered to by pests) and greater resistance to rain washing. On the other hand, MC-L demonstrated an excellent later-stage insecticidal activity (maintained from 3 to 10 days after application) mainly because of its outstanding light stability. The study demonstrated that by simply adjusting the particle size, the transfer and release behavior of pesticide MCs in the field can be regulated. This finding is of great value to the application of pesticide MCs in agriculture and could provide a new approach for the efficient utilization of pesticide MC formulation.

Stejskal and coworkers [42] have explored how microcapsule size and brief exposure affected the bioavailability of microencapsulated pesticide on porous and nonporous surfaces of *Blattella germanica* L. The authors tested five microcapsule formulations containing one of the following insecticides: chlorpyrifos 23.1 g L(-1) CS (Detmol-PRO), chlorpyrifos 20 g L(-1) CS (Empire 20), fenitrothion 20 g L(-1) CS (Detmol-Mic), cyphenothrin 10 g L(-1) CS (Detmol-CAP), and diazinon 30 g L(-1) CS (Diacap). The authors found main differences in bioavailability on the porous and the nonporous surfaces. The largest difference was observed in Empire 20 and Detmol CAP, while the bioavailability of Detmol MIC did not differ on porous and nonporous surfaces. Comparison of their microcapsule size spectra revealed that formulations containing larger microcapsules had higher efficacy on porous surfaces than formulations with smaller microcapsules. In order to explain this difference in efficacy, the variance of microcapsule sizes was regressed on the efficacy ratio on porous versus nonporous surfaces. Although negative correlation was evident between the size of capsules and the efficacy ratio on porous and nonporous surfaces, the difference in the slope parameter was not statistically significant.

Recently, Wu and coworkers [43] proposed an efficient synthesis of starch-regulated porous calcium carbonate microspheres as the carrier for slow-release herbicide (Figure 11.5).

Calcium carbonate ($CaCO_3$) is attracting more and more attention due to its low cost, environmental friendliness, and high stability. Therefore, fabricating porous $CaCO_3$ microspheres (PCMs) has become not only a hot research subject but also a big

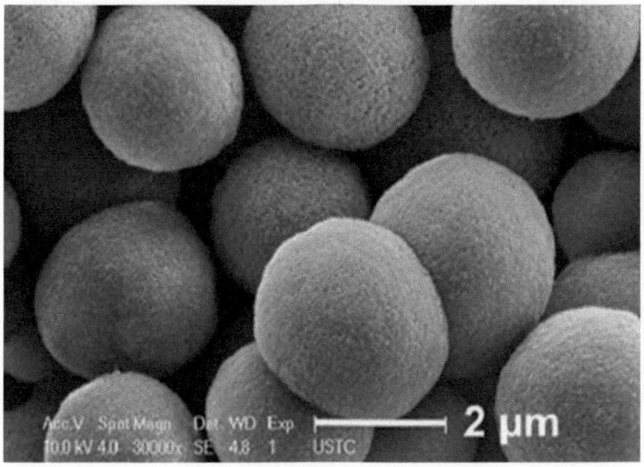

Fig. 11.5: SEM micrograph of starch-regulated porous calcium carbonate microspheres. Adapted with permission from [43] Copyright 2018 American Chemical Society.

challenge for researchers in the material field. It is known that under thermal treatment, soluble starch (SS) molecules are able to self-organize into nanoaggregates that bind Ca^{2+} through chelation and electrostatic interaction. It means that CO_3^{2-} could be introduced to the system producing $CaCO_3$ nanoparticles through heterogeneous nucleation regulated by these starch aggregates, as it was designed by Wu and coworkers. The authors selected prometryn (PMT) as a model pesticide and *Cynodon dactylon* as the model weed plant to evaluate the control efficacy of optimal-PCMs-SS/PMT formulation. Figure 11.6 shows digital photographs of *Cynodon dactylon* after 1-week treatments with (a) deionized water, (b) ethanol/deionized water solution (vethanol/vdeionized water = 1:4), (c) optimal-PCM-SS suspension, (d) PMT aqueous solution, and (e) optimal-PCMs-SS/PMT suspension, wherein PMT and optimal-PCM-SS/PMT were applied with equal PMT amounts (14 mg).

As shown in Figure 11.6, after treatment with PMT and encapsulated PMT (optimal-PCM-SS/PMT), the plant height dramatically decreased, suggesting that PMT and encapsulated PMT could effectively inhibit the growth of *Cynodon dactylon* compared with deionized water and ethanol/deionized water. Moreover, encapsulated PMT displayed a significantly higher control efficacy than PMT alone. Reported data suggest that the optimal-PCMs-SS could efficiently increase the utilization efficacy of PMT because it could not only promote the slow release but also improve the adhesion ability of PMT on the rough surfaces of the seed and root of *Cynodon dactylon*.

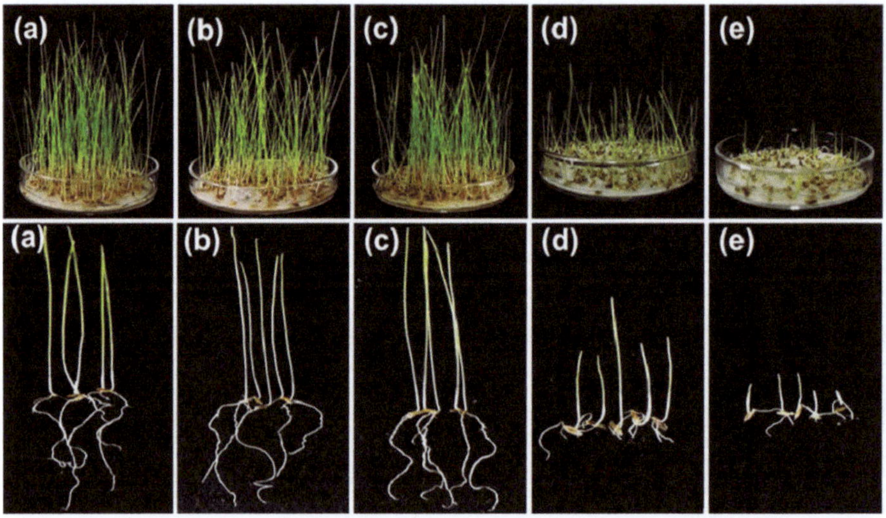

Fig. 11.6: Digital photographs of *Cynodon dactylon* after 1-week treatments with (a) deionized water, (b) ethanol/deionized water solution (vethanol/vdeionized water = 1:4), (c) optimal-PCM-SS suspension, (d) PMT aqueous solution, and (e) optimal-PCM-SS/PMT suspension; PMT and optimal-PCM-SS/PMT were applied with equal PMT amounts (14 mg). Adapted with permission from [43]. Copyright 2018 American Chemical Society.

11.4 Conclusions

Application of encapsulation technology in agriculture may significantly increase product efficiency because the encapsulated particles are designed to release their contents when exposed to a particular trigger. This system allows growers to precisely control the conditions under which the active ingredient is released.

Furthermore, encapsulation technology of agrochemicals minimizes their unwanted loss in the environment, thus reducing their impact on the environment. The encapsulated product is protected until appropriate conditions for release are met, possibly matching the plant nutrient and protection requirements. Such condition results in a reduced need of product and frequency of application, which increases the use efficiency of the substance. Therefore, as an important side effect, encapsulation also positively impacts on the environmental fate of pesticides and fertilizers: it decreases leaching and pollution of groundwater and surface water, as well as ensures that the encapsulated compounds remain in the site where they are needed.

In other words, the use of encapsulation in agriculture increases product efficiency, minimizes the potential for negative effects associated with overdosage, and reduces the frequency of the applications in accordance with normal crop requirements. Thus, by applying encapsulation technology, the agriculture will be not only more productive but also more sustainable and able to yield safe and healthy food.

11.5 References

[1] Corsi, C., Lamberth, C., New paradigms in crop protection research: registrability and cost of goods. In: Peter Maienfisch and Thomas M. Stevenson (eds.) Discovery and Synthesis of Crop Protection Products. American Chemical Society: 2015. Vol. 1204, pp. 25–37.

[2] Mullan, B., Haqq-Misra, J., Population growth, energy use, and the implications for the search for extraterrestrial intelligence. Futures **106** (2019) 4–17. doi:https://doi.org/10.1016/j.futures.2018.06.009.

[3] Keating, B. A., Herrero, M., Carberry, P. S., Gardner, J., Cole, M. B., Food wedges: framing the global food demand and supply challenge towards 2050. Global Food Security **3** (2014) 125–132, doi:https://doi.org/10.1016/j.gfs.2014.08.004.

[4] Chen, L., Redmile-Gordon, M., Li, J., Zhang, J., Xin, X., Zhang, C., Ma, D., Zhou, Y., Linking cropland ecosystem services to microbiome taxonomic composition and functional composition in a sandy loam soil with 28-year organic and inorganic fertilizer regimes. Appl Soil Ecol **139** (2019) 1–9. doi:https://doi.org/10.1016/j.apsoil.2019.03.011.

[5] Cobo, S., Levis, J. W., Dominguez-Ramos, A., Irabien, A., Economics of enhancing nutrient circularity in an organic waste valorization system. Environ Sci Technol **53** (2019) 6123–6132. doi:10.1021/acs.est.8b06035.

[6] Galloway, J. N., Townsend, A. R., Erisman, J. W., Bekunda, M., Cai, Z., Freney, J. R., Martinelli, L. A., Seitzinger, S. P., Sutton, M. A. Transformation of the nitrogen cycle: recent trends, questions, and potential solutions. Science **320** (2008) 889–892. doi:10.1126/science.1136674.

[7] Von Blottnitz, H., Rabl, A., Boiadjiev, D., Taylor, T., Arnold, S. Damage costs of nitrogen fertilizer in Europe and their internalization. J Environ Plan Manage **49** (2006) 413–433. doi:10.1080/09640560600601587.

[8] Bateman, A. S., Kelly, S. D., Fertilizer nitrogen isotope signatures. Isotopes Environ Health Stud **43** (2007) 237–247, doi:10.1080/10256010701550732.

[9] Yang, M., Fang, Y., Sun, D., Shi, Y. Efficiency of two nitrification inhibitors (dicyandiamide and 3, 4-dimethypyrazole phosphate) on soil nitrogen transformations and plant productivity: a meta-analysis. Sci Rep **6** (2016) 22075. doi:10.1038/srep22075. https://www.nature.com/articles/srep22075#supplementary-information.

[10] Roy, R. N., Finck, A., Blair, G. J., Tandon, H. L. S., Plant Nutrition for Food Security. A Guide for Integrated Nutrient Management. FAO, Rome, 2006.

[11] Pesticide residue in foods. In: Cheremisinoff, N. P., Rosenfeld, P. E. (eds.) Handbook of Pollution Prevention and Cleaner Production: Best Practices in the Agrochemical Industry. William Andrew Publishing, Oxford, 2011; pp. 269–279.

[12] Alavanja, M. C. R. Introduction: pesticides use and exposure extensive worldwide. Rev Environ Health **24** (2009) 303–309.

[13] Zhao, X., Cui, H., Wang, Y., Sun, C., Cui, B., Zeng, Z. Development strategies and prospects of nano-based smart pesticide formulation. J Agric Food Chem **66** (2018) 6504–6512. doi:10.1021/acs.jafc.7b02004.

[14] Massinon, M., De Cock, N., Forster, W. A., Nairn, J. J., McCue, S. W., Zabkiewicz, J. A., Lebeau, F. Spray droplet impaction outcomes for different plant species and spray formulations. Crop Protection **99** (2017) 65–75. doi:https://doi.org/10.1016/j.cropro.2017.05.003.

[15] Sinclair, R. G. Slow-release pesticide system. Polymers of lactic and glycolic acids as ecologically beneficial, cost-effective encapsulating materials. Environ Sci Technol **7** (**1973**) 955–956. doi:10.1021/es60082a011.

[16] Quaglia, F., Barbato, F., De Rosa, G., Granata, E., Miro, A., La Rotonda, M. I. Reduction of the Environmental Impact of Pesticides: waxy microspheres encapsulating the insecticide carbaryl. J Agric Food Chem **49** (2001) 4808–4812. doi:10.1021/jf010418w.

[17] Roy, A., Singh, S. K., Bajpai, J., Bajpai, A. K. Controlled pesticide release from biodegradable polymers. Central Eur J Chem **12** (2014) 453–469. doi:10.2478/s11532-013-0405-2.

[18] Walker, G. W., Kookana, R. S., Smith, N. E., Kah, M., Doolette, C. L., Reeves, P. T., Lovell, W., Anderson, D. J., Turney, T. W., Navarro, D. A. Ecological risk assessment of nano-enabled pesticides: a perspective on problem formulation. J Agric Food Chem **66** (2018) 6480–6486. doi:10.1021/acs.jafc.7b02373.

[19] Calabi-Floody, M., Medina, J., Rumpel, C., Condron, L. M., Hernandez, M., Dumont, M., Mora, M.d. l. L. Smart fertilizers as a strategy for sustainable agriculture. In: Donald L. Sparks (eds.), Advances in Agronomy. Academic Press, Elsevier, Amsterdam, The Netherlands, 2018. Vol. 147, pp. 119–157.

[20] Xiang, Y., Ru, X., Shi, J., Song, J., Zhao, H., Liu, Y., Zhao, G. Granular, Slow-release fertilizer from urea-formaldehyde, ammonium polyphosphate, and amorphous silica gel: a new strategy using cold extrusion. J Agric Food Chem **66** (2018) 7606–7615. doi:10.1021/acs.jafc.8b02349.

[21] Elhassani, C. E., Essamlali, Y., Aqlil, M., Nzenguet, A. M., Ganetri, I., Zahouily, M. Urea-impregnated HAP encapsulated by lignocellulosic biomass-extruded composites: a novel slow-release fertilizer. Environ Technol Inno. **15** (2019) 100403, doi:https://doi.org/10.1016/j.eti.2019.100403.

[22] Wang, J., Liu, S., Qin, Y., Chen, X., Xing, R.e., Yu, H., Li, K., Li, P. Preparation and characterization of controlled-release fertilizers coated with marine polysaccharide derivatives. Chin J Oceanol Limn. **35** (2017) 1086–1093, doi:10.1007/s00343-017-6074-9.

[23] Jarosiewicz, A., Tomaszewska, M. Controlled-release NPK fertilizer encapsulated by polymeric membranes. J Agric Food Chem **51** (2003) 413–417, doi:10.1021/jf020800o.

[24] Han, W., Dong, S., Li, B., Ge, L. Preparation of polyacrylonitrile-based porous hollow carbon microspheres. Colloid Surface A. **520** (2017) 467–476, doi:https://doi.org/10.1016/j.colsurfa.2017.02.009.

[25] Abedi-Koupai, J., Varshosaz, J., Mesforoosh, M., Khoshgoftarmanesh, A. H. Controlled release of fertilizer microcapsules using ethylene vinyl acetate polymer to enhance micronutrient and water use efficiency. J. Plant Nutr. **35** (2012) 1130–1138, doi:10.1080/01904167.2012.676126.

[26] Malusá E., Sas-Paszt L., and J. Ciesielska. Technologies for beneficial microorganisms inocula used as biofertilizers. Sci. World J (2012), Article ID 491206, 12 pages. doi:10.1100/2012/491206

[27] Yeung, R. A., Kennedy, R. A. A comparison of selected physico-chemical properties of calcium alginate fibers produced using two different types of sodium alginate. J Mech Behav Biomed **90** (2019) 155–164, doi:https://doi.org/10.1016/j.jmbbm.2018.10.011.

[28] Wang, X., Liu, F., Gao, Y., Xue, C.-h., Li, R. W., Tang, Q.-j. Transcriptome analysis revealed anti-obesity effects of the Sodium Alginate in high-fat diet -induced obese mice. Int. J. Biol. Macromol. **115** (2018) 861–870, doi:https://doi.org/10.1016/j.ijbiomac.2018.04.042.

[29] Kim, Y. S., Kim, H. W., Lee, S. H., Shin, K. S., Hur, H. W., Rhee, Y. H. Preparation of alginate-quaternary ammonium complex beads and evaluation of their antimicrobial activity. Int J Biol Macromol **41** (2007) 36–41.

[30] Wu, Z., Zhao, Y., Kaleem, I., Li, C. Preparation of calcium-alginate microcapsuled microbial fertilizer coating Klebsiella oxytoca Rs-5 and its performance under salinity stress. Eur J Soil Biol. **47** (2011) 152–159, doi:https://doi.org/10.1016/j.ejsobi.2010.11.008.

[31] Kay, A. N. An Alginate-based fertilizer to reduce eutrophication of aquatic environments. Trans Kans Acad Sci. (1903-) **107** (2004) 155–164.

[32] Wang, Y., Liu, M., Ni, B., Xie, L. κ-Carrageenan-sodium alginate beads and superabsorbent coated nitrogen fertilizer with slow-release, water-retention, and anticompaction properties. Ind. Eng. Chem. Res. **51** (2012) 1413–1422, doi:10.1021/ie2020526.

[33] Muxika, A., Etxabide, A., Uranga, J., Guerrero, P., de la Caba, K. Chitosan as a bioactive polymer: Processing, properties and applications. Int. J. Biol. Macromol. **105** (2017) 1358–1368, doi:https://doi.org/10.1016/j.ijbiomac.2017.07.087.

[34] Santos, B. R. d., Bacalhau, F. B., Pereira, T. d. S., Souza, C. F., Faez, R. Chitosan-montmorillonite microspheres: a sustainable fertilizer delivery system. Carbohyd Polym. **127** (2015) 340–346, doi:https://doi.org/10.1016/j.carbpol.2015.03.064.

[35] França, D. Medina, Â. F. Messa, L. L. Souza, C. F. Faez, R. Chitosan spray-dried microcapsule and microsphere as fertilizer host for swellable – controlled release materials. Carbohyd Polym. **196** (2018) 47–55, doi:https://doi.org/10.1016/j.carbpol.2018.05.014.

[36] Feng, D., Bai, B., Wang, H., Suo, Y. Novel fabrication of biodegradable superabsorbent microspheres with diffusion barrier through thermo-chemical modification and their potential agriculture applications for water holding and sustained release of fertilizer. J Agric Food Chem **65** (2017) 5896–5907, doi:10.1021/acs.jafc.7b01849.

[37] Liu, B., Wang, Y., Yang, F., Cui, H., Wu, D. Development of a chlorantraniliprole microcapsule formulation with a high loading content and controlled-release property. J Agric Food Chem **66** (2018) 6561–6568, doi:10.1021/acs.jafc.7b01295.

[38] 川島, 操., 大坪, 敏.; 津田, 重., 伊藤, 高., 辻, 孝. フェニトロチオンマイクロカプセルの防虫合板への適用性. 日本農薬学会誌 (J Pestic Sci) **16** (1991) 41–46, doi:10.1584/jpestics.16.41.

[39] Scher, H. B. Microencapsulation of pesticides by interfacial polymerization: process and performance considerations. In *Pesticide Chemistry: Human Welfare and the Environment*, Greenhalgh, R., Drescher, N., Eds. Pergamon: 1983; pp. 295–300.

[40] HASHEMI, S. A., ZANDI, M. Encapsulation process in synthesizing polyurea microcapsules containing pesticide. Iran Polym J. (English) **10** (2001).

[41] Luo, J., Huang, X-p., Jing, T-f., Zhang, D-x., Li, B., Liu, F. Analysis of particle size regulating the insecticidal efficacy of phoxim polyurethane microcapsules on leaves. ACS Sustain Chem Eng **6** **(12)** (2018) 17194–17203. DOI: 10.1021/acssuschemeng.8b04567.
[42] Stejskal, V., Aulicky, R., Pekar, S. Brief exposure of Blattella germanica (Blattodea) to insecticides formulated in various microcapsule sizes and applied on porous and non-porous surfaces. Pest Manag Sci **65** (2009) 93–98.
[43] Xiang, Y., Han, J., Zhang, G., Zhan, F., Cai, D., Wu, Z. Efficient synthesis of starch-regulated porous calcium carbonate microspheres as a carrier for slow-release herbicide. ACS Sustain Chem Eng **6** (2018) 3649–3658, doi:10.1021/acssuschemeng.7b03973.

Index

https://doi.org/10.1515/9783110642070-012

Bei Fragen zur Produktsicherheit wenden Sie sich bitte an:
If you have any questions regarding product safety,
please contact:

Walter de Gruyter GmbH
Genthiner Straße 13
10785 Berlin
productsafety@degruyterbrill.com